IN MENDEL'S MIRROR

In Mendel's Mirror

Philosophical Reflections on Biology

PHILIP KITCHER

UNIVERSITY PRESS

2003

OXFORD
UNIVERSITY PRESS

Oxford   New York
Auckland   Bangkok   Buenos Aires   Cape Town   Chennai
Dar es Salaam   Delhi   Hong Kong   Istanbul   Karachi   Kolkata
Kuala Lumpur   Madrid   Melbourne   Mexico City   Mumbai   Nairobi
São Paulo   Shanghai   Taipei   Tokyo   Toronto

Published by Oxford University Press, Inc.
198 Madison Avenue, New York, New York 10016

www.oup.com

Oxford is a registered trademark of Oxford University Press

Library of Congress Cataloging-in-Publication Data
Kitcher, Philip, 1947–
In Mendel's mirror : philosophical reflections on biology / Philip Kitcher.
    p. cm.
ISBN 0-19-515178-X; 0-19-515179-8 (pbk.)
1. Biology—Philosophy.   I. Title.
QH331 K6145 2002
570′.1—dc21   2002025759

9 8 7 6 5 4 3 2 1

Printed in the United States of America
on acid-free paper

For Dick Lewontin and in memory of Steve Gould

i maestri di color che sanno

Acknowledgments

I would like to thank the following people and institutions for permission to reprint the essays included here.

The *Philosophical Review* for "1953 and All That: A Tale of Two Sciences"

Kluwer publications for "The Hegemony of Molecular Biology" and "Some Puzzles about Species"

Nicholas Rescher for "Darwin's Achievement"

The *Journal of Philosophy* for "The Return of the Gene" and "The Evolution of Human Altruism"

George Gale and the *Philosophy of Science Association* for "Species" and Development Decomposition and the Future of Human Behavioral Ecology"

Howard Wettstein and *Midwest Studies in Philosophy* for "Function and Design"

The *Journal of Theoretical Biology* for "Evolution of Altruism in Optional and Compulsory Games"

Barry Smith and *The Monist* for "Infectious Ideas"

Leonard Harris and Prometheus Books for "Race, Ethnicity, Biology, Culture"

Phillip Sloan and University of Notre Dame Press for "Utopian Eugenics and Social Inequality"

Cambridge University Press for "Battling the Undead: How (and How Not) to Resist Genetic Determinism"

Kurt Bayertz and MIT Press for "Four Ways of 'Biologicizing' Ethics"

Cheryl Travis and MIT Press for "Pop Sociobiology Reborn: The Evolutionary Psychology of Sex and Violence"

Robert Pennock and MIT Press for "Born-Again Creationism"

In addition, I am grateful to my coauthors John Batali ("Evolution of Altruism in Optional and Compulsory Games"), Kim Sterelny ("The Return of the Gene"), and A. Leah Vickers ("Pop Sociobiology Reborn: The Evolutionary Psychology of Sex and Violence") for allowing me to reprint our joint work.

Contents

Introduction

When I was a graduate student in the early 1970s, philosophy of science clearly meant philosophy of physical science. Novices in the field were expected to know some mathematical physics, particularly parts of relativity and quantum mechanics, and to be acquainted with bits and pieces of optics, thermodynamics, and elementary chemistry. Biological ignorance was rampant, and tolerated. The general ways in which we thought about theory, explanation, and the growth of scientific knowledge were developed in relation to a handful of examples from physical science, and nobody seemed to worry that they wouldn't apply in biology or psychology.

When I began teaching, I came to realize how partial a perspective this was (see chapter 6). Even in the mid-1970s, biology was becoming the science of choice for a large number of the brightest American undergraduates. (Today the number of biology majors on a campus typically dwarfs the number of physics majors.) Quite reasonably, my students wanted me to relate philosophical discussions to the areas of science they were studying. I discovered that a handful of philosophers had already been broaching issues in the philosophy of biology and that their discussions were fascinating. Besides the intrinsic interest of the scientific material, I quickly learned that the standard models from the general philosophy of science were remarkably ill-adapted to an area of science, biology, which was emerging as *the* dominant field of inquiry at the end of the twentieth century.

The essays that follow reflect twenty years of pondering the significance of various parts of biological practice. In some of them I'm concerned with the implications of biological discoveries for philosophical questions, sometimes questions within the philosophy of science, at other times broader issues in metaphysics, epistemology, ethics, or social philosophy. On other occasions, I'm chiefly interested in a dispute that has emerged within biology itself, and the work undertaken is the activity of philosophical clarification that generations of philosophers have carried out so well with respect to areas within physics. Sometimes, then, the arrow of illumination is supposed to run from biology to philosophy and, at other times, in the opposite direction. But, quite frequently, it would be overly neat to assign a single direction to an essay; even when I'm primarily concerned with one direction, I often can't resist some suggestions about the other.

The opening essay is principally focused on the inadequacies of a general philosophical account of intertheoretic relations and with the provision of a substitute. Many philosophers have believed that sciences in some fields can be reduced to sciences in other fields—biology can be reduced to physics and chemistry, psychology to neuroscience, the social sciences to psychology. General philosophy of science has tried to explain the reduction relation. I try to show how the traditional views about reduction fail, and how we can use a different approach to appreciate the connections between one part of biology (genetics) and the physical sciences. The second essay, written fifteen years later, looks at the same material from a different angle, trying to understand what kinds of ventures within biology might be needed to supplement the biochemical research that has enjoyed such explosive success in recent decades.

In the third essay, I return to the inadequacies of older philosophical perspectives and try to present an interpretation of Darwin's scientific achievement that will enable us to recognize its magnitude and that will expose the structure of the argument Darwin offered. From the 1860s to the present, Darwin's evolutionary theory (or theories) have invited trivialization at the hands of hostile critics, including philosophers. I try to set the record straight.

The fourth essay takes up an important theoretical issue in contemporary Darwinism. In the 1960s and 1970s, George Williams and Richard Dawkins offered accounts of natural selection that viewed the Darwinian struggle as occurring among genes (rather than individual organisms), and their proposals led to fierce debates about the proper "unit(s) of selection." Kim Sterelny and I attempted to resolve these debates by suggesting that the processes biologists mark out as instances of natural selection can typically be approached from many perspectives. (This position was also developed independently by Kenneth Waters, who joined Sterelny and me in a reply to criticisms.) As I'd now prefer to put it (and Sterelny and Waters might disagree), there's a constant danger of overinterpreting Darwin's metaphor, so that there's an alleged causal fact about the world concerning the exact place at which natural selection "acts." Better, I believe, to see that selection is just our metaphorical way of approaching the complex facts of birth, mating, and death; we can organize those facts in various Darwinian ways, and our choices among styles of organization should be thoroughly pragmatic.

Chapters 5 and 6 take up one of the thorniest issues in theoretical biology, the character of biological species. In them, I'm concerned both to use ideas from metaphysics and the philosophy of science to clarify the biological material, and also to use findings from biology to illuminate philosophical discussions (philosophers writing about natural kinds often seem remarkably unaware of how badly their ideas fit biological species). But I also swim against a major current. Most writers on biological species tend to assume that divisions of organisms into species must serve the purposes of evolutionary biology, as if other fields of biology had no competing claim. A general theme of the essays in this collection, one that emerges particularly in the essays on species, is my belief that different classifications are appropriate to different areas of biology and that none has priority. Most philosophical discussions of biology are centered on evolutionary biology, and place evolution *über alles*; ironically, the vast majority of contemporary biologists are pursuing molecular studies that have little direct tie to evolution.

That theme is continued in the next essay, which explores the concept of function. Here, once again, I'm concerned with a traditional philosophical problem,

"How should we think about functions after Darwin?" but also with clarifying the biological usages. And, once again, my approach is pluralistic, recognizing one notion of function that has intimate connections with evolution and another, typically figuring in molecular and cell-biological studies, where the link to evolutionary concerns is much more remote.

The eighth and ninth essays focus on a problem in theoretical biology that has, for obvious reasons, caught philosophical attention. Biologists think of altruistic behavior as behavior that promotes the reproductive success of the beneficiary at reproductive cost to the agent. They then wonder how altruism can emerge and be maintained in a Darwinian world. Thanks to W. D. Hamilton, John Maynard Smith, Robert Trivers, and Robert Axelrod, there are theoretical ways of addressing the worry. Philosophers have two residual problems. First, can these theoretical devices be applied to the evolutionary history of our own species? Second, can we give a clear account of the kind of altruism that is of interest to philosophy (particularly to moral philosophy) and account for the evolution of that? Chapter 8 tries to approach both questions; chapter 9 (coauthored with John Batali) offers a deeper version of my answer to the first.

For the past three decades, behavioral biologists have been wrestling with problems about culture and cultural transmission. The successes of the mathematical models introduced in connection with the problems of altruism (and elsewhere) have fostered hopes of giving a mathematical account of "cultural inheritance." In chapter 10, I try to lay the groundwork for a formal treatment of cultural transmission, one that favors an "infection" model rather than an "inheritance" approach. Here, as in some of the earlier essays, my main concern is to clear up confusions that have muddied the biological literature.

The first ten essays might easily run parallel to a collection in philosophy of physics. Philosophy of biology, however, has an important dimension that the philosophy of physical science typically lacks. Biological ideas are often introduced into discussions of social issues. With unfortunate frequency, the result is often to generate or to sustain conceptual confusions or mistakes in reasoning that can be socially costly. In my view, the philosophy of biology has the task of attempting to expose such errors. In the last seven essays, although I'm often doing the same kind of philosophical work that figured earlier, I do so in connection with problems and discussions that bear on social issues.

Chapter 11 offers a sober look at the notions of race and ethnicity, trying to understand the extent to which these notions can be reconstructed using the tools of contemporary biology (including the idea of cultural transmission discussed in chapter 10).

In chapter 12, I turn my attention to the Human Genome Project and its social implications, attempting simultaneously to identify the scientific and medical consequences we can expect and apply ideas from contemporary social and political philosophy to appraise them. Chapter 13 considers the idea of genetic determinism—a staple of popular reports about biological findings. Here I endeavor to show why the attack on unwarranted determinist claims must proceed piecemeal, why, in other words, biologists and philosophers of biology have to be continually vigilant in opposing a socially damaging family of myths.

The next three essays extend the critique of my 1985 book on sociobiology, *Vaulting Ambition*. At the end of the 1980s, it seemed that researchers interested in applying Darwinian ideas to human behavior had learned restraint, and that they would no longer repeat the errors I and others had diagnosed. Chapter 14 was my attempt to understand how their enterprise might be pursued carefully and rigorously—although at least one reader suggested that it might be a *reductio* of the venture. In chapter 15, I consider the popular idea of relating biology to morality, distinguishing those views of the relation that are legitimate from those that seem flawed. Although I continue to believe that my critique is correct with respect to the positions actually considered in it, I would now read the conclusion slightly differently: to say that there is much work to be done in relating biology to ethics is not to say that the enterprise is in principle impossible, but only that it must avoid the pitfalls I have tried to expose.

In the 1990s, human sociobiology came back. It returned with a new name—"evolutionary psychology"—and, like the old sociobiology, it came in various grades of respectability. Chapter 16, coauthored with A. Leah Vickers, is a response to some of the most disreputable "work" in evolutionary psychology. We aim to show how recent (quite popular) books by David Buss and by Randy Thornhill and Craig Palmer repeat the old errors, and why the provocative conclusions about sex and violence are far from established.

Like human sociobiology, "creation science" metamorphosed between the 1980s and the 1990s, as savvy theists mined the possibilities of "intelligent design creation." The last essay goes beyond my 1982 book, *Abusing Science*, in considering this Born-Again Creationism. To use a biblical image, it is weighed in the balance and found wanting.

In collecting the essays here, I've avoided making substantive changes. This means that there will occasionally be some anachronisms—as, for example, when I write about the status of the sequencing work in the Human Genome Project. Moreover, attentive readers will note that I sometimes change my mind about various things (for example, note 38 of chapter 1 is belied by chapter 4). It has seemed to me better to let the originals stand, because modifying them would often make the perspective of the essay less easily intelligible. In the case of the anachronisms, simply noting the date of publication makes the judgments comprehensible; for my changes of heart, it may turn out that some people prefer my younger self.

Like much work in the philosophy of biology, this collection contains plenty of references to Darwin. In my title, however, I've chosen to honor the other great nineteenth-century pioneer of contemporary biological theory. I have two reasons for this, both of which have just been bruited. First, just as philosophy of science was the poorer for its exclusion of biology, so too, philosophy of biology has been diminished by insufficient attention to issues in molecular biology, cell biology, and developmental biology. I pay some attention to some of those issues here, and my title serves as a way of reminding myself (as well as my colleagues) to do better. Second, the social impact of biological discussions, central to the themes of the last seven chapters, almost always involves Mendel's great innovation, the notion of the "hereditary factor" or in our terminology, "the gene." Insofar as philosophy of biology is socially relevant, it's because of our tendency to view ourselves in Mendel's mirror.

In writing many of these essays, I've been aided by large numbers of people. My gratitude is recorded in footnotes. Here, I'd like to single out a few who have helped me greatly. I want to thank my three coauthors, John Batali, Kim Sterelny, and Leah Vickers, for their permission to reprint our joint work—and even more for the pleasure of our collaborations. Like other philosophers of biology of my generation and subsequent generations, I'm indebted to the early pioneers; when the subject wasn't even on the radar screen of philosophy of science, Marjorie Grene, Michael Ruse, William Wimsatt, and especially David Hull, showed the way. In my own work, I've appreciated the generous suggestions and wise advice of John Dupré, Alex Rosenberg, and Elliott Sober; their conversations have left an even greater imprint than is recorded in my footnotes. My thanks also to Peter Ohlin for his help, advice, and support.

Finally, many biologists, in many places, have helped me enormously. Among them two in particular stand out. Steve Gould and Dick Lewontin offered me counsel, encouragement, and criticism for more than two decades. When this collection was conceived, I asked them if I might dedicate it to them both, and I was delighted when they agreed. In the intervening year, many people—friends, colleagues, scientists from numerous fields, general readers who thought they hated science, and lots of others—have lost a great inspiration to thinking about and reveling in the many-sidedness of life. So, with a mixture of pleasure and sadness, I offer the following essays to two great mentors, whose influence will readily be visible throughout—to Dick and in memory of Steve.

IN MENDEL'S MIRROR

1

1953 and All That

A Tale of Two Sciences (1984)

Must we geneticists become bacteriologists, physiological chemists and physicists, simultaneously with being zoologists and botanists? Let us hope so.

—H. J. Muller, "Variation Due to Change in the Individual Gene"

1. The Problem

Toward the end of their paper announcing the molecular structure of DNA, James Watson and Francis Crick remark, somewhat laconically, that their proposed structure might illuminate some central questions of genetics.[1] Thirty years have passed since Watson and Crick published their famous discovery. Molecular biology has indeed transformed our understanding of heredity. The recognition of the structure of DNA; the understanding of gene replication, transcription, and translation; the cracking of the genetic code; the study of gene regulation; these and other breakthroughs have combined to answer many of the questions that baffled classical geneticists. Muller's hope—expressed in the early days of classical genetics—has been amply fulfilled.

Yet the success of molecular biology and the transformation of classical genetics into molecular genetics bequeath a philosophical problem. There are two recent theories that have addressed the phenomena of heredity. One, *classical genetics*, stemming from the studies of T. H. Morgan, his colleagues, and students, is the successful outgrowth of the Mendelian theory of heredity rediscovered at the beginning of this century. The other, *molecular genetics*, descends from the work of Watson and Crick. What is the relationship between these two theories? How does the molecular theory illuminate the classical theory? How exactly has Muller's hope been fulfilled?

There used to be a popular philosophical answer to the problem posed in these three connected questions: classical genetics has been reduced to molecular genetics. Philosophers of biology inherited the notion of reduction from general discussions in philosophy of science, discussions that usually center on examples from physics. Unfortunately attempts to apply this notion in the case of genetics have been vulnerable to cogent criticism. Even after considerable tinkering with the concept of reduction, one cannot claim that classical genetics has been (or is being) reduced to molecular genetics.[2] However, the antireductionist point is typically negative.[3] It denies the adequacy of a particular solution to the problem of characterizing the relation between classical genetics and molecular genetics. It does not offer an alternative solution.

My aim in this paper is to offer a different perspective on intertheoretic relations. The plan is to invert the usual strategy. Instead of trying to force the case of genetics into a mold, which is alleged to capture important features of examples in physics, or resting content with denying that the material can be forced, I shall try to arrive at a view of the theories involved and the relations between them that will account for the almost universal idea that molecular biology has done something important for classical genetics. In so doing, I hope to shed some light on the general questions of the structure of scientific theories and the relations which may hold between successive theories. Since my positive account presupposes that something is wrong with the reductionist treatment of the case of genetics, I shall begin with a diagnosis of the foibles of reductionism.

2. What's Wrong with Reductionism?

Ernest Nagel's classic treatment of reduction[4] can be simplified for our purposes. Scientific theories are regarded as sets of statements.[5] To reduce a theory T_2 to a theory T_1, is to deduce the statements of T_2 from the statements of T_1. If there are nonlogical expressions which appear in the statements of T_2, but do not appear in the statements of T_1, then we are allowed to supplement the statements of T_1 with some extra premises connecting the vocabulary of T_1 with the distinctive vocabulary of T_2 (so-called *bridge principles*). Intertheoretic reduction is taken to be important because the statements which are deduced from the reducing theory are supposed to be explained by this deduction.

Yet, as everyone who has struggled with the paradigm cases from physics knows all too well, the reductions of Galileo's law to Newtonian mechanics and of the ideal gas laws to the kinetic theory do not exactly fit Nagel's model. Study of these examples suggests that, to reduce a theory T_2 to a theory T_1, it suffices to deduce the laws of T_2 from a suitably modified version of T_1, possibly augmented with appropriate extra premises.[6] Plainly, this sufficient condition is dangerously vague.[7] I shall tolerate its vagueness, proposing that we understand the issue of reduction in genetics by using the examples from physics as paradigms of what "suitable modifications" and "appropriate extra premises" are like. Reductionists claim that the relation between classical genetics and molecular biology is sufficiently similar to the intertheoretical relations exemplified in the examples from physics to count as the same type of thing: to wit, as intertheoretical reduction.

It may seem that the reductionist thesis has now become so amorphous that it will be immune to refutation. But this is incorrect. Even when we have amended the classical model of reduction so that it can accommodate the examples that originally motivated it, the reductionist claim about genetics requires us to accept three theses:

(R1)   Classical genetics contains general laws about the transmission of genes which can serve as the conclusions of reductive derivations.

(R2)   The distinctive vocabulary of classical genetics (predicates like '① is a gene', '① is dominant with respect to ②') can be linked to the vocabulary of molecular biology by bridge principles.

(R3)   A derivation of general principles about the transmission of genes from principles of molecular biology would explain why the laws of gene transmission hold (to the extent that they do).

I shall argue that each of the theses is false, offering this as my diagnosis of the ills of reductionism.

Before offering my criticisms, it may help to explain why reductionism presupposes (R1)–(R3). If the relation between classical genetics and molecular biology is to be like that between the theory of ideal gases and the kinetic theory (say), then we are going to need to find general principles, identifiable as the central laws of classical genetics, that can serve as the conclusions of reductive derivations. (We need counterparts for the Boyle-Charles law.) These will be general principles about genes, and because classical genetics seems to be a theory about the inheritance of characteristics, the only likely candidates are laws describing the transmission of genes between generations. [So reductionism leads to (R1).] If we are to derive such laws from molecular biology, then there must be bridge principles connecting the distinctive vocabulary figuring in the laws of gene transmission (presumably expressions like '① is a gene', and perhaps '① is dominant with respect to ②') with the vocabulary of molecular biology. [Hence (R2).] Finally, if the derivations are to achieve the goal of intertheoretical reduction then they must explain the laws of gene transmission [(R3)].

Philosophers often identify theories as small sets of general laws. However, in the case of classical genetics, the identification is difficult and those who debate the reducibility of classical genetics to molecular biology often proceed differently. David Hull uses a characterization drawn from Dobzhansky: classical genetics is "concerned with gene differences; the operation employed to discover a gene is hybridization: parents differing in some trait are crossed and the distribution of the trait in hybrid progeny is observed."[8] This is not unusual in discussions of reduction in genetics. It is much easier to identify classical genetics by referring to the subject matter and to the methods of investigation, than it is to provide a few sentences that encapsulate the content of the theory.

Why is this? Because when we read the major papers of the great classical geneticists or when we read the textbooks in which their work is summarized, we find it hard to pick out *any* laws about genes. These documents are full of informative statements. Together, they tell us an enormous amount about the chromosomal arrangement of particular genes in particular organisms, about the effect on the phenotype of various mutations, about frequencies of recombination, and so forth.[9] In some cases, we might explain the absence of formulations of general laws about genes (and even of reference to such laws) by suggesting that these things are common knowledge. Yet that hardly accounts for the nature of the textbooks or of the papers that forged the tools of classical genetics.

If we look back to the pre-Morgan era, we do find two general statements about genes, namely Mendel's Laws (or "Rules"). Mendel's second law states that, in a diploid organism which produces haploid gametes, genes at differenct loci will be transmitted independently; so, for example, if A, a and B, b are pairs of alleles at different loci, and if an organism is heterozygous at both loci, then the probabilities that a gamete will receive any of the four possible genetic combinations, AB, Ab, aB, ab, are all equal.[10] Once it was recognized that genes are (mostly) chromosomal segments (as biologists discovered soon after the rediscovery of Mendel's laws), we understand that the law will not hold in general: alleles which are on the same chromosome (or, more exactly, close together on the same chromosome) will tend to be transmitted together because (ignoring recombination)[11] one member of each homologous pair is distributed to a gamete.[12]

Now it might seem that this is not very important. We could surely find a correct substitute for Mendel's second law by restricting the law so that it only talks about genes on nonhomologous chromosomes. Unfortunately, this will not quite do. There can be interference with normal cytological processes so that segregation of nonhomologous chromosomes need not be independent.[13] However, my complaint about Mendel's second law is not that it is incorrect: many sciences use laws that are clearly recognized as approximations. Mendel's second law, amended or unamended, simply becomes irrelevant to subsequent research in classical genetics.

We envisaged amending Mendel's second law by using elementary principles of cytology, together with the identification of genes as chromosomal segments, to correct what was faulty in the unamended law. It is the fact that the application is so easy and that it can be carried out far more generally that makes the "law" it generates irrelevant. We can understand the transmission of genes by analyzing the cases that interest us from a cytological perspective—by proceeding from "first principles," as it were. Moreover, we can adopt this approach whether the organism is haploid, diploid, or polyploid, whether it reproduces sexually or asexually, whether the genes with which we are concerned are or are not on homologous chromosomes, whether or not there is distortion of independent chromosomal segregation at meiosis. Cytology not only teaches us that the second law is false; it also tells us how to tackle the problem at which the second law was directed (the problem of determining frequencies for pairs of genes in gametes). The amended second law is a restricted statement of results obtainable using a general technique. What figures largely in genetics after Morgan is the technique, and this is hardly surprising when we realize that one of the major research problems of classical genetics has been the problem of discovering the distribution of genes *on the same chromosome*, a problem which is beyond the scope of the amended law.

Let us now turn from (R1) to (R2), assuming, contrary to what has just been argued, that we can identify the content of classical genetics with general principles about gene transmission. (Let us even suppose, for the sake of concreteness, that the principles in question are Mendel's laws—amended in whatever way the reductionist prefers.) To derive these principles from molecular biology, we need a bridge principle. I shall consider first statements of the form

(*) (x) (x is a gene $\leftrightarrow$ Mx)

where 'Mx' is an open sentence (possibly complex) in the language of molecular biology. Molecular biologists do not offer any appropriate statement. Nor do they seem interested in providing one. I claim that no appropriate bridge principle can be found.

Most genes are segments of DNA. (There are some organisms — viruses — whose genetic material is RNA; I shall henceforth ignore them.) Thanks to Watson and Crick, we know the molecular structure of DNA. Hence the problem of providing a statement of the above form becomes that of saying, in molecular terms, which segments of DNA count as genes.

Genes come in different sizes, and for any given size, we can find segments of DNA of that size that are not genes. Therefore genes cannot be identified as segments of DNA containing a particular number of nucleotide pairs. Nor will it do to give a molecular characterization of those codons (triplets of nucleotides) that initiate and terminate transcription, and take a gene to be a segment of DNA between successive initiating and terminating codons. In the first place, mutation might produce a *single* allele containing within it codons for stopping and restarting transcription.[14] Secondly, and much more importantly, the criterion is not general since not every gene is transcribed on mRNA.

The latter point is worth developing. Molecular geneticists recognize regulatory genes as well as structural genes. To cite a classic example, the operator region in the *lac* operon of E. *coli* serves as a site for the attachment of protein molecules, thereby inhibiting transcription of mRNA and regulating enzyme production.[15] Moreover, it is becoming increasingly obvious that genes are not always transcribed, but play a variety of roles in the economy of the cell.[16]

At this point, the reductionist may try to produce a bridge principle by brute force. Trivially, there are only a finite number of terrestrial organisms (past, present and future) and only a finite number of genes. Each gene is a segment of DNA with a particular structure and it would be possible, in principle, to provide a detailed molecular description of that structure. We can now give a molecular specification of the gene by enumerating the genes and disjoining the molecular descriptions.[17] The point made above, that the segments which we count as genes do not share any structural property can now be put more precisely: any instantiation of (*) which replaces 'M' by a structural predicate from the language of molecular biology will insert a predicate that is essentially disjunctive.

Why does this matter? Let us imagine a reductionist using the enumerative strategy to deduce a general principle about gene transmission. After great labor, it is revealed that all actual genes satisfy the principle. I claim that more than this is needed to reduce a *law* about gene transmission. We envisage laws as sustaining counterfactuals, as applying to examples that might have been but which did not actually arise. To reduce the law it is necessary to show how possible but nonactual genes would have satisfied it. Nor can we achieve the reductionist's goal by adding further disjuncts to the envisaged bridge principle. For although there are only finitely many *actual* genes, there are indefinitely many genes which *might* have arisen.

At this point, the reductionist may protest that the deck has been stacked. There is no need to produce a bridge principle of the form (*). Recall that we are trying to derive a general law about the transmission of genes, whose paradigm is Mendel's second law. Now the gross logical form of Mendel's second law is:

(1) $(x) (y) ((Gx \ \& \ Gy) \rightarrow Axy)$.

We might hope to obtain this from statements of the forms

(2) $(x) (Gx \rightarrow Mx)$

(3) $(x) (y) ((Mx \ \& \ My) \rightarrow Axy)$

where 'Mx' is an open sentence in the language of molecular biology. Now there will certainly be true statements of the form (2): for example, we can take 'Mx' as 'x is composed of DNA v.x is composed of RNA'. The question is whether we can combine some such statement with other appropriate premises—for example, some instance of (3)—so as to derive, and thereby explain (1). No geneticist or molecular biologist has advanced any suitable premises, and with good reason. We discover true statements of the form (2) by hunting for weak necessary conditions on genes, conditions that have to be met by genes but which are met by hordes of other biological entities as well. We can only hope to obtain *weak* necessary conditions because of the phenomenon that occupied us previously: from the molecular standpoint, genes are not distinguished by any common structure. Trouble will now arise when we try to show that the weak necessary condition is jointly sufficient for the satisfaction of the property (independent assortment at meiosis) that we ascribe to genes. The difficulty is illustrated by the example given above. If we take 'Mx' to be 'x is composed of DNA v.x is composed of RNA' then the challenge will be to find a general law governing the distribution of all segments of DNA and RNA!

I conclude that (R2) is false. Reductionists cannot find the bridge principles they need, and the tactic of abandoning the form (*) for something weaker is of no avail. I shall now consider (R3). Let us concede both of the points that I have denied, allowing that there are general laws about the transmission of genes and that bridge principles are forthcoming. I claim that exhibiting derivations of the transmission laws from principles of molecular biology and bridge principles would not explain the laws, and, therefore, would not fulfill the major goal of reduction.

As an illustration, I shall use the envisaged amended version of Mendel's second law. Why do genes on nonhomologous chromosomes assort independently? Cytology provides the answer. At meiosis, chromosomes line up with their homologues. It is then possible for homologous chromosomes to exchange some genetic material, producing pairs of recombinant chromosomes. In the meiotic division, one member of each recombinant pair goes to each gamete, and the assignment of one member of one pair to a gamete is probabilistically independent of the assignment of a member of another pair to that gamete. Genes that occur close on the same chromosome are likely to be transmitted together (recombination is not likely to occur between them), but genes on nonhomologous chromosomes will assort independently.

This account is a perfectly satisfactory explanation of why our envisaged law is true to the extent that it is. (We recognize how the law could fail if there were some unusual mechanism linking particular nonhomologous chromosomes.) To emphasize the adequacy of the explanation is not to deny that it could be extended in certain ways. For example, we might want to know more about the mechanics of the process by which the chromosomes are passed on to the gametes. In fact, cytology provides such information. However, appeal to molecular biology would not deepen our understanding of the transmission law. Imagine a successful derivation of the law from principles of chemistry and a bridge principle of the form (*). In charting the details of the molecular rearrangements the derivation would only blur the outline of a simple cytological story, adding a welter of irrelevant detail. Genes on nonhomologous chromosomes assort independently because nonhomologous chromosomes are transmitted independently at meiosis, and so long as we recognize this, we do not need to know what the chromosomes are made of.

In explaining a scientific law, L, one often provides a deduction of L from other principles. Sometimes it is possible to explain some of the principles used in the deduction by deducing them, in turn, from further laws. Recognizing the possibility of a sequence of deductions tempts us to suppose that we could produce a better explanation of L by combining them, producing a more elaborate derivation in the language of our ultimate premises. But this is incorrect. What is relevant for the purposes of giving one explanation may be quite different from what is relevant for the purposes of explaining a law used in giving that original explanation. This general point is illustrated by the case at hand. We begin by asking why genes on nonhomologous chromosomes assort independently. The simple cytological story rehearsed above answers the question. That story generates *further* questions. For example, we might inquire why nonhomologous chromosomes are distributed independently at meiosis. To answer this question we would describe the formation of the spindle and the migration of chromosomes to the poles of the spindle just before meiotic division.[18] Once again, the narrative would generate yet further questions. Why do the chromosomes "condense" at prophase? How is the spindle formed? Perhaps in answering these questions we would begin to introduce the chemical details of the process. Yet simply plugging a molecular account into the narratives offered at the previous stages would *decrease* the explanatory power of those narratives. What is relevant to answering our original question is the fact that nonhomologous chromosomes assort independently. What is relevant to the issue of why nonhomologous chromosomes assort independently is the fact that the chromosomes are not selectively oriented toward the poles of the spindle. (We need to eliminate the doubt that, for example, the paternal and maternal chromosomes become separated and aligned toward opposite poles of the spindle.) In neither case are the molecular details relevant. Indeed, adding those details would only disguise the relevant factor.

There is a natural reductionist response. The considerations of the last paragraphs presuppose far too subjective a view of scientific explanation. After all, even

if *we* become lost in the molecular details, beings who are cognitively more powerful than we could surely recognize the explanatory force of the envisaged molecular derivation. However, this response misses a crucial point. The molecular
derivation forfeits something important.

Recall the original cytological explanation. It accounted for the transmission
of genes by identifying meiosis as a process of a particular kind: a process in which
paired entities (in this case, homologous chromosomes) are separated by a force
so that one member of each pair is assigned to a descendant entity (in this case, a
gamete). Let us call processes of this kind *PS-processes*. I claim first that explaining
the transmission law requires identifying PS-processes as forming a natural kind to
which processes of meiosis belong, and second that PS-processes cannot be identified as a kind from the molecular point of view.

If we adopt the familiar covering law account of explanation, then we shall view
the cytological narrative as invoking a law to the effect that processes of meiosis are
PS-processes and as applying elementary principles of probability to compute the
distribution of genes to gametes from the laws that govern PS-processes. If the illumination provided by the narrative is to be preserved in a molecular derivation,
then we shall have to be able to express the relevant laws as laws in the language
of molecular biology, and this will require that we be able to characterize PS-
processes as a natural kind from the molecular point of view. The same conclusion,
to wit that the explanatory power of the cytological account can be preserved only
if we can identify PS-processes as a natural kind in molecular terms, can be reached
in analogous ways if we adopt quite different approaches to scientific explanation—
for example, if we conceive of explanation as specifying causally relevant properties or as fitting phenomena into a unified account of nature.

However, PS-processes are heterogeneous from the molecular point of view.
There are no constraints on the molecular structures of the entities that are paired
or on the ways in which the fundamental forces combine to pair them and to separate them. The bonds can be forged and broken in innumerable ways: all that
matters is that there be bonds that initially pair the entities in question and that are
subsequently (somehow) broken. In some cases, bonds may be formed directly
between constituent molecules of the entities in question; in others, hordes of accessory molecules may be involved. In some cases, the separation may occur because
of the action of electromagnetic forces or even of nuclear forces; but it is easy to
think of examples in which the separation is effected by the action of gravity. I claim,
therefore, that PS-processes are realized in a motley of molecular ways. (I should
note explicitly that this conclusion is independent of the issue of whether the reductionist can find bridge principles for the concepts of classical genetics.)

We thus obtain a reply to the reductionist charge that we reject the explanatory power of the molecular derivation simply because we anticipate that our brains
will prove too feeble to cope with its complexities.[19] The molecular account objectively fails to explain because it cannot bring out that feature of the situation that
is highlighted in the cytological story. It cannot show us that genes are transmitted
in the ways that we find them to be because meiosis is a PS-process and because
any PS-process would give rise to analogous distributions. Thus (R3)—like (R1) and
(R2)—is false.

3. The Root of the Trouble

Where did we go wrong? Here is a natural suggestion. The most fundamental failure of reductionism is the falsity of (R1). Lacking an account of theories that could readily be applied to the cases of classical genetics and molecular genetics, the attempt to chart the relations between these theories was doomed from the start. If we are to do better, we must begin by asking a preliminary question: what is the structure of classical genetics?

I shall follow this natural suggestion, endeavoring to present a picture of the structure of classical genetics that can be used to understand the intertheoretic relations between classical and molecular genetics.[20] As we have seen, the main difficulty in trying to axiomatize classical genetics is to decide what body of statements one is attempting to axiomatize. The history of genetics makes it clear that Morgan, Muller, Sturtevant, Beadle, McClintock, and others have made important contributions to genetic theory. But the statements occurring in the writings of these workers seem to be far too specific to serve as parts of a general theory. They concern the genes of particular kinds of organisms—primarily paradigm organisms, like fruit flies, bread molds, and maize. The idea that classical genetics is simply a heterogeneous set of statements about dominance, recessiveness, position effect, nondisjunction and so forth, in *Drosophila*, *Zea mays*, *E. coli*, *Neurospora*, etc. flies in the face of our intuitions. The statements advanced by the great classical geneticists seem more like *illustrations* of the theory than *components* of it. (To know classical genetics it is not necessary to know the genetics of any particular organism, not even *Drosophila melanogaster*.) But the only alternative seems to be to suppose that there are general laws in genetics, never enunciated by geneticists but reconstructible by philosophers. At the very least, this supposition should induce the worry that the founders of the field, and those who write the textbooks of today, do a singularly bad job.

Our predicament provokes two main questions. First, if we focus on a particular time in the history of classical genetics, it appears that there will be a set of statements about inheritance in particular organisms, which constitutes the corpus that geneticists of that time accept: what is the relationship between this corpus and the version of classical genetic theory in force at the time? (In posing this question, I assume, contrary to fact, that the community of geneticists was always distinguished by unusual harmony of opinion; it is not hard to relax this simplifying assumption.) Second, we think of genetic theory as something that persisted through various versions: what is the relation among the versions of classical genetic theory accepted at different times (the versions of 1910, 1930, and 1950, for example) that makes us want to count them as versions of the same theory?

We can answer these questions by amending a prevalent conception of the way in which we should characterize the state of a science at a time. The corpus of statements about the inheritance of characteristics accepted at a given time is only one component of a much more complicated entity that I shall call the *practice* of classical genetics at that time. There is a common language used to talk about hereditary phenomena, a set of accepted statements in that language (the corpus of beliefs about inheritance mentioned above), a set of questions taken to be the appropriate questions to ask about hereditary phenomena, and a set of patterns of reasoning which are instantiated in answering some of the accepted questions; (also: sets of experimental procedures and methodological rules, both designed for use in evaluating proposed answers; these may be ignored for present purposes). The practice of classical genetics at a time is completely specified by identifying each of the components just listed.[21]

A pattern of reasoning is a sequence of *schematic sentences*, that is, sentences in which certain items of nonlogical vocabulary have been replaced by dummy letters, together with a set of *filling instructions* which specify how substitutions are to be made in the schemata to produce reasoning that instantiates the pattern.[22] This notion of pattern is intended to explicate the idea of the common structure that underlies a group of problem-solutions.

The foregoing definitions enable us to answer the two main questions I posed above. Beliefs about the particular genetic features of particular organisms illustrate or exemplify the version of genetic theory in force at the time in the sense that these beliefs figure in particular particular problem-solutions generated by the current practice. Certain patterns of reasoning are applied to give the answers to accepted questions, and in making the application, one puts forward claims about inheritance in particular organisms. Classical genetics persists as a single theory with different versions at different times in the sense that different practices are linked by a chain of practices along which there are relatively small modifications in language, in accepted questions, and in the patterns for answering questions. In addition to this condition of historical connection, versions of classical genetic theory are bound by a common structure: each version uses certain expressions to characterize hereditary phenomena, accepts as important questions of a particular form, and offers a general style of reasoning for answering those questions. Specifically, throughout the career of classical genetics, the theory is directed toward answering questions about the distribution of characteristics in successive generations of a genealogy, and it proposes to answer those questions by using the probabilities of chromosome distribution to compute the probabilities of descendant genotypes.

The approach to classical genetics embodied in these answers is supported by reflection on what beginning students learn. Neophytes are not taught (and never have been taught) a few fundamental theoretical laws from which genetic "theorems" are to be deduced. They are introduced to some technical terminology, which is used to advance a large amount of information about special organisms. Certain questions about heredity in these organisms are posed and answered. Those who understand the theory are those who know what questions are to be asked about hitherto unstudied examples, who know how to apply the technical language to the organisms involved in these examples, and who can apply the patterns of reasoning that are to be instantiated in constructing answers. More simply, successful students grasp general patterns of reasoning which they can use to resolve new cases.

I shall now add some detail to my sketch of the structure of classical genetics, and thereby prepare the way for an investigation of the relations between classical genetics and molecular genetics. The initial family of problems in classical genetics, the family from which the field began, is the family of *pedigree problems*. Such problems arise when we confront several generations of organisms, related by specified connections of descent, with a given distribution of one or more characteristics. The question that arises may be to understand the given distribution of phenotypes, or to predict the distribution of phenotypes in the next generation, or to specify the probability that a particular phenotype will result from a particular mating. In general, classical genetic theory answers such questions by making hypotheses about the relevant genes, their phenotypic effects, and their distribution among the individuals in the pedigree. Each version of classical genetic theory contains one or more problem-solving patterns exemplifying this general idea, but the detailed character of the pattern is refined in later versions, so that previously recalcitrant cases of the problem can be accommodated.

Each case of a pedigree problem can be characterized by a set of *data*, a set of *constraints*, and a question. In any example, the data are statements describing the distribution of phenotypes among the organisms in a particular pedigree, or a diagram conveying the same information. The level of detail in the data may vary widely: at one extreme we may be given a full description of the interrelationships among all individuals and the sexes of all those involved; or the data may only provide the numbers of individuals with specific phenotypes in each generation; or, with minimal detail, we may simply be told that from crosses among individuals with specified phenotypes a certain range of phenotypes is found.

The constraints on the problem consist of general cytological information and descriptions of the chromosomal constitution of members of the species. The former will include the thesis that genes are (almost always)[23] chromosomal segments and the principles that govern meiosis. The latter may contain a variety of statements. It may be pertinent to know how the species under study reproduces, how sexual dimorphism is reflected at the chromosomal level, the chromosome number typical of the species, what loci are linked, what the recombination frequencies are, and so forth. As in the case of the data, the level of detail (and thus of stringency) in the constraints can vary widely.

Lastly, each problem contains a question that refers to the organisms described in the data. The question may take several forms: "What is the expected distribution of phenotypes from a cross between *a* and *b*?" (where *a*, *b* are specified individuals belonging to the pedigree described by the data), "What is the probability that a cross between *a* and *b* will produce an individual having P?" (where *a*, *b* are specified individuals of the pedigree described by the data and P is a phenotypic property manifested in this pedigree), "Why do we find the distribution of phenotypes described in the data?" and others.

Pedigree problems are solved by advancing pieces of reasoning that instantiate a small number of related patterns. In all cases the reasoning begins from a *genetic hypothesis*. The function of a genetic hypothesis is to specify the alleles that are relevant, their phenotypic expression, and their transmission through the pedigree. From that part of the genetic hypothesis that specifies the genotypes of the parents in any mating that occurs in the pedigree, together with the constraints on the problem, one computes the expected distribution of genotypes among the offspring. Finally, for any mating occurring in the pedigree, one shows that the expected distribution of genotypes among the offspring is consistent with the assignment of genotypes given by the genetic hypothesis.

The form of the reasoning can easily be recognized in examples—examples that are familiar to anyone who has ever looked at a textbook or a research report in genetics.[24] What interests me is the style of reasoning itself. The reasoning begins with a genetic hypothesis that offers four kinds of information: (a) specification of the number of relevant loci and the number of alleles at each locus; (b) specification of the relationships between genotypes and phenotypes; (c) specification of the relations between genes and chromosomes, of facts about the transmission of chromosomes to gametes (for example, resolution of the question whether there is disruption of normal segregation) and about the details of zygote formation; (d) assignment of genotypes to individuals in the pedigree. After showing that the genetic hypothesis is consistent with the data and constraints of the problem, the principles of cytology and the laws of probability are used to compute expected distributions of genotypes from crosses. The expected distributions are then compared with those assigned in part (d) of the genetic hypothesis.[25]

Throughout the career of classical genetics, pedigree problems are addressed and solved by carrying out reasoning of the general type just indicated. Each version of classical genetic theory contains a pattern for solving pedigree problems with a method for computing expected genotypes that is adjusted to reflect the particular form of the genetic hypotheses that it sanctions. Thus one way to focus the differences among successive versions of classical genetic theory is to compare their conceptions of the possibilities for genetic hypotheses. As genetic theory develops, there is a changing set of conditions on admissible genetic hypotheses. Prior to the discovery of polygeny and pleiotropy (for example), part (a) of any adequate genetic hypothesis was viewed as governed by the requirement that there would be a one-one correspondence between loci and phenotypic traits.[26] After the discovery of incomplete dominance and epistasis, it was recognized that part (b) of an adequate hypothesis might take a form that had not previously been allowed: one is not compelled to assign to the heterozygote a phenotype assigned to one of the homozygotes, and one is also permitted to relativize the phenotypic effect of a gene to its genetic environment.[27] Similarly, the appreciation of phenomena of linkage, recombination, nondisjunction, segregation distortion, meiotic drive, unequal crossing over, and crossover suppression, modify conditions previously imposed on part (c) of any genetic hypothesis. In general, we can take each version of classical genetic theory to be associated with a set of conditions (usually not formulated explicitly) that govern admissible genetic hypotheses. While a general form of reasoning persists through the development of classical genetics, the patterns of reasoning used to resolve cases of the pedigree problem are constantly fine-tuned as geneticists modify their views about what forms of genetic hypothesis are allowable.

So far I have concentrated exclusively on classical genetic theory as a family of related patterns of reasoning for solving the pedigree problem. It is natural to ask if versions of the theory contain patterns of reasoning for addressing other questions. I believe that they do. The heart of the theory is the theory of *gene transmission*, the family of reasoning patterns directed at the pedigree problem. Out of this theory grow other subtheories. The theory of *gene mapping* offers a pattern of reasoning that addresses questions about the relative positions of loci on chromosomes. It is a direct result of Sturtevant's insight that one can systematically investigate the set of pedigree problems associated with a particular species. In turn, the theory of gene mapping raises the question of how to identify mutations, issues that are to be tackled by the *theory of mutation*. Thus we can think of classical genetics as having a central theory, the theory of gene transmission, which develops in the ways I have described above, surrounded by a number of satellite theories that are directed at questions arising from the pursuit of the central theory. Some of these satellite theories (for example, the theory of gene mapping) develop in the same continuous fashion. Others, like the theory of mutation, are subject to rather dramatic shifts in approach.

4. Molecular Genetics and Classical Genetics

Armed with some understanding of the structure and evolution of classical genetics, we can finally return to the question with which we began. What is the relation between classical genetics and molecular genetics? When we look at textbook presentations and the pioneering research articles that they cite, it is not hard to discern major ways in which molecular biology has advanced our understanding of hereditary phenomena. We can readily identify particular molecular explanations which illuminate issues that were treated incompletely, if at all, from the classical perspective. What proves puzzling is the connection of these explanations to the theory of classical genetics. I hope that the account of the last section will enable us to make the connection.

I shall consider three of the most celebrated achievements of molecular genetics. Consider first the question of *replication*. Classical geneticists believed that genes can replicate themselves. Even before the experimental demonstration that all genes are transmitted to all the somatic cells of a developing embryo, geneticists agreed that normal processes of mitosis and meiosis must involve gene replication. Muller's suggestion that the central problem of genetics is to understand how mutant alleles, incapable of performing wild-type functions in producing the phenotype, are nonetheless able to replicate themselves, embodies this consensus. Yet classical genetics had no account of gene replication. A molecular account was an almost immediate dividend of the Watson-Crick model of DNA.

Watson and Crick suggested that the two strands of the double helix unwind and each strand serves as the template for the formation of a complementary strand. Because of the specificity of the pairing of nucleotides, reconstruction of DNA can be unambiguously directed by a single strand. This suggestion has been confirmed and articulated by subsequent research in molecular biology.[28] The details are more intricate than Watson and Crick may originally have believed, but the outline of their story stands.

A second major illumination produced by molecular genetics concerns the characterization of mutation. When we understand the gene as a segment of DNA we recognize the ways in which mutant alleles can be produced. "Copying errors" during replication can cause nucleotides to be added, deleted, or substituted. These changes will often lead to alleles that code for different proteins, and which are readily recognizable as mutants through their production of deviant phenotypes. However, molecular biology makes it clear that there can be *hidden* mutations, mutations that arise through nucleotide substitutions that do not change the protein produced by a structural gene (the genetic code is redundant) or through substitutions that alter the form of the protein in trivial ways. The molecular perspective provides us with a general answer to the question, "What is a mutation?" namely that a mutation is the modification of a gene through insertion, deletion, or substitution of nucleotides. This general answer yields a basic method for tackling (in principle) questions of form, "Is *a* a mutant allele?" namely a demonstration that *a* arose through nucleotide changes from alleles that persist in the present population. The method is frequently used in studies of the genetics of bacteria and bacteriophage, and can sometimes be employed even in inquiries about more complicated organisms. So, for example, there is good biochemical evidence for believing that some alleles that produce resistance to pesticides in various species of insects arose through nucleotide changes in the alleles naturally predominating in the population.[29]

I have indicated two general ways in which molecular biology answers questions that were not adequately resolved by classical genetics. Equally obvious are a large number of more specific achievements. Identification of the molecular structures of particular genes in particular organisms has enabled us to understand why those genes combine to produce the phenotypes they do. One of the most celebrated cases is that of the normal allele for the synthesis of human hemoglobin and the mutant allele that is responsible for sickle-cell anemia.[30] The hemoglobin molecule—whose structure is known in detail—is built up from four amino-acid chains (two "α-chains" and two "β-chains"). The mutant allele results from substitution of a single nucleotide with the result that one amino acid is different (the sixth amino acid in the β-chains). This slight modification causes a change in the interactions of hemoglobin molecules: deoxygenated mutant hemoglobin molecules combine to form long fibres. Cells containing the abnormal molecule become deformed after they have given up their oxygen, and because they become rigid, they can become stuck in narrow capillaries, if they give up their oxygen too soon. Individuals who are homozygous for the mutant gene are vulnerable to experience blockages of blood flow. However, in heterozygous individuals, there is enough normal hemoglobin in blood cells to delay the time of formation of the distorting fibres, so that the individual is physiologically normal.

This example is typical of a broad range of cases, among which are some of the most outstanding achievements of molecular genetics. In all of the cases, we replace a simple assertion about the existence of certain alleles which give rise to various phenotypes with a molecular characterization of those alleles from which we can derive descriptions of the phenotypes previously attributed.

I claim that the successes of molecular genetics that I have just briefly described—and which are among the accomplishments most emphasized in the biological literature—can be understood from the perspective on theories that I have developed above. The three examples reflect three different relations among successive theories, all of which are different from the classical notion of reduction (and the usual modifications of it). Let us consider them in turn.

The claim that genes can replicate does not have the status of a central law of classical genetic theory.[31] It is not something that figures prominently in the explanations provided by the theory (as, for example, the Boyle-Charles law is a prominent premise in some of the explanations yielded by phenomenological thermodynamics). Rather, it is a claim that classical geneticists took for granted, a claim presupposed by explanations, rather than an explicit part of them. Prior to the development of molecular genetics that claim had come to seem increasingly problematic. If genes can replicate, how do they manage to do it? Molecular genetics answered the worrying question. It provided a theoretical demonstration of the possibility of an antecedently problematic presupposition of classical genetics.

We can say that a theory presupposes a statement p if there is some problem-solving pattern of the theory, such that every instantiation of the pattern contains statements that jointly imply the truth of p. Suppose that, at a given stage in the development of a theory, scientists recognize an argument from otherwise acceptable premises which concludes that it is impossible that p. Then the presupposition p is problematic for those scientists. What they would like would be an argument showing that it is possible that p and explaining what is wrong with the line of reasoning which appears to threaten the possibility of p. If a new theory generates an argument of this sort, then we can say that the new theory gives a theoretical demonstration of the possibility of an antecedently problematic presupposition of the old theory.

A less abstract account will help us to see what is going on in the case of gene replication. Very frequently, scientists take for granted in their explanations some general property of entities that they invoke. Their assumption can come to seem problematic if the entities in question are supposed to belong to a kind, and there arises a legitimate doubt about whether members of the kind can have the property attributed. A milder version of the problem arises if, in all cases in which the question of whether things of the general kind have the property can be settled by appealing to background theory, it turns out that the answer is negative. Under these circumstances, the scientists are committed to regarding their favored entities as unlike those things of the kind which are amenable to theoretical study with respect to the property under discussion. The situation is worse if background theory provides an argument for thinking that *no* things of the kind can have the property.

Consider now the case of gene replication. For any problem-solution offered by any version of the theory of gene transmission (the central subtheory of classical genetic theory), that problem-solution will contain sentences implying that the alleles which it discusses are able to replicate. Classical genetics presupposes that a large number of identifiable genes can replicate. This presupposition was always weakly problematic because genes were taken to be complicated molecules and, in all cases in which appeal to biochemistry could be made to settle the issue of whether a molecular structure was capable of replication, the issue was decided in the negative. Muller exacerbated the problem by suggesting that mutant alleles are damaged molecules (after all, many of them were produced through x-ray bombardment, an extreme form of molecular torture). So there appeared to be a strong argument against the possibility that any mutant allele can replicate. After the work of Watson, Crick, Kornberg, and others, there was a theoretical demonstration of the allegedly problematic possibility. One can show that genes can replicate by showing that any segment of DNA (or RNA) can replicate. (DNA and RNA are the genetic materials. Establishing the power of the genetic material to replicate bypasses the problem of deciding which segments are genes. Thus the difficulties posed by the falsity of [R2] are avoided.) The Watson-Crick model provides a characterization of the (principal) genetic material, and when this description is inserted into standard patterns of chemical reasoning one can generate an argument whose conclusion asserts that, under specified conditions, DNA replicates. Moreover, given the molecular characterization of DNA and of mutation, it is possible to see that although mutant alleles are "damaged" molecules, the kind of damage (insertion, deletion, or substitution of nucleotides) does not affect the ability of the resultant molecule to replicate.

Because theoretical demonstrations of the possibility of antecedently problematic presuppositions involve derivation of conclusions of one theory from the premises supplied by a background theory, it is easy to assimilate them to the classical notion of reduction. However, on the account I have offered, there are two important differences. First, there is no commitment to the thesis that genetic theory can be formulated as (the deductive closure of) a conjunction of laws. Second, it is not assumed that all general statements about genes are equally in need of molecular derivation. Instead, one particular thesis, a thesis that underlies all the explanations provided by classical genetic theory, is seen as especially problematic, and the molecular derivation is viewed as addressing a specific problem that classical geneticists had already perceived. Where the reductionist identifies a general benefit in deriving all the axioms of the reduced theory, I focus on a particular derivation of a claim that has no title as an axiom of classical genetics, a derivation that responds to a particular explanatory difficulty of which classical geneticists were acutely aware. The reductionist's global relation between theories does not obtain between classical and molecular genetics, but something akin to it does hold between special fragments of these theories.[32]

The second principal achievement of molecular genetics, the account of mutation, involves a conceptual refinement of prior theory. Later theories can be said to provide conceptual refinements of earlier theories when the later theory yields a specification of entities that belong to the extensions of predicates in the language of the earlier theory, with the result that the ways in which the referents of these predicates are fixed are altered in accordance with the new specifications. Conceptual refinement may occur in a number of ways. A new theory may supply a descriptive characterization of the extension of a predicate for which no descriptive characterization was previously available; or it may offer a new description which makes it reasonable to amend characterizations that had previously been accepted.[33] In the case at hand, the referent of many tokens of 'mutant allele' was initially fixed through the description "chromosomal segment producing a heritable deviant phenotype." After Bridges's discovery of unequal crossing-over at the *Bar* locus in *Drosophila*, it was evident to classical geneticists that this descriptive specification covered cases in which the internal structure of a gene was altered and cases in which neighboring genes were transposed. Thus it was necessary to retreat to the less applicable description "chromosomal segment producing a heritable deviant phenotype as the result of an internal change within an allele." Molecular genetics offers a precise account of the internal changes, with the result that the description can be made more informative: mutant alleles are segments of DNA that result from prior alleles through deletion, insertion, or substitution of nucleotides. This refixing of the referent of *mutant allele* makes it possible in principle to distinguish cases of mutation from cases of recombination, and thus to resolve those controversies that frequently arose from the use of *mutant allele* in the later days of classical genetics.[34]

Finally, let us consider the use of molecular genetics to illuminate the action of particular genes. Here we again seem to find a relationship that initially appears close to the reductionist's ideal. Statements that are invoked as premises in particular problem-solutions—statements that ascribe particular phenotypes to particular genotypes—are derived from molecular characterizations of the alleles involved. On the account of classical genetics offered in section 3 of this chapter, each version of classical genetic theory includes in its schema for genetic hypotheses a clause which relates genotypes to phenotypes (clause [b] in the description of a genetic hypothesis on p. 14). Generalizing from the hemoglobin example, we might hope to discover a pattern of reasoning within molecular genetics that would generate as its conclusion the schema for assigning phenotypes to genotypes.

It is not hard to characterize the relation just envisioned. Let us say that a theory T' provides an *explanatory extension* of a theory T just in case there is some problem-solving pattern of T one of whose schematic premises can be generated as the conclusion of a problem-solving pattern of T'. When a new theory provides an explanatory extension of an old theory, then particular premises occurring in explanatory derivations given by the old theory can themselves be explained by using arguments furnished by the new theory. However, it does not follow that the explanations provided by the old theory can be improved by replacing the premises in question with the pertinent derivations. What is relevant for the purposes of explaining some statement S may not be relevant for the purposes of explaining a statement S' which figures in an explanatory derivation of S.

Even though reductionism fails, it may appear that we can capture part of the spirit of reductionism by deploying the notion of explanatory extension. The thesis that molecular genetics provides an explanatory extension of classical genetics embodies the idea of a global relationship between the two theories, while avoiding two of the three troubles that were found to beset reductionism. That thesis does not simply assert that some specific presupposition of classical genetics (for example, the claim that genes are able to replicate) can be derived as the conclusion of a molecular argument, but offers a general connection between premises of explanatory derivations in classical genetics and explanatory arguments from molecular genetics. It is formulated so as to accommodate the failure of (R1) and to honor the picture of classical genetics developed in section 3. Moreover, the failure of (R2) does not affect it. If we take the hemoglobin example as a paradigm, we can justifiably contend that the explanatory extension does not require any general characterization of genes in molecular terms. All that is needed is the possibility of deriving phenotypic descriptions from molecular characterizations of the structures of *particular* genes. Thus, having surmounted two hurdles, our modified reductionist thesis is apparently within sight of success.

Nevertheless, even Born-Again Reductionism is doomed to fall short of salvation. Although it is true that molecular genetics belongs to a cluster of theories which, taken together, provide an explanatory extension of classical genetics, molecular genetics, on its own, cannot deliver the goods. There are some cases in which the ancillary theories do not contribute to the explanation of a classical claim about gene action. In such cases, the classical claim can be derived and explained by instantiating a pattern drawn from molecular genetics. The example of human hemoglobin provides one such case. But this example is atypical.

Consider the way in which the hemoglobin example works. Specification of the molecular structures of the normal and mutant alleles, together with a description of the genetic code, enables us to derive the composition of normal and mutant hemoglobin. Application of chemistry then yields descriptions of the interactions of the proteins. With the aid of some facts about human blood cells, one can then deduce that the sickling effect will occur in abnormal cells, and given some facts about human physiology, it is possible to derive the descriptions of the phenotypes. There is a clear analogy here with some cases from physics. The assumptions about blood cells and physiological needs seem to play the same role as the boundary conditions about shapes, relative positions, and velocities of planets that occur in Newtonian derivations of Kepler's laws. In the Newtonian explanation we can see the application of a general pattern of reasoning—the derivation of explicit equations of motion from specifications of the forces acting—which yields the general result that a body under the influence of a centrally directed inverse square force will travel in a conic section; the general result is then applied to the motions of the planets by incorporating pieces of astronomical information. Similarly, the derivation of the classical claims about the action of the normal and mutant hemoglobin genes can be seen as a purely chemical derivation of the generation of certain molecular structures and of the interactions among them. The chemical conclusions are then applied to the biological system under consideration by introducing three "boundary conditions": first, the claim that the altered molecular structures only affect development to the extent of substituting a different molecule in the erythrocytes (the blood cells that transport hemoglobin); second, a description of the chemical conditions in the capillaries; and third, a description of the effects upon the organism of capillary blockage.

The example is able to lend comfort to reductionism precisely because of an atypical feature. In effect, one concentrates on the *differences* among the phenotypes, takes for granted the fact that in all cases development will proceed normally to the extent of manufacturing erythrocytes—which are, to all intents and purposes, simply sacks for containing hemoglobin molecules—and compares the difference in chemical effect of the cases in which the erythrocytes contain different molecules. *The details of the process of development can be ignored.* However, it is rare for the effect of a mutation to be so simple. Most structural genes code for mole-

cules whose presence or absence make subtle differences. Thus, typically, a mutation will affect the distribution of chemicals in the cells of a developing embryo. A
likely result is a change in the timing of intracellular reactions, a change that may,
in turn, alter the shape of the cell. Because of the change of shape, the geometry
of the embryonic cells may be modified. Cells that usually come into contact may
fail to touch. Because of this, some cells may not receive the molecules necessary
to switch on certain batteries of genes. Hence the chemical composition of these
cells will be altered. And so it goes.[35]

Quite evidently, in examples like this (which include most of the cases in
which molecular considerations can be introduced into embryology) the reasoning
that leads us to a description of the phenotype associated with a genotype will be
much more complicated than that found in the hemoglobin case. It will not simply
consist in a chemical derivation adapted with the help of a few boundary conditions furnished by biology. Instead, we shall encounter a sequence of subarguments:
molecular descriptions lead to specifications of cellular properties, from these
specifications we draw conclusions about cellular interactions, and from these conclusions we arrive at further molecular descriptions. There is clearly a pattern of
reasoning here that involves molecular biology and which extends the explanations
furnished by classical genetics by showing how phenotypes depend upon genotypes—but I think it would be folly to suggest that the extension is provided by
molecular genetics alone.

In section 2, we discovered that the traditional answer to the philosophical question of understanding the relation that holds between molecular genetics and classical genetics, the reductionist's answer, will not do. Section 3 attempted to build
on the diagnosis of the ills of reductionism, offering an account of the structure and
evolution of classical genetics that would improve on the picture offered by those
who favor traditional approaches to the nature of scientific theories. In the present
section, I have tried to use the framework of section 3 to understand the relations
between molecular genetics and classical genetics. Molecular genetics has done
something important for classical genetics, and its achievements can be recognized
by seeing them as instances of the intertheoretic relations that I have characterized.
Thus I claim that the problem from which we began is solved.

So what? Do we have here simply a study of a particular case—a case which
has, to be sure, proved puzzling for the usual accounts of scientific theories and
scientific change? I hope not. Although the traditional approaches may have
proved helpful in understanding some of the well-worn examples that have
been the stockin-trade of twentieth-century philosophy of science, I believe that
the notion of scientific practice sketched in section 3 and the intertheoretic
relations briefly characterized here will prove helpful in analyzing the structure
of science and the growth of scientific knowledge *even in those areas of science
where traditional views have seemed most successful.*[36] Hence the tale of two sciences
which I have been telling is not merely intended as a piece of local history that fills
a small but troublesome gap in the orthodox chronicles. I hope that it introduces
concepts of general significance in the project of understanding the growth of
science.

5. Antireductionism and the Organization of Nature

One loose thread remains. The history of biology is marked by continuing opposition between reductionists and antireductionists. Reductionism thrives on exploiting the charge that it provides the only alternative to the mushy incomprehensibility of vitalism. Antireductionists reply that their opponents have ignored the organismic complexity of nature. Given the picture painted above, where does this traditional dispute now stand?

I suggest that the account of genetics that I have offered will enable reductionists to provide a more exact account of what they claim, and will thereby enable antireductionists to be more specific about what they are denying. Reductionists and antireductionists agree in a certain minimal physicalism. To my knowledge, there are no major figures in contemporary biology who dispute the claim that each biological event, state, or process is a complex physical event, state, or process. The most intricate part of ontogeny or phylogeny involves countless changes of physical state. What antireductionists emphasize is the organization of nature and the "interactions among phenomena at different levels." The appeal to organization takes two different forms. When the subject of controversy is the proper form of evolutionary theory, then antireductionists contend that it is impossible to regard all selection as operating at the level of the gene.[37] What concerns me here is not this area of conflict between reductionists and their adversaries, but the attempt to block claims for the hegemony of molecular studies in understanding the physiology, genetics, and development of organisms.[38]

A sophisticated reductionist ought to allow that, in the current practice of biology, nature is divided into levels that form the proper provinces of areas of biological study: molecular biology, cytology, histology, physiology, and so forth. Each of these sciences can be thought of as using certain language to formulate the questions it deems important and as supplying patterns of reasoning for resolving those questions. Reductionists can now set forth one of two main claims. The stronger thesis is that the explanations provided by any biological theories can be reformulated in the language of molecular biology and be recast so as to instantiate the patterns of reasoning supplied by molecular biology. The weaker thesis is that molecular biology provides explanatory extension of the other biological sciences.

Strong reductionism falls victim to the considerations that were advanced against (R3). The distribution of genes to gametes is to be explained, not by rehearsing the gory details of the reshuffling of the molecules, but through the observation that chromosomes are aligned in pairs just prior to the meiotic division, and that one chromosome from each matched pair is transmitted to each gamete. We may formulate this point in the biologists' preferred idiom by saying that the assortment of alleles is to be understood at the cytological level. What is meant by this description is that there is a pattern of reasoning that is applied to derive the description of the assortment of alleles and which involves predicates that characterize cells and their large-scale internal structures. That pattern of reasoning is to be objectively preferred to the molecular pattern which would be instantiated by the derivation that charts that complicated rearrangements of individual molecules because it can be applied across a range of cases which would look heterogeneous from a molecular perspective. Intuitively, the cytological pattern makes connections that are lost at the molecular level, and it is thus to be preferred.

So far, antireductionism emerges as the thesis that there are *autonomous levels of biological explanation*. Antireductionism construes the current division of biology not simply as a temporary feature of our science stemming from our cognitive imperfections but as the reflection of levels of organization in nature. Explanatory patterns that deploy the concepts of cytology will endure in our science because we would foreswear significant unification (or fail to employ the relevant laws, or fail to identify the causally relevant properties) by attempting to derive the conclusions to which they are applied using the vocabulary and reasoning patterns of molecular biology. But the autonomy thesis is only the beginning of antireductionism. A stronger doctrine can be generated by opposing the weaker version of sophisticated reductionism.

In section 4, I raised the possibility that molecular genetics may be viewed as providing an explanatory extension of classical genetics through deriving the schematic sentence that assigns phenotypes to genotypes from a molecular pattern of reasoning. This apparent possibility fails in an instructive way. Antireductionists are not only able to contend that there are autonomous levels of biological explanation. They can also resist the weaker reductionist view that explanation always flows from the molecular level up. Even if reductionists retreat to the modest claim that, while there are autonomous levels of explanation, descriptions of cells and their constituents are always explained in terms of descriptions about genes, descriptions of tissue geometry are always explained in terms of descriptions of cells, and so forth, antireductionists can resist the picture of a unidirectional flow of explanation. Understanding the phenotypic manifestation of a gene, they will maintain, requires constant shifting back and forth across levels. Because developmental processes are complex and because changes in the timing of embryological events may produce a cascade of effects at several different levels, one sometimes uses descriptions at higher levels to explain what goes on at a more fundamental level.

For example, to understand the phenotype associated with a mutant limb-bud allele, one may begin by tracing the tissue geometry to an underlying molecular structure. The molecular constitution of the mutant allele gives rise to a nonfunctional protein, causing some abnormality in the internal structures of cells. The abnormality is reflected in peculiarities of cell shape, which, in turn, affects the spatial relations among the cells of the embryo. So far we have the unidirectional flow of explanation which the reductionist envisages. However, the subsequent course of the explanation is different. Because of the abnormal tissue geometry, cells that are normally in contact fail to touch; because they do not touch, certain important molecules, which activate some batteries of genes, do not reach crucial cells; because the genes in question are not "switched on" a needed morphogen is not produced; the result is an abnormal morphology in the limb.

Reductionists may point out, quite correctly, that there is some very complex molecular description of the entire situation. The tissue geometry is, after all, a configuration of molecules. But this point is no more relevant than the comparable claim about the process of meiotic division in which alleles are distributed to gametes. Certain genes are not expressed because of the geometrical structure of the cells in the tissue: *the pertinent cells are too far apart*. However this is realized at the molecular level, our explanation must bring out the salient fact that it is the presence of a gap between cells that are normally adjacent that explains the non-expression of the genes. As in the example of allele transmission at meiosis, we lose sight of the important connections by attempting to treat the situation from a molecular point of view. As before, the point can be sharpened by considering situations in which radically different molecular configurations realize the crucial feature of the tissue geometry: situations in which heterogeneous molecular structures realize the breakdown of communication between the cells.

Hence, embryology provides support for the stronger antireductionist claim. Not only is there a case for the thesis of autonomous levels of explanation, but we find examples in which claims at a more fundamental level (specifically, claims about gene expression) are to be explained in terms of claims at a less fundamental level (specifically, descriptions of the relative positions of pertinent cells). Two antireductionist biologists put the point succinctly:

> A developmental program is not to be viewed as a linearly organized causal chain from genome to phenotype. Rather, morphology emerges as a consequence of an increasingly complex dialogue between cell populations, characterized by their geometric continuities, and the cells' genomes, characterized by their states of gene activity.[39]

A corollary is that the explanations provided by the "less fundamental" biological sciences are not extended by molecular biology alone.

It would be premature to claim that I have shown how to reformulate the antireductionist appeals to the organization of nature in a completely precise way. My conclusion is that, to the extent that we can make sense of the present explanatory structure within biology—that division of the field into subfields corresponding to levels of organization in nature—we can also understand the antireductionist doctrine. In its minimal form, it is the claim that the commitment to several explanatory levels does not simply reflect our cognitive limitations; in its stronger form, it is the thesis that some explanations oppose the direction of preferred reductionistic explanation. Reductionists should not dismiss these doctrines as incomprehensible mush unless they are prepared to reject as unintelligible the biological strategy of dividing the field (a strategy which seems to me well understood, even if unanalyzed).

The examples I have given seem to support both antireductionist doctrines. To clinch the case, further analysis is needed. The notion of explanatory levels obviously cries out for explication, and it would be illuminating to replace the informal argument that the unification of our beliefs is best achieved by preserving multiple explanatory levels with an argument based on a more exact criterion for unification. Nevertheless, I hope that I have said enough to make plausible the view that, despite the immense value of the molecular biology that Watson and Crick launched in 1953, molecular studies cannot cannibalize the rest of biology. Even if geneticists must become "physiological chemists" they should not give up being embryologists, physiologists, and cytologists.

Notes

Earlier versions of this essay were read at Johns Hopkins University and at the University of Minnesota, and I am very grateful to a number of people for comments and suggestions. In particular, I would like to thank Peter Achinstein, John Beatty, Barbara Horan, Patricia Kitcher, Richard Lewontin, Kenneth Schaffner, William Wimsatt, an anonymous reader, and the editors of the *Philosophical Review*, all of whom have had an important influence on the final version. Needless to say, these people should not be held responsible for residual errors. I am also grateful to the American Council of Learned Societies and the Museum of Comparative Zoology at Harvard University for support and hospitality while I was engaged in research on the topics of this essay.

1. "Molecular Structure of Nucleic Acids," *Nature* 171 (1953): 737–738; reprinted in Peters, *Classic Papers*, 241–243. Watson and Crick amplified their suggestion in "Genetic Implications of the Structure of Deoxyribonucleic Acid," *Nature* 171 (1953): 934–937.

2. The most sophisticated attempts to work out a defensible version of reductionism occur in articles by Kenneth Schaffner. See, in particular, Schaffner, "Approaches to Reduction," *Philosophy of Science* 34 (1967): 137–147; Schaffner, "The Watson-Crick Model and Reductionism," *British Journal for the Philosophy of Science* 20 (1969): 325–348; Schaffner, "The Peripherality of Reductionism in the Development of Molecular Biology," *Journal of the History of Biology* 7 (1974): 111–139; and Schaffner, "Reductionism in Biology: Prospects and Problems," in R. S. Cohen et al. eds., *PSA 1974* (Boston: Reidel, 1976), 613–632. See also Michael Ruse, "Reduction, Replacement, and Molecular Biology," *Dialectica* 25 (1971): 38–72; and William K. Goosens, "Reduction by Molecular Genetics," *Philosophy of Science* 45 (1978): 78–95. A variety of antireductionist points are made in David Hull, "Reduction in Genetics—Biology or Philosophy?" *Philosophy of Science* 39 (1972): 491–499; Hull, chapter 1 of *Philosophy of Biological Science* (Englewood Cliffs, N.J.: Prentice Hall, 1974); Steven Orla Kimbrough, "On the Reduction of Genetics to Molecular Biology," *Philosophy of Science* 46 (1979): 389–406; and Ernst Mayr, *The Growth of Biological Thought* (Cambridge, Mass.: Harvard University Press, 1982), 59–63.

3. Typically, though not invariably. In a suggestive essay, "Reductive Explanation: A Functional Account" (in Cohen et al., *PSA 1974*, 671–710), William Wimsatt offers a number of interesting ideas about intertheoretic relations and the case of genetics. Also provocative are Nancy Maull's "Unifying Science without Reduction," *Studies in the History and Philosophy of Science* 8 (1977): 143–171; and Lindley Darden and Nancy Maull, "Interfield Theories," *Philosophy of Science* 44 (1977): 43–64. My chief complaint about the works I have cited is that unexplained technical notions—"mechanisms," "levels," "domain," "field," "theory"—are invoked (sometimes in apparently inconsistent ways), so that no precise answer to the philosophical problem posed in the text is ever given. Nevertheless, I hope that the discussion of the later sections of this essay will help to articulate more fully some of the genuine insights of these authors, especially those contained in Wimsatt's rich essay.

4. E. Nagel, *The Structure of Science* (New York: Harcourt Brace, 1961), chapter 11. A simplified presentation can be found in chapter 8 of C. G. Hempel, *Philosophy of Natural Science* (Englewood Cliffs, N.J.: Prentice Hall, 1966).

5. Quite evidently, this is a weak version of what was once the "received view" of scientific theories, articulated in the works of Nagel and Hempel cited in the previous note. A sustained presentation and critique of the view is given in the introduction to F. Suppe, ed., *The Structure of Scientific Theories* (Urbana: University of Illinois Press, 1973). The fact that the standard model of reduction presupposes the thesis that theories are reasonably regarded as sets of statements has been noted by Clark Glymour, "On Some Patterns of Reduction,"

Philosophy of Science 36 (1969): 340–353, 342; and by Jerry Fodor, *The Language of Thought* (New York: Crowell, 1975), 11, footnote 10. Glymour endorses the thesis; Fodor is skeptical about it.

6. Philosophers often suggest that, in reduction, one derives *corrected* laws of the reduced theory from an *unmodified* reducing theory. But this is not the way things go in the paradigm cases: one doesn't correct Galileo's law by using Newtonian mechanics; instead, one neglects "insignificant terms" in the Newtonian equation of motion for a body falling under the influence of gravity; similarly, in deriving the Boyle-Charles law from kinetic theory (or statistical mechanics), it is standard to make idealizing assumptions about molecules, and so obtain the exact version of the Boyle-Charles law; subsequently, corrected versions are generated by "subtracting" the idealizing procedures. Although he usually views reduction as deriving a corrected version of the reduced theory, Schaffner notes that reduction might sometimes proceed by modifying the reducing theory ("Approaches to Reduction," 138; "The Watson-Crick Model and Reductionism," 322). In fact, the point was already made by Nagel, *Structure of Science*.

7. In part, because modification might produce an inconsistent theory that would permit the derivation of anything. In part, because of the traditionally vexing problem of the proper form for bridge principles in heterogeneous reductions in physics. The former problem is discussed in Glymour, "Patterns of Reduction," 352 and in Dudley Shapere, "Notes towards a Post-Positivistic Interpretation of Science," in P. Achinstein and S. Barker, eds., *The Legacy of Logical Positivism* (Baltimore: Johns Hopkins University Press, 1971). For discussion of the latter issue, see Larry Sklar, "Types of Inter-Theoretic Reduction," *British Journal for the Philosophy of Science* 18 (1967): 109–124; Robert Causey, "Attribute Identities in Microreductions," *Journal of Philosophy* 69 (1972): 407–422; and Berent Enc, "Identity Statements and Micro-reductions," *Journal of Philosophy* 73 (1976): 285–306. The concerns that I shall raise are orthogonal to these familiar areas of dispute.

8. Hull, *Philosophy of Biological Science*, 23, adapted from Theodosius Dobzhansky, *Genetics of the Evolutionary Process* (New York: Columbia University Press, 1970), 167. Similarly molecular genetics is said to have the task of "discovering how molecularly characterized genes produce proteins which in turn combine to form gross phenotypic traits" (Hull, *Philosophy of Biological Science*; see also James D. Watson, *Molecular Biology of the Gene* [Menlo Park, Calif.: Benjamin, 1976], 54).

9. The phenotype/genotype distinction was introduced to differentiate the observable characteristics of an organism from the underlying genetic factors. In subsequent discussions the notion of phenotype has been extended to include properties which are not readily observable (for example, the capacity of an organism to metabolize a particular amino acid). The expansion of the concept of phenotype is discussed in Philip Kitcher, "Genes," *British Journal for the Philosophy of Science* 33 (1982): 337–359.

10. A *locus* is the place on a chromosome occupied by a gene. Different genes which can occur at the same locus are said to be *alleles*. In diploid organisms, chromosomes line up in pairs just before the meiotic division that gives rise to gametes. The matched pairs are pairs of *homologous chromosomes*. If different alleles occur at corresponding loci on a pair of homologous chromosomes, the organism is said to be *heterozygous* at these loci.

11. *Recombination* is the process (which occurs before meiotic division) in which a chromosome exchanges material with the chromosome homologous with it. Alleles which occur on one chromosome may thus be transferred to the other chromosome, so that new genetic combinations can arise.

12. Other central Mendelian claims also turn out to be false. The Mendelian principle that if an organism is heterozygous at a locus then the probabilities of either allele being transmitted to a gamete are equal falls afoul of cases of meiotic drive. (A notorious example

is the *t*-allele in the house mouse, which is transmitted to 95% of the sperm of males who are heterozygous for it and the wild-type allele; see R. C. Lewontin and L. C. Dunn, "The Evolutionary Dynamics of a Polymorphism in the House Mouse," *Genetics* 45 (1960): 705–722. Even the idea that genes are transmitted across the generations, unaffected by their presence in intermediate organisms, must be given up once we recognize that intra-allelic recombination can occur.

13. To the best of my knowledge, the mechanisms of this interference are not well understood. For a brief discussion, see J. Sybenga, *General Cytogenetics* (New York: American Elsevier, 1972), 313–314. In this essay, I shall use "segregation distortion" to refer to cases in which there is a propensity for nonhomologous chromosomes to assort together. "Meiotic drive" will refer to examples in which one member of a pair of homologous chromosomes has a greater probability of being transmitted to a gamete. The literature in genetics exhibits some variation in the use of these terms. Let me note explicitly that, on these construals, both segregation distortion and meiotic drive will be different from *nondisjunction*, the process in which a chromosome together with the whole (or a part) of the homologous chromosome is transmitted to a gamete.

14. This point raises some interesting issues. It is common practice in genetics to count a segment of DNA as a single gene if it was produced by mutation from a gene. Thus many mutant alleles are viewed as DNA segments in which modification of the sequence of bases has halted transcription too soon, with the result that the gene product is truncated and nonfunctional. My envisaged case simply assumes that a second mutation occurs further down the segment so that transcription starts and stops in two places, generating two useless gene products. The historical connection with the original allele serves to identify the segment as one gene.

Conversely, where there is no historical connection to any organism, one may have qualms about counting a DNA segment as a gene. Suppose that, in some region of space, a quirk of nature brings together the constituent atoms for the white eye mutant in *Drosophila melanogaster*, and that the atoms become arranged in the right way. Do we have here a *Drosophila* gene? If the right answer is "No" then it would seem that a molecular structure only counts as a gene given an appropriate history. I hasten to add that "appropriate histories" need not simply involve the usual biological ways in which organisms transmit, replicate, and modify genes: one can reasonably hope to synthesize genes in the laboratory. The case seems analogous to questions that arise about personal identity. If a person's psychological features are replicated by a process that sets up the "right sort of causal connection" between person and product, then we are tempted to count the product as the surviving person. Similarly, if a molecular structure is generated in a way that sets up "the right sort of causal connection" between the structure and some prior gene then it counts as a gene. In both cases, causal connections of "the right sort" may be set up in everyday biological ways and by means of deliberate attempts to replicate a prior structure.

15. So-called *structural genes* direct the formation of proteins by coding for RNA molecules. They are "transcribed" to produce *messenger* RNA (mRNA) which serves as a more immediate "blueprint" for the construction of the protein. Transcription is started and stopped through the action of regulatory genes. In the simplest regulatory system (that of the *lac* operon) an area adjacent to the structural gene serves as a "dumping ground" for a molecule. When concentration of the protein product becomes too high, the molecule attaches to this site and transcription halts; when more protein is required, the cell produces a molecule that removes the inhibiting molecule from the neighborhood of the structural gene, and transcription begins again. For much more detail, see Watson, *Molecular Biology*, chapter 14, and M. W. Strickberger, *Genetics* (New York: Macmillan, 1976), chapter 29.

16. The situation is complicated by the existence of "introns"—segments within genes whose products under transcription are later excised—and by the enormous amount of repetitive DNA that most organisms seem to contain. Moreover, the regulatory systems in eukaryotes appear to be much more complicated than the prokaryote systems (of which the *lac* operon is *one* paradigm). For a review of the situation, as of a few years ago, see Eric H. Davidson, *Gene Expression in Early Development* (New York: Academic Press, 1976).

17. The account will be even more complicated if we honor the suggestion of note 14 and suppose that, for a molecular structure to count as a gene, it must be produced in the right way.

18. Early in the process preceding meiotic division the chromosomes become more compact. As meiosis proceeds, the nucleus comes to contain a system of threads that resembles a spindle. Homologous chromosomes line up together near the center of the spindle, and they are oriented so that one member of each pair is slightly closer to one pole of the spindle, while the other is slightly closer to the opposite pole.

19. The point I have been making is related to an observation of Hilary Putnam's. Discussing a similar example, Putnam writes: "The same explanation will go in any world (whatever the microstructure) in which those *higher level structural features* are present"; he goes on to claim that "explanation is superior not just subjectively but methodologically, . . . if it brings out relevant laws" (Putnam, "Philosophy and Our Mental Life," in *Mind, Language, and Reality* [Cambridge: Cambridge University Press, 1975], 291–303, 296). The point is articulated by Alan Garfinkel, *Forms of Explanation* (New Haven: Yale University Press, 1981), and William Wimsatt has also raised analogous considerations about explanation in genetics.

It is tempting to think that the independence of the "higher level structural features" in Putnam's example and in my own can be easily established: one need only note that there are worlds in which the same feature is present without any molecular realization. So, in the case discussed in the text, PS-processes might go on in worlds where all objects were perfect continua. But although this shows that PS-processes form a kind which could be realized without molecular reshufflings, we know that all *actual* PS-processes do involve such reshufflings. The reductionist can plausibly argue that if the set of PS-processes with molecular realizations is itself a natural kind, then the explanatory power of the cytological account can be preserved by identifying meiosis as a process of this narrower kind. Thus the crucial issue is not whether PS-processes form a kind with nonmolecular realizations, but whether those PS-processes which have molecular realizations form a kind that can be characterized from the molecular point of view. Hence, the easy strategy of responding to the reductionist must give way to the approach adopted in the text. (I am grateful to the editors of the *Philosophical Review* for helping me to see this point.)

20. It would be impossible in the scope of this essay to do justice to the various conceptions of scientific theory that have emerged from the demise of the "received view." Detailed comparison of the perspective I favor with more traditional approaches (both those that remain faithful to core ideas of the "received view" and those that adopt the "semantic view" of theories) must await another occasion.

21. My notion of a practice owes much to some neglected ideas of Sylvain Bromberger and Thomas Kuhn. See, in particular, Bromberger, "A Theory about the Theory of Theory and about the Theory of Theories," in W. L. Reese ed., *Philosophy of Science: The Delaware Seminar* (New York: University of Delaware Press, 1963); Bromberger, "Questions," *Journal of Philosophy* 63 (1966): 597–606; and Kuhn, *The Structure of Scientific Revolutions* (Chicago: University of Chicago Press, 1962), chapters ii–v. The relation between the notion of a practice and Kuhn's conception of a paradigm is discussed in chapter 7 of Kitcher, *The Nature of Mathematical Knowledge* (New York: Oxford University Press, 1983).

22. More exactly, a general argument pattern is a triple consisting of a sequence of schematic sentences (a *schematic argument*), a set of filling instructions (directions as to how dummy letters are to be replaced), and a set of sentences describing the inferential characteristics of the schematic argument (a *classification* for the schematic argument). A sequence of sentences instantiates the general argument pattern just in case it meets the following conditions: (i) the sequence has the same number of members as the schematic argument of the general argument pattern; (ii) each sentence in the sequence is obtained from the corresponding schematic sentence in accordance with the appropriate filling instructions; (iii) it is possible to construct a chain of reasoning which assigns to each sentence the status accorded to the corresponding schematic sentence by the classification. For some efforts at explanation and motivation, see Kitcher, "Explanatory Unification," *Philosophy of Science* 48 (1981): 507–531.

23. Sometimes particles in the cytoplasm account for hereditary traits. See Strickberger, *Genetics*, 257–265.

24. For examples, see Strickberger, *Genetics*, chapters 6–12, 14–17, especially chapter 11; Peters, *Classic Papers*; and H. L. K. Whitehouse, *Towards an Understanding of the Mechanism of Heredity* (London: Arnold, 1965).

25. The comparison will make use of standard statistical techniques, such as the chi-square test.

26. *Polygeny* occurs when many genes affect one characteristic; *pleiotropy* occurs when one gene affects more than one characteristic.

27. *Incomplete dominance* occurs when the phenotype of the heterozygote is intermediate between that of the homozygotes; *epistasis* occurs when the effect of a particular combination of alleles at one locus depends on what alleles are present at another locus.

28. See Watson, *Molecular Biology*, chapter 9; and Arthur Kornberg, DNA *Synthesis* (San Francisco: Freeman, 1974).

29. See G. P. Georghiou, "The Evolution of Resistance to Pesticides," *Annual Review of Ecology and Systematics* 3 (1972): 133–168.

30. See Watson, *Molecular Biology*, 189–193; and T. H. Maugh II, "A New Understanding of Sickle Cell Emerges," *Science* 211 (1981): 265–267.

31. However, one might claim that "Genes can replicate" is a law of genetics, in that it is general, lawlike, and true. This does not vitiate my claim that the structure of classical genetics is not to be sought by looking for a set of general laws, for the law in question is so weak that there is little prospect of finding supplementary principles which can be conjoined with it to yield a representation of genetic theory. I suggest that "Genes can replicate" is analogous to the thermodynamic "law," "Gases can expand," or to the Newtonian "law," "Forces can be combined." If the only laws that we could find in thermodynamics and mechanics were weak statements of this kind we would hardly be tempted to conceive of these sciences as sets of laws. I think that the same point goes for genetics.

32. A similar point is made by Kenneth Schaffner in a forthcoming book on theory structure in the biomedical sciences [*Discovery and Explanation in Biology and Medicine* (Chicago: University of Chicago Press, 1993)]. Schaffner's terminology is different from my own, and he continues to be interested in the prospects of global reduction, but there is considerable convergence between the conclusions that he reaches and those that I argue for in the present section.

33. There are numerous examples of such modifications from the history of chemistry. I try to do justice to this type of case in Philip Kitcher, "Theories, Theorists, and Theoretical Change," *Philosophical Review* 87 (1978): 519–547, and in "Genes."

34. Molecular biology also provided significant refinement of the terms *gene* and *allele*. See Kitcher, "Genes."

35. For examples, see N. K. Wessels, *Tissue Interactions and Development* (Menlo Park, Calif.: Benjamin, 1977), especially chapters 6, 7, 13–15; and Donald Ede, *An Introduction to Developmental Biology* (London: Blackie, 1978), especially chapter 13.

36. I attempt to show how the same perspective can be fruitfully applied to other examples in Philip Kitcher, "Explanatory Unification," sections 3 and 4; Kitcher, *Abusing Science* (Cambridge, Mass.: MIT Press, 1982), chapter 2; Kitcher, "Darwin's Achievement" (chapter 3 of this book).

37. The extreme version of reductionism is defended by Richard Dawkins in *The Selfish Gene* (New York: Oxford University Press, 1976) and *The Extended Phenotype* (San Francisco: Freeman, 1982). For an excellent critique, see Elliott Sober and Richard C. Lewontin, "Artifact, Cause, and Genic Selection," *Philosophy of Science* 49 (1982): 157–180. More ambitious forms of antireductionism with respect to evolutionary theory are advanced in S. J. Gould, "Is a New and General Theory of Evolution Emerging?" *Paleobiology* 6 (1980): 119–130; N. Eldredge and J. Cracraft, *Phylogenetic Patterns and the Evolutionary Process* (New York: Columbia University Press, 1980); and Steven M. Stanley, *Macroevolution* (San Francisco: Freeman, 1979). A classic early source of some (but not all) later antireductionist themes is Ernst Mayr's *Animal Species and Evolution* (Cambridge, Mass.: Harvard University Press, 1963), especially chapter 10.

38. Gould's *Ontogeny and Phylogeny* (Cambridge, Mass.: Harvard University Press, 1977) provides historical illumination of both areas of debate about reductionism. Contemporary antireductionist arguments about embryology are expressed by Wessels, *Tissue Interactions*, and Ede, *Developmental Biology*. See also G. Oster and P. Alberch, "Evolution and Bifurcation of Developmental Programs," *Evolution* 36 (1982): 444–459.

39. Oster and Alberch, "Evolution and Bifurcation," 454. The diagram on p. 452 provides an equally straightforward account of their antireductionist position.

2

The Hegemony of Molecular Biology (1999)

I

Dick Lewontin has probably had more influence on contemporary philosophy of biology than any other living biologist—partly because of his brilliant and wide-ranging contributions to genetics and evolutionary theory, partly because of the warmth and kindness he has extended to philosophers, and especially because of his own major philosophical contributions. For nearly two decades, my own writings have been substantially indebted to Dick's insights, and I've found myself fighting on the same side in many of the same battles. But there have been important differences. As Dick has often noted, his own opposition to various popular doctrines, especially in sociobiology and other forms of genetic determinism, has been more radical than my own. Where I have accepted the ground rules of a particular enterprise and argued that the alleged conclusions don't follow, Dick has often wanted to sweep away the enterprise as misguided. In effect, we've replayed the relationship between early twentieth-century British socialists and their more revolutionary counterparts in continental Europe—my Keir Hardie to Dick's Lenin. I can think of no better way to honor his legacy to philosophy of biology than to play it again.

Although his principal concern about the contemporary practice of molecular biology has centered on ideas about genetic causation, Lewontin has a broader interest in debunking what he sees as a "Cartesian" strategy of explanation by dissection.[1] After opposing the "ideology" that we can study all the nature by breaking the world up into independent parts, and after condemning "obscurantist holism," Lewontin continues: "The problem is to construct a third view, one that sees the entire world neither as an indissoluble whole nor with the equally incorrect, but currently dominant, view that at every level the world is made up of bits and pieces that can be isolated and that have properties that can be studied in isolation."[2] In effect, Lewontin wants to resist the hegemony of molecular biology without lapsing into mysticism. So do I. In what follows I shall try to articulate an antireductionist view that sees molecular studies as an important part of, but not the whole of, contemporary biology. I suspect that this view will assign molecular biology a more important role than that which Lewontin would favor.

II

Although the idea of the hegemony of molecular biology is often presented by philosophers in terms of the notion of intertheoretic reduction, a more common formulation in biology discussions would emphasize two themes.

(H1)   All organisms are composed of molecules.

(H2)   Real (rigorous, complete) explanations of the properties of living things trace those properties to interactions among molecules ("life is to be explained at the molecular level").

(H1) is a truism. The real debate centers on (H2).

Proponents of (H2) envisage a reformulated biology in which the properties of organisms are described in a language that allows for application of biochemical principles to derive biological consequences. The first objection is that the envisaged derivations are unobtainable because we can't produce the appropriate language. The second is that, even if we had such derivations, they would not always be explanatory. One very obvious way to pose the first is to ask how we could ever hope to provide a biochemical explication of such notions as *species*, *predator*, and *ecosystem*. But the issue can be more sharply posed if we focus on what seems to be a much more promising case for the hegemonist, to wit genetics.

Consider two statements from classical genetics.

(G1)   Human beings who are homozygous for the sickling allele experience crises at low levels of oxygen.

(G2)   Genes on different chromosomes, or sufficiently far apart on the same chromosome, assort independently.

Hegemonists can point to (G1) as a partial success, but, as I'll argue, (G2) represents total failure.

Since the late 1940s, biologists have known how the hemoglobin transcribed and translated from the sickling allele differs from that translated from the normal allele. Ignoring complexities of development, they can treat erythrocytes as sacs containing hemoglobin, and using principles of chemistry, they can then show that, under conditions of low oxygen, the mutant hemoglobin would tend to clump in ways that produce the characteristic rigid crescents that give sickle-cell anemia its name. Once it's recognized that these crescents would tend to block narrow capillaries, we have an explanation for (G1). Although that explanation isn't fully molecular—recall that we've ignored the developmental process entirely and have taken a very macrolevel view of the pertinent physiology—it's a start.

The envisaged explanation would start with the derivation of the normal and mutant sequences of amino acids from the specifications of DNA sequences and the genetic code. Of course, to achieve that, we only need the sequence specification of particular alleles. By contrast, when we turn to (G2), the reformulation in biochemical terms would require a specification of the general property *being a gene*. That is, what is needed is completion of the open sentence.

x is a gene if and only if *x* is . . .

Now surely we know *something* about how to complete this. An important necessary condition on genes is that they be segments of DNA or RNA; but of course there are lots of segments of DNA and RNA (most of them, in fact) that are not genes. The task is thus to identify the property that distinguishes the right segments of nucleic acid from the wrong ones.

There is an important constraint on doing this, a constraint that's sometimes unrecognized. If the principles of chemistry are to be employed in deriving the reformulated biological conclusion, then we'll need a characterization of the pertinent entities—in this instance genes—that will mesh with standard ways of drawing chemical consequences. That meant that the characterization will have to be *structural*, identifying genes in terms of their constituent molecules. Hence a proposal to specify genes as functional entities, for example those nucleic acid segments that are transcribed and translated to produce polypeptides, won't serve the hegemonist's turn.[3] (I should note that this proposal is also inadequate because it wrongly excludes segments that happen to lose their regulatory regions.)

No structural specification of the general notion of a gene is currently available. That's not because the project of finding one wouldn't be important to contemporary molecular biology. On the contrary, as masses of sequence data pour in, investigators hunting for genes would welcome a systematic method of searching the long string of A's, C's, G's, and T's. The best they can do is to pick out Open Reading Frames (ORFs)—relatively long stretches bounded by start and stop codons—treating these as candidates and then checking to see if they can discover corresponding mRNAs.

Further, as the intricacies of genomes become more evident, the possibilities of split genes, overlapping genes, truncated sequences that are still associated with regulatory regions, sequences that have lost their regulatory regions, embedded genes, and so forth make any structural and functional criteria, with ORFs coupled with functional mRNAs as the central instances and with peripheral examples settled by conventions that sometimes vary from study to study.

The first trouble with (G2) is thus that the required cross-science identifications aren't available. I'll now argue that, even if they were, a derivation of (G2) from principles of chemistry wouldn't be explanatory.

III

I'll begin indirectly with a motivational story. In 1710, John Arbuthnot, a physician, pointed out that the previous eighty-two years in London were all "male"—that is, in each of these years, there was a preponderance of male births in London. Publishing his finding in the *Philosophical Transactions of the Royal Society*, Arbuthnot calculated the probability that this occurred by chance, and, finding that probability to be minute, he chalked up the phenomenon to Divine Providence. Let's imagine two secular characters who try to give a better explanation.

First is the Mad Mechanist (MM). His guiding principle is that "life is to be explained at the molecular level," and he puts this to work in offering an explanation. More exactly, he provides a recipe for an explanation, admitting that, because of his ignorance of trillions of details, he can't go further. The recipe runs as follows:

> Start with the first birth of 1628. Go back to the copulatory act that began the pregnancy resulting in this birth. Give a molecular characterization of the circumstances that preceded fertilization. From this characterization derive a conclusion about the sperm that was incorporated into the zygote. Continue with the molecular account of the course of the pregnancy and birth. You have now explained the sex of the first infant of 1628.
>
> Continue in the same fashion with the second birth, the third birth and so on. When you are done, add up the totals for both sexes. You now have a complete explanation of why 1628 was a male year.
>
> Repeat the same procedure for subsequent years until you reach 1709. Stop. You now have a complete explanation for why all 82 years are male.

Actually you don't.

To see why, consider our next character, the Sensible Sex-Ratio Theorist (SST). She proceeds from R. A. Fisher's insight about the evolution of sex ratio.

> In species without special conditions of mating (including *Homo sapiens*) if the sex ratio at sexual maturity departs from $1:1$, there will be a selective advantage to a tendency to produce members of the underrepresented sex (this will show up in terms of increased numbers of expected grand offspring). In human populations that are sufficiently large, we should thus expect the sex ratio at sexual maturity to approximate $1:1$ (the more closely the larger the population; even in the seventeenth century London had a large population).
>
> If one sex is more vulnerable than the other to mortality between birth and sexual maturity, then that sex will have to be produced in greater numbers if the sex ratio at sexual maturity is to be $1:1$. In human beings, males are more vulnerable to prepubertal death. Thus the birth sex ratio is skewed toward males.

I claim that the SST would give a better explanation than the MM, even if the latter could actually deliver the details. Part of the reason is that the SST's account shows that Arbuthnot's data are no fluke. The significant point for our purposes is that we don't need the masses of accidental molecular minutiae: we want to see how a regularity in nature is part of a broad general pattern.

A Not-So-Mad Mechanist (NSMM) would see the point and modify his position. Recognizing the fact that the best explanation of the phenomenon doesn't grub through the molecular details, he might ask whether there are different facets of this situation that molecular research might illuminate. Indeed there are. SST tells us why years are male (or, more exactly, likely to be male for large populations). But that leaves it open how various populations of *Homo sapiens* find their ways to (rough) equilibrium. NSMM will propose a division of explanatory labor. After SST has shown the shape of the explanation, physiologists can delve into the mechanisms of Y-biased fertilization (are Y-bearing sperm faster? are vaginal conditions more suited to the voyages of Y-bearing sperm? are there polymorphisms in human populations?), leading eventually to a molecular understanding of the most important processes. I'll return to the significance of this point below.

Now back to (G2). We can envisage a counterpart to MM, bravely trying to show how gory chemical details yield the independent assortment of genes (provided that the genes are on different chromosomes or are sufficiently far apart on the same chromosome). But there's no reason to think that these efforts would be any more illuminating than MM's. For there's also a counterpart to SST, whose explanation goes as follows.

> Consider the following kind of process, a PS-process (for *pairing* and *separation*). There are some basic entities which come in pairs. For each pair, there's a correspondence relation between the parts of one member of the pair and the parts of the other member. At the first stage of the process, the entities are placed in an *arena*. While they are in the arena, they can exchange segments, so that the parts of one member of a pair are replaced by the corresponding parts of the other member, and conversely. After exactly one round of exchanges, one and only one member of each pair is drawn from the arena and placed in the *winner's box*.
>
> In any PS-process, the chances that small segments that belong to members of different pairs or that are sufficiently far apart on members of the same pair will be found in the winner's box are independent of one another. (G2) holds because the distribution of chromosomes to gametes at meiosis is a PS-process.

This, I submit, is a full explanation of (G2), an explanation that prescinds entirely from the stuff that genes are made of. Understanding the probabilistic regularities that govern the transmission of genes is a matter of seeing that transmission is a PS-process, and it's irrelevant whether the genes are made of nucleic acid or of swiss cheese.

The conclusion we ought to draw is that some important biological regularities cannot be captured in the language of molecular biology—or, more strictly, in a molecular biological language that restricts itself to structural notions[4]—and that these regularities are fully explained without grinding out molecular detail. An Enlightened Hegemonist (EH) ought to appreciate the point, recognizing the need to absorb functional concepts, and claims involving those concepts, from traditional areas of biology. EH will insist, however, that there are important molecular issues about the functionally characterized regularities—question concerning the mechanisms of Y-biased fertilization or the molecular underpinnings of the pairing of homologous chromosomes at meiosis. That point, I'll argue later, is correct. If EH is ambitious, however, there may be a further proposal: Although it is right for molecular biology to absorb functional insights from the classical areas of biology, further investigations in these areas are unnecessary; from now on, molecular biology is all the new biology we need. I now want to suggest that we ought to resist such hegemonist yearnings.

IV

In 1917, D'Arcy Wentworth Thompson published a remarkable book. Like Tom Stoppard's Lady Thomasina, Thompson yearned for the mathematics of the animate world.[5] In recent years, mathematical biologists have begun to realize Thompson's program, and the result, I'll suggest, is a view of developmental biology that both assigns an important place to molecular studies and deepens the challenge to the hegemony of molecular biology.

For present purposes we'll only need to consider the most elementary parts of a few major approaches, and I want to emphasize that the simple models I'll describe are elaborated in much more subtle versions. I begin with the use of *Lindenmeyer systems*—L-systems—to characterize the growth of plants. A *string OL-system* is a triple $\langle V, I, P \rangle$ where V is a vocabulary, I is an initial string, and P is a set of production rules. A *developmental sequence* in an OL-system is a sequence of strings whose first member is the initial string and such that the $n + 1$st member is obtained from the nth member by applying all the production rules that can be applied to the nth string. So, for example, consider the L-system

$$I: \quad a_r$$
$$P_1: \quad a_r \rightarrow a_l b_r$$
$$P_2: \quad a_l \rightarrow b_l a_r$$
$$P_3: \quad b_r \rightarrow a_r$$
$$P_4: \quad b_l \rightarrow a_l$$

Within this system, we can obtain the following developmental sequence:

$$a_r$$
$$a_l b_r$$
$$b_l a_r a_r$$
$$a_l a_l b_r a_l b_r$$
$$b_l a_r b_l a_r a_r b_l a_r a_r$$
$$\ldots$$

This formalism can be used to model the development of a multicellular filament found in the blue-green bacteria *Anabaena catenula* (the *a*s and *b*s represent different types of cell and the suffixes show the polarity; see (figure 2.1).[6]

In general, L-systems model the development of plants by supposing that there are elementary biological processes that are applied recursively to certain kinds of structures: intuitively, in a growing plant, a particular kind of structure gives way to a different kind of structure, and the process of replacement is represented by a production rule. Note that this treats the development of plants in an extremely abstract way, prescinding from the details of the types of processes involved. Thus the growth of two quite different plants could be represented by the same L-system, if in the one instance a production rule called for the replacement of a particular kind of nodal cell with a branch and in the other it specified that a very different type of cell should be surrounded by a specific geometrical cluster of certain kinds of cells.

For a less abstract treatment of issues in development, we can turn to the mathematics of diffusion equations. Inspired by work of Alan Turing,[7] a number of

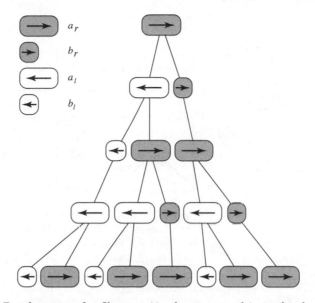

FIGURE 2.1 Development of a filament (*Anabaena catenula*) simulated using a DOL-system. From A. Lindenmayer and P. Prusinkiewicz, *The Algorithmic Beauty of Plants* (New York: Springer, 1990).

biologists have explored the possibility that various kinds of patterns could be generated through "activator-inhibitor" systems. Basic to this approach is the thought that a cell might start to produce greater concentrations of a particular molecule, that this molecule could then diffuse into adjacent cells, with the activation and diffusion prompting the production of an inhibitor. In the boring case, an entire tissue of cells reaches a uniform steady state. Much more interesting is the possibility that a local departure from uniformity gives rise to a stable pattern.

Once again, I'll focus on the very simplest system. We imagine an interaction among two molecules, the activator (whose concentration is a) and the inhibitor (whose concentration is b). The inhibitor is assumed to diffuse much more rapidly than the activator and also to reach equilibrium almost instantaneously. The concentrations are governed by the differential equations:

$$\partial a / \partial t = a^2 - a$$
$$\partial b / \partial t = a^2 - b.$$

It's not hard to see that there's a steady state at $a = b = 1$. It's possible to generate a pattern, however, if one cell in an array has a slightly increased activator concentration. Because of the assumption that the inhibitor diffuses rapidly, it responds to the average concentration of the activator, and thus remains virtually constant. Hence the activator will continue to increase (since, by the first equation, the time-derivative of its concentration will be positive). Once the increase becomes sufficiently large, there will be an effect on the average sufficient to produce

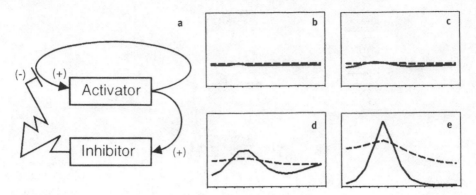

FIGURE 2.2 Pattern formation by autocatalysis and long-range inhibition. (a) Reaction scheme. An activator catalyses its own production and that of its highly diffusing antagonist, the inhibitor. (b–e) Stages in pattern formation after a local perturbation. Computer simulation in a linear array of cells. A homogeneous distribution of both substances is unstable. A minute local increase of the activator (—) grows further until a steady state is reached in which self-activation and the surrounding cloud of inhibitor (----) are balanced [S22]. From H. Meinhardt, *The Algorithmic Beauty of Sea Shells* (New York: Springer, 1998).

inhibitor to stop the process. We thus obtain a steady state with a locally high concentration of activator and relatively elevated levels of inhibitor elsewhere (see figure 2.2).[8]

Suppose, then, that the growth of seashells is a process in which concentrations of activator molecules and of inhibitor molecules are governed by a coupled set of partial differential equations that allow for nontrivial steady states. If the difference between these concentrations is associated with pigmentation (or possibly with differentially directed cell growth and division), then it is possible to understand how patterns of various kinds emerge. Hans Meinhardt has explored a wide range of growth processes, showing how the patterns found in a diverse class of seashells can be generated from particular sets of equations. His analysis, while less abstract than the Lindenmayer-Prusinkiewicz treatment of plants, continues to prescind from the molecular details. Two shells might result from the same growth process—accretion of new material at the margin—and might conform to the same set of differential equations, even though the molecules that play the roles of activator and inhibitor are different in the two cases. It might even turn out that, in the one instance, the relationship between the molecules produces a pigmented pattern while, in the other, that relationship yields a pattern of relief (ridges and valleys on the shell surface).

My last example inches further in the direction of diminished abstraction. Meinhardt's attempt to find a general set of models for shell pattern ranges more widely than an endeavor of James D. Murray to explain the diversity of mammalian coat patterns.[9] Following Turing, Murray considers a reaction diffusion system governed by the following equations:

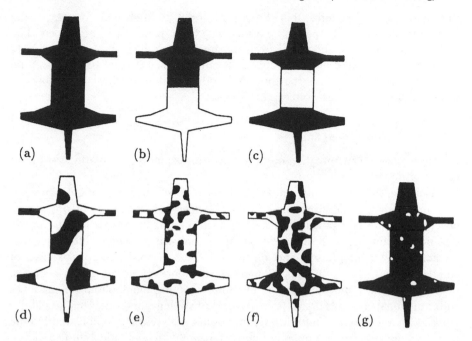

FIGURE 2.3 Effect of body surface scale on the spatial patterns formed by the reaction diffusion mechanism (Mammalian Coat Patterns—"How the Leopard Got Its Spots") with parameter values $\alpha = 1.5$, $K = 0.125$, $p = 13$, $a = 103$, $b = 77$ (steady state $u_s = 23$, $v_s = 24$), $d = 7$. Domain dimension is related directly to γ. (a) $\gamma < 0.1$; (b) $\gamma = 0.5$; (c) $\gamma = 25$; (d) $\gamma = 250$; (e) $\gamma = 1250$; (f) $\gamma = 3000$; (g) $\gamma = 5000$. From Murray, *Mathematical Biology* (New York: Springer, 1989).

$$\partial u / \partial t = \gamma f(u, v) + \nabla^2 u$$
$$\partial v / \partial t = \gamma g(u, v) + d \nabla^2 v$$
$$f(u, v) = a - u - h(u, v)$$
$$g(u, v) = \alpha(b - v) - h(u, v)$$
$$h(u, v) = \rho u v / (1 + u + K u^2)$$

Here u and v are molecular concentrations, a, b, d. and γ arc dimensionless parameters, d being the ratio of diffusion coefficients and γ a scaling parameter (γ varies as the area of the surface on which the pattern is being laid down).[10] Murray shows that, when $d > 1$, processes conforming to these equations can give rise to spatially inhomogeneous patterns. Whether such a pattern occurs, and what form it takes, depends on the value of γ. As this value increases, the character of the pigmentation pattern changes from uniform to bicolored to blotched to striped to spotted (see figure 2.3).

It's now possible to arrive at a clever "theorem." Assume that mammalian coat markings are generated from reactions among chemicals that satisfy the given system of equations. For a given value of $d > 1$, provided that it allows for both striped and spotted patterns, there'll be a threshold γ^*, such that, for $\gamma \geq \gamma^*$, the resultant pattern will be spotted, and for $\gamma < \gamma^*$, the pattern will be striped. The value of γ for an animal body will always be greater than the value for that animal's tails (bodies are always bigger in area than tails). Hence it can't happen that the value for the tail lies above the threshold and the value for the body below the threshold. In other words we have the "theorem":

> Although there can be spotted animals with striped tails, there can't be striped animals with spotted tails.

Murray's model of mammalian coat patterns thus explains a regularity we find in nature.

Consider now three different proposals for research in developmental biology. The first, the original hegemonist position, suggests that studies of organismic development are best pursued by starting with a complete understanding of the genetics, continuing with an investigation of the ways in which different genes are activated and suppressed, and, on this basis, exploring the molecular bases of cellular differentiation. In light of the considerations raised in earlier sections, hegemonists may concede the need for supplementing the "bottom up" analyses with functional concepts drawn from classical physiology (and other traditional disciplines), but see no reason for further functional analyses that do not attend to the molecular details. The examples I've chosen from the mathematical study of development are intended to show that this concession is too limited. A third, and more enlightened, approach would view the mathematical and molecular programs as working in tandem.

At the most concrete level of mathematical analysis, theorists may try to formulate differential equations that govern the interactions of molecules whose identities they don't know, seeking in this way to understand a general pattern of development in some group of organisms—as in Murray's treatment of mammalian coat patterns, with its pretty result about spots and stripes. Their research then poses the problem of trying to find the hypothetical molecules and, quite possibly, of rebuilding the model to accommodate the complexities that emerge.

Above this is a level of analysis represented by Meinhardt's work on seashells, where the emphasis is on a family of related models. Here theorists attempt to discover more general regularities, consisting in conformity to a family of sets of differential equations. For this enterprise to succeed it will be important to show how particular phenomena in particular organisms are governed by particular members of the family and, in consequence, to supplement the mathematical details with the identification of the pertinent molecules.

Even more abstract is a style of analysis that focuses on formal features of growth without reference to a specific interpretation of the biological processes represented by formal transformation, and without specification of equations that are to be satisfied. The study of plant growth in terms of Lindenmayer systems allows for various physiological "readings" of the production rules, interpretations that might be given in terms of macroscopic plant physiology, in terms of some mathematically characterized process, or in terms of molecular interactions. It's easy to see how there could be a nested sequence of abstract accounts, subsumed at the most formal level in claims about L-systems, with intermediate levels of mathematical biology that eventually are instantiated in detailed studies of heterogeneous molecules in very different organisms.

If the hegemonist's likely mistake consists in the loss of understanding through immersion in detail, with concomitant failure to represent general regularities that are important to "growth and form," the mathematical analyst can easily lose touch with the biological realities. Hegemonic grumbling about the ease with which one can make up pleasing mathematical models is unfair—it isn't so easy—but it has a point. The multileveled picture of theorizing about development that I have recommended needs its molecular base. To admit that is to recognize that the questions about individual mechanisms that excite molecular biologists (partly, of course, because they have powerful tools for addressing them)[11] are important, both for their confirmation of the more abstract models and for their uncovering of constraints on model-building. The mistake is to think that these are the *only* important questions, that once we have PCR there is no further need of classical "whole organism" biology. D'Arcy Thompson's vision should be integrated with the achievements and programs of contemporary molecular biology to generate a multileveled study of development.

V

I return, in conclusion, to Lewontin's concerns about the hegemony of molecular biology, expressed in his critique of the "Dream of the Human Genome."[12] Because of much of the propaganda that has surrounded it, the Human Genome Project (HGP) has become a symbol of the hegemony of molecular biology. In part, Lewontin's critique focuses on different questions from those that have concerned me here. He sees, quite correctly, that the basis on which the HGP was advertised consists of a dubious set of claims about the causal roles of genes and about the existence of a royal road to a future molecular medicine of enormous power. There's no reason to believe that, when we've mapped and sequenced the human genome, we will "understand who we are."[13] Similarly, merely knowing the sequences of lots of human genes isn't going to tell us very much directly, given the difficulties of the protein-folding problem, the uncertainties of methods of tracing protein function, our ignorance of developmental pathways, and so forth. The immediate bio-medical upshot of the HGP will be an enormously enhanced ability to give genetic tests, and both Lewontin and I have doubts about whether this is likely to be socially beneficial. Even if it is in principle possible to apply the new means of testing to promote the welfare of citizens (as I have argued at some length), it is becoming depressingly clear that the needed safeguards are not likely to be in place by the time the technologies flood the marketplace.[14]

But I want to distinguish the status of the HGP as part of a sociomedical agenda from its role in contemporary biological research. Lewontin's critique, and the kindred remarks of historians and philosophers of biology,[15] convey the message that the HGP is biologically misguided, either because the mass of sequence data it will generate is useless or because it is inextricably entwined with a reductionistic research program. These reactions reflect a disposition to accept the propaganda for the HGP at face value. It is quite right to point out that there is nothing biologically special about the genome of our own species and to question the hegemonist suggestion that we can proceed from knowledge of sequence data to knowledge of genes and thence to all manner of biological understandings. Yet the research actually conducted under the auspices of the HGP fully absorbs these points.

From a biological point of view, the most important work being conducted with HGP funding (or the parallel research carried out with private support, most notably that of Craig Venter and his colleagues) consists in fine-grained mapping and sequencing of nonhuman organisms, from bacteria to yeast, to nematode worms, and still in progress, flies. The fruits of this research are likely to make any number of research projects in physiology and developmental biology enormously easier in coming decades (as well as paving the way for evolutionary insights obtained from comparisons of the genomes of closely related species). Specifically, molecular biologists working on *Caenorhabditis elegans*, *Drosophila melanogaster*, and *Dictyostelium discoideum* already envisage the possibility of identifying major developmental pathways, possibly pathways that have been highly conserved in the evolutionary process. There's no automatic route to picking out such pathways but the ability to discover which genes are activated in which cells (which will flow from complete genome sequencing) is likely to offer important clues.

There should be no illusions that this molecular work can proceed by ignoring macrolevel studies of development and physiology. On the contrary, the full exploitation of the sequence data generated by the genome project will require just the kinds of functional studies—including mathematical modeling—that I have emphasized throughout this essay.[16] Critics of the HGP may be correct in thinking that the current balance of research in biology has tipped too far toward this particular molecular endeavor, that it is not the *only* project of biological value. It is wrong, however, to overstate the claim by taking the project to be devoid of biological significance and to accuse it of commitment to the hegemonist manifesto. Provided we have a rich enough repertoire of visions, the dream of (say) the fruit-fly genome is a dream worth having.

Notes

1. This opposition is evident in many of the contributions to *The Dialectical Biologist* (Cambridge, Mass.: Harvard University Press, 1986), a volume of essays, some of which are by Lewontin alone, some by his colleague Richard Levins, and some that are jointly written. The attack on genetic determinism also permeates this volume, as well as surfacing in *Not in Our Genes* (written with Leon Kamin and Steven Rose [New York: Pantheon, 1984]) and the more recent *Biology as Ideology* (New York: Harper, 1992). I discuss the varieties of Lewontin's critique of genetic determinism in Philip Kitcher, "Battling the Undead: How (and How Not) to Resist Genetic Determinism," in a *Festschrift* for Dick Lewontin; that essay, reprinted as chapter 13 of this book, should be seen as a companion piece to my efforts here.

2. Lewontin, *Biology as Ideology*, 15.

3. This problem affects the suggestion made in a provocative essay by C. Kenneth Waters, "Genes Made Molecular," *Philosophy of Science* 61 (1994): 163–185.

4. In fact, contemporary molecular biology is permeated by language that can't be replaced with an austere physicochemical idiom. Consider the standard account of transcription. One talks of RNA polymerases "associating" with DNA. The suggestion, of course, is that the RNA polymerases come close—but how close is close enough? Well, that's going to depend on the conformation of the DNA, and there's no general structural criterion. In effect, molecular biologists, here and elsewhere, quietly take over functional concepts. This moved Sylvia Culp and me to suggest that molecular biology turns out not to be reducible to molecular biology (see Culp and Kitcher, "Theory Structure and Theory Change in Contemporary Molecular Biology," *British Journal for the Philosophy of Science*, 40 [1989]: 459–483.

5. D'Arcy Thompson, *On Growth and Form* (Cambridge: Cambridge University Press, 1917); an abridged version, edited by the distinguished developmental biologist John Tyler Bonner, was published by Cambridge University Press in 1961. Thompson's wish for a mathematical account of development and morphology is echoed in several speeches by the heroine of Stoppard's *Arcadia*.

6. This example is used as the first (simplest) illustration by Przemyslaw Prusinkiewicz and Aristid Lindenmayer in *The Algorithmic Beauty of Plants* (New York: Springer, 1990). The example comes from p. 5 (I have slightly modified the notation). As Prusinkiewicz and Lindenmayer note, L-systems are related to Chomskyan grammars. Subsequent examples reveal the possibilities of far more complex relations between symbols and biological entities (particularly through processes that draw out shapes dependent on the symbols), context-dependence, three-dimensionality, probabilistic systems, and so forth. The resultant systems can simulate the growth of flowers and trees, generate the Fibonacci spirals found in sunflowers, and model compound leaves, among other achievements.

7. Specifically a paper, Alan Turing, "The Chemical Basis of Morphogenesis," *Philosophical Transactions of the Royal Society* B, 237 (1952): 37–72. This essay was written shortly before Turing's tragic suicide. It is interesting to ponder whether, if he had lived, the pace of work in mathematical developmental biology would have accelerated, producing a very different distribution of work in the current field.

8. This extremely elementary example is from Hans Meinhardt, *The Algorithmic Beauty of Sea Shells* (New York: Springer, 1998). As Meinhardt shows, much more complex sets of partial differential equations give rise to a wide variety of patterns, including the elaborate branching and meshwork found in some shells. The case in the text is the foundation of a system that will yield regular stripes.

9. In fact, Murray has a very broad program of trying to understand pattern formation, but I'll only consider one aspect of it here. The discussion is drawn from chapter 15 of J. D. Murray, *Mathematical Biology* (New York: Springer, 1989), although the essentials were already given in Murray's "On Pattern Formation Mechanisms for Lepidopteran Wing Patterns and Mammalian Coat Markings," *Philosophical Transactions of the Royal Society* B, 295 (1961), 473–496. Murray provided an accessible overview in "How the Leopard Gets Its Spots," *Scientific American* 258 (1988): 80–87.

10. Murray provides a lucid discussion of these equations in chapter 14 of *Mathematical Biology*. He notes there that the equations describe the chemical kinetics of a substrate-inhibition system, which has been studied experimentally. Real instantiations of the system are thus known.

11. As in other parts of science, techniques in molecular biology have a life of their own, sometimes inspiring people to pursue questions because they can be addressed. For an illuminating study of the ways in which instruments and experimental skills possess an inertia

that shapes the course of research, see Peter Galison's *Image and Logic* (Chicago: University of Chicago Press, 1998), which pursues this theme in the context of particle physics.

12. Originally published in the *New York Review of Books* and reprinted in *Biology as Ideology*.

13. Lewontin's attacks on this specific form of genetic determinism are quite devastating. I've tried to argue similar points in Philip Kitcher, *The Lives to Come* (New York: Simon and Schuster, 1996), especially chapter 11. In general, however, my critique of genetic determinism differs from that which has featured most prominently in Lewontin's recent writings. See my essay "Battling the Undead" (chapter 13 of this book).

14. I argued for the in principle possibility in *The Lives to Come*. Since I finished writing that book, there has been virtually no progress in addressing the problems of the proliferation of genetic tests, not only in the United States but also in other affluent nations. Of course, the United States is especially backward because of its notable lack of commitment to universal health care coverage. My current position is thus much closer to Lewontin's pessimistic view of the likely social effects of the HGP.

15. See, for example, Alexander Rosenberg, "Subversive Reflections on the Human Genome Project," *PSA 1994* (East Lansing, Mich.: Philosophy of Science Association, 1995), Volume 2, 329–335, and A. Tauber and S. Sarkar, "The Human Genome Project: Has Blind Reductionism Gone Too Far?" *Perspectives in Biology and Medicine* 35 (1992): 222–235.

16. In a forthcoming essay, Kenneth Schaffner argues for similar themes [Genes, Behavior, and Emergentism," *Philosophy of Science* 65 (1998): 209–252]. Schaffner's lucid analysis of investigations of behavioral genetics in the nematode *C. elegans* reveals exactly the need for multileveled studies that I've been emphasizing. It seems to me also to show the fruitful possibilities of combining molecular work with mathematical studies of the properties of networks. Interestingly, the same cross-fertilization of intellectual disciplines is already envisaged in work on the development of the soil ameba *Dictyostelium discoideum* (in the work of William Loomis and his colleagues).

3

Darwin's Achievement (1985)

By 1844 Charles Darwin had begun to feel that his growing friendship with Joseph Hooker was strong enough to be tested by the revelation of his unorthodox ideas on "the species question." Darwin's disclosure cost him some misgivings: "it is like confessing a murder," he wrote.[1] Yet, little more than a quarter of a century later, Darwin's heresy had won endorsements from many prominent scientists in Britain, Europe, and the United States. By 1871, Thomas Henry Huxley was prepared to declare that "in a dozen years the 'Origin of Species' has worked as complete a revolution in Biological Science as the 'Principia' did in Astronomy."[2]

How was so swift a victory accomplished? Part of the answer must give credit to Darwin's political skills. We should not be beguiled by the picture of the unworldly invalid of Down, whose quiet walks in his beloved garden were the occasion only for lofty musings on points of natural philosophy. Darwin's study was the headquarters of a brilliant campaign (which he sometimes saw in explicitly military terms),[3] directed with enormous energy and insight. His letters are beautifully designed to make each of his eminent correspondents—Hooker and Huxley, Lyell, Wallace, and Asa Gray—feel that he is the crucial lieutenant, the man on whose talents and dedication the cause depends.[4] Morale is kept up, and the troops are deployed with skill.

Yet Darwin's brilliant use of the social structure of British (and American) science is not the entire secret of his success. Those who fought on his behalf were initially recruited through Darwin's careful presentation of the arguments for his theory,[5] and in their public defenses of that theory, they explained and amplified the reasoning distilled in *The Origin of Species*.[6] As Darwin himself clearly saw[7] the recruitment of eminent allies was necessary to secure a hearing for his ideas. Despite the suggestions of his opponents to the contrary, Darwin's adroit politicking did not dictate the verdict.[8]

In what follows I shall defend an old-fashioned view. The *Origin* is what Darwin advertised it as being—"one long argument" for the theory of evolution.[9] Ultimately, the Darwinian revolution was resolved by reason and evidence, and the reasons and the evidence are crystallized in the *Origin*. We would do well to remember that, for several of Darwin's closest friends and staunchest supporters, it was the reading of the *Origin* that stiffened their convictions and fired their enthusiasm.[10]

Nevertheless, if my claim that Darwin's heresy triumphed because of the reasons he provided is traditional, my defense of that claim will break with the usual views about how those reasons work. I believe that historians and philosophers of science have brought to the study of Darwin a conception of theory and evidence that distorts his achievement.[11] I shall offer a different approach to the theory advanced in the *Origin*, an approach which will, I hope, enable us to see clearly why Darwin's "long argument" was so successful.

II

Virtually everyone would agree that the *Origin* offers a new theory, the theory of evolution by natural selection.[12] When one attempts to specify exactly what this novel theory is, the result is inevitably influenced by general ideas about scientific theories. Once, there was a well-articulated philosophical view on this topic. Scientific theories were held to be axiomatized sets of statements, among whose axioms occurred statements formulated in a special vocabulary, the "theoretical vocabulary" of the theory in question. Expressions in this theoretical vocabulary were supposed to apply to unobservable entities, and because philosophers harbored worries about how they could do so, the account required that there be special statements ("correspondence rules") whose function was to fix the meaning of the theoretical terms. In general, it was supposed that the axioms of the theory would include laws, that these laws would be used in conjunction with particular statements ("boundary conditions" or "initial conditions") to derive previously unaccepted statements whose truth or falsity could be determined by observation, and that theories were confirmed by yielding a large number of such observational consequences which investigation revealed to be true.

There are a number of excellent reasons why this account of scientific theories is no longer *aptly* called the "received view."[13] But a battered and truncated version of it lingers on. Even those who are skeptical about the need for distinctive theoretical vocabulary, or correspondence rules, or axiomatizability, are likely to suppose that any scientific theory worthy of the name must consist of a set of statements, among which are some general laws (laws that set forth the most fundamental regularities in the domain of natural phenomena under investigation), and that such laws should be used to derive previously unaccepted statements whose truth values are subject to empirical determination. When this residual thesis about theories is applied to the case of Darwin, we are led to expect that the *Origin* advances some collection of new general principles about organisms. After all, what else could Darwin's theory be?

The expectation is fostered when we turn to the opening chapters of the *Origin*, where we seem to discover exactly the kind of principles that were anticipated. Darwin's theory apparently rests on four fundamental claims.

1. At any stage in the history of a species, there will be variation among the members of the species; different organisms belonging to the species will have different properties (*Principle of Variation*; *Origin*, chapters 1–2, passim).

2. At any stage in the history of a species, more organisms are born than can survive to reproduce (*Principle of the Struggle for Existence*; *Origin*, chapter 3).

3. At any stage in the history of a species, some of the variation among members of the species is variation with respect to properties that affect the ability to survive and reproduce; some organisms have characteristics that better dispose them to survive and reproduce (*Principle of Variation in Fitness*; *Origin*, 80).

4. Heritability is the norm; most properties of an organism are inherited by its descendants (*Strong Principle of Inheritance*; *Origin*, 5, 13).

From these principles—more exactly, from (2), (3), and (4)—one can obtain by a plausible argument

5. Typically, the history of a species will show the modification of that species in the direction of those characteristics which better dispose their bearers to survive and reproduce; properties which dispose their bearers to survive and reproduce are likely to become more prevalent in successive generations of the species (*Principle of Natural Selection*; *Origin*, chapter 4).

The justification for reconstructing Darwin's theory in this way is relatively straight-forward. The first four principles are assembled and defended at the beginning of the *Origin*, and the main theoretical work then appears to be the derivation of the principle of natural selection from them.

Expositors of Darwin from T. H. Huxley to Richard Lewontin have recon-structed the "heart" of his theory in the way that I have done.[14] Nor will my own account of the theory entirely forsake this great tradition. But it should trouble us that the suggested reconstruction is at odds with an assumption that historians and philosophers of science often tacitly and legitimately make. We expect that the fun-damental principles of a novel scientific theory should be those statements intro-duced by the theory that most stand in need of defense and confirmation, and that the arguments assembled by the innovative theorist should be directed at the fundamental principles of the new theory. My reconstruction of Darwin's theory is crucially inadequate not so much for refined scientific reasons (for example, concerns about the need for additional assumptions in the passage to [5]), nor because of esoteric philosophical scruples (my failure to make plain the role of Darwin's key theoretical concept, the concept of fitness). The trouble is that the theory I have ascribed to Darwin is uncontroversial—so uncontroversial as to border on triviality.

Virtually all of Darwin's opponents would have accepted (1)–(4). None of the great scientists of the mid-nineteenth century would have denied—and none should have denied—that species vary, that the increase of a species is checked, that some variation affects characters relevant to the ability to survive and reproduce, that many properties are heritable. Moreover, they would have seen the force of the argument for (5), assenting to the idea that natural selection has the power to adjust the properties of a species, eliminating variants whose properties render them less able to compete in a struggle for limited resources. What was in dispute in the

Darwinian revolution was not so much the *truth* of (1)–(5), as their *significance*.[15] For the committed Darwinian, these principles were the key to understanding a vast range of biological phenomena, and the principal theoretical and argumentative work of the *Origin* consists in showing how the seemingly banal observations about variation, competition, and inheritance might answer questions that had previously seemed to be beyond the scope of scientific treatment.

Acceptance of (1)–(5) is compatible with the doctrine of the fixity of species (which states that species are closed under reproduction). But Darwin did not simply accept (1)–(5) and add the historical claim that lineages (ancestor-descendant sequences of organisms) have been modified. Something like that view was defended by one of his most bitter critics, the disappointed dean of British biology, Richard Owen.[16] Nor did Darwin simply conjoin (1)–(4) with the historical thesis of evolution and the vague declaration that natural selection has been the primary force of evolution. The *Origin* contains a novel and well-articulated theory precisely because it fuses (1)–(5) with the suggestion that species are mutable to fashion powerful techniques of biological explanation. Darwin, Darwinians, and critics of Darwin agreed on what was at stake.

> Nothing is easier than to admit in words the truth of the universal struggle for life, or more difficult—at least I have found it so—than constantly to bear this conclusion in mind. Yet unless it be thoroughly engrained in the mind, I am convinced that the whole economy of nature, with every fact on distribution, rarity, abundance, extinction, and variation, will be dimly seen or quite misunderstood.[17]

> Few can deny the reality of this struggle for existence, and few can dispute the method of its action and the tendency of its results. The main ground of controversy is this, Will this constant accumulation of inherited variations ever constitute a specific difference?[18]

The major theoretical work of the *Origin* lies in displaying the unanticipated significance of (1)–(5).

Before I make this suggestion more precise, let me respond to an obvious objection. The triviality of the reconstructions that I have envisaged might be thought to stem from the fact that I have remained at a very informal level. Perhaps a more significant version of Darwin's theory could be obtained by disambiguating (5), presenting rigorous derivations of the disambiguated principles, and thus exposing the precise conditions that are needed for the Darwinian argument to go through. Quite evidently, if one imports the ideas of contemporary mathematical population genetics, then it is possible to replace (5) with precise claims about the sequential frequencies of phenotypes (or genotypes) found in successive generations of abstractly characterized populations, and to derive them from precise versions of (1)–(4). Equally evidently, this approach ascribes to Darwin a theory of heredity that he did not have, and it is appropriate to point out that one can hardly claim to have found a precise and nontrivial theory in Darwin by inserting such a theory in a field about which he confessed his own ignorance.[19]

A much more promising approach, pursued by Mary Williams,[20] is to attempt to formalize (1)–(5) without making use of any specific theory of heredity. Williams succeeds in showing that a formal version of (5) can be stated using only primitive notions that are arguably Darwinian, and that, given certain extra assumptions, this version can be obtained from formalizations of (1)–(4). Moreover, her approach could easily be extended to provide explicit commitment to Darwin's gradualism and to incorporate stochastic elements in a more satisfactory way than her original version.[21] However, none of this is of any avail in meeting the complaint that I have been developing. Given a little training in logic, Owen, Sedgwick, and Agassiz would all have endorsed Williams's axioms and her derivations. Their objections would not have concerned the truth of the statements put forward, but the fuss that was being made about matters of so little biological importance.

My brief survey of attempts to find a small set of general principles about organisms which can be hailed as Darwin's theory hardly shows that the enterprise is inevitably doomed. However, I hope that it provides some motivation for a different approach to the nature of scientific theories in general, and of Darwin's theory in particular. In the next few sections I shall offer a picture of Darwin's theory that is explicitly designed to focus the main claims and arguments of the *Origin*. Once this has been done, I shall return to the broader issues concerning the nature and confirmation of scientific theories.

III

Darwin's theory of evolution by natural selection is an explanatory device, aimed at answering some general families of questions, questions which Darwin made central to biology, by presenting and applying what I shall call *Darwinian histories*. To fix ideas, I shall characterize a Darwinian history in a preliminary way as a narrative that traces the successive modifications of a group of organisms from generation to generation in terms of various factors, most notably that of natural selection. The main claim of the *Origin* is that we can understand numerous biological phenomena in terms of the Darwinian histories of the organisms involved.

Consider first issues of biogeographical distribution. For any group G of organisms, characterized, perhaps, on the grounds of similar morphology, similar behavior, or a propensity to interbreed in nature, we can identify the range of that group. With respect to any such group we can envisage a complete description of its history. From the Darwinian perspective, this historical description will trace the modification of the current group from its ancestors, revealing how properties change along the ancestral-descendant line, and how, as these changes occur, the area occupied by members of the group alters. Darwinian histories provide the basis for answers to biogeographical questions.

One general form of biogeographical question concerns the distribution of particular groups. Thus, for any group G with range R, we may inquire why the range of G is exactly R. Let us call questions of this form *pure questions about particular features of organismic distribution*.[22] Obvious examples of pure biogeographical questions about particular groups are the question of why pangolins are found in southern Africa and Southeast Asia, and in these regions alone, and why koala bears are confined to Australia. Darwin's suggestion is that we can answer such questions by relating Darwinian histories for the pangolins and the koalas respectively.

We can envisage an ideal Darwinian answer to pure biogeographical questions about the distribution of particular groups. That answer would trace the modification of the ancestral-descendant sequence from generation to generation, showing how, at each stage, the range of the existing group of organisms resulted from their properties and the local environment. Quite evidently, the gory details would not only be irrelevant, but also confusing. As Darwin conceded, he could "in no one instance explain the course of modification in any particular instance."[23] However, he suggests that, despite the difficulty of the "descent to details," his theory owes its "chief" support to its ability to connect "under an intelligible point of view a host of facts."[24] I understand his point as follows: complete Darwinian histories would provide ideal answers to pure biogeographical questions, and it is the fact that the same *form* of answer is always to be given that constitutes the unifying power of the theory; but, in our practical study of biogeography we do not need (nor could we use) such detailed narratives; our explanation-seeking questions are answered by noting certain major features of the Darwinian history of a group of organisms.[25]

Darwinian histories provide the *basis* for acts of explanation, and, confronted with a practical question of biogeographical distribution, *incomplete* knowledge of a Darwinian history will suffice to enable us to offer an answer. When we ask why a group G occupies a range R, we typically have a more particular puzzle in mind. For example, someone who wonders why many marsupial species are found in Australia is likely to be puzzled by the fact that so few are found elsewhere. That puzzlement is relieved by outlining the Darwinian history of the marsupials—describing how they were able to reach Australia before the evolution of successful placental competitors, how the placentals were able to invade many marsupial strongholds, and how the placentals were prevented from reaching Australia. Similarly, people who inquire why the birds known as "Darwin's finches" are confined to the Galapagos are typically concerned to know what accounts for the presence on these islands of forms similar to, but specifically different from, mainland American birds. That concern is adequately addressed by pointing out that the Galapagos finches are the evolved descendants of South American birds who managed to reach the islands and successfully colonized them. In both cases, a general, unfocused, explanation-seeking question is determined in context as a more precise request for information. The request can be honored by abstracting from the Darwinian history, so that the needed information can be given despite considerable ignorance about the details.

Darwin's new proposal thus consists, in part, of two general claims: first, complete Darwinian histories provide ideal answers to pure biogeographical questions; second, incomplete knowledge about Darwinian histories can be used to answer the biogeographical questions that arise in practice. Darwin's theory displays the way in which such questions will be answered. The theory is not simply an *assertion that* certain questions are important and that they can be answered by a description of the history of a lineage, but a *demonstration of the form* that the answers are to take.

It is helpful to contrast the biogeographical part of Darwin's theory with the corresponding parts of the rival theories available in Darwin's day. One (Creationist) approach would take the history of a group of organisms to be irrelevant to their current distribution: because each group of organisms was created to inhabit a particular region, and because it has always inhabited that region, our understanding of biogeographical distribution is advanced by recognizing those features of the organisms in the group that fit them to live where they do. By 1859, this approach had fallen into well-deserved disrepute. It was well known that organisms transported by humans could thrive in areas that they had previously been unable to reach, and naturalists knew of other cases in which organisms seem ill-suited to their natural habitat (a popular example is that of those woodpeckers who inhabit treeless terrain). More promising was a Creationist view that provides some scope for history. On this view, it is suggested that the current range of an organismic group is the result of a process in which an unmodified (or relatively unmodified) sequence of organisms has dispersed from an original "center of creation." Unless this approach is supplemented with a scheme for explaining the distribution of original centers of creation, then it is evident that it will terminate our biogeographical inquiries more rapidly than Darwin's proposal does. Although we may sometimes be able to understand the current distribution of a species in terms of its dispersal from an original center of creation, there will be all too many cases in which this only postpones our puzzlement. Darwin makes the point forcefully:

> But if the same species can be produced at two separate points, why do we not find a single mammal common to Europe and Australia or South America? The conditions of life are nearly the same, so that a multitude of European plants and animals have become naturalised in America and Australia; and some of the aboriginal plants are identically the same at these distant points of the northern and southern hemispheres?[26]

Darwin's challenge is to provide a comprehensible distribution of centers of creation that will allow for the disconnected distribution of the plants common to the hemispheres, while explaining the failure of the mammals to radiate into regions for which they are well suited. The thrust is that Creationists will ultimately be forced into conceding that the distribution of original centers of creation is inexplicable. By contrast, as Darwin will emphasize, the theory of evolution claims for scientific investigation questions which rival theories dismiss as unanswerable.

I have begun with the example of biogeography because it is the case on which Darwin often lays the greatest stress, suggesting that it was reflection on biogeography that originally led him to the theory of evolution.[27] However, biogeography is only part of the story. The rest is more of the same kind of thing. With respect to comparative anatomy, embryology, and adaptation, Darwin also provided strategies for answering major families of questions.

Consider comparative anatomy. Here, the task is to provide answers to question of the general form

Why do organisms belonging to the groups G, G′ share the property P?

where G and G′ will typically be acknowledged taxa (e.g. species, genera, families, etc.) and P will be some structural property (such as bone structure in a forelimb, for example). A Darwinian answer to these questions will take one of two common forms. In cases where P is recognized as a homology (perhaps on the grounds that it is one element in a rich collection of common properties), the presence of P in both G and G′ will be ideally explained by relating the history of descent of G and G′ from a common ancestor which also possessed P. In cases where P is a "mere analogy" (perhaps recognized as such because it is only an isolated example of a shared property), its common presence will be understood by tracing the history of the emergence of P in the groups G, G′, showing how ancestors of the present members of those groups were modified so that they came to possess P, perhaps as the result of similar environmental pressures. Classic cases of both types were already described by Darwin: Similarities in the bone structure of the forelimbs in various mammalian groups—moles, seals, bats, ruminants—are to be understood in terms of descent from a common ancestor. By contrast, the existence of wings in birds, bats, flying reptiles, and insects, is understood by recognizing the paths which these groups have followed in evolving the ability to fly.[28]

As in the case of biogeography, while relating the complete Darwinian histories of the groups involved would provide an ideal answer to a question about the relationships among them, our practical questions about the similarities among organisms do not require such detail. Quite frequently, the question of why two groups of organisms agree in a morphological property stems from puzzlement that organisms so different in other respects should share the morphological property in question. In the case where the property is a homology, the puzzle is resolved by outlining enough of the Darwinian histories of the organisms to reveal the main lines of their modifications from a common ancestor. Similarly, in the case of analogy, we need to tell enough of the Darwinian history to recognize how a similar feature has been produced in unrelated lineages.

Moreover, like Darwin's treatment of biogeography, the approach to comparative anatomy is easily contrasted with potential Creationist accounts. Appeals to common design for common environments are difficult to defend when the Creationist comes to details:

> It is difficult to imagine conditions of life more nearly the same than deep limestone caverns under a nearly similar climate; so that on the common view of the blind animals having been separately created for the American and European caverns, close similarity in their organization and affinities might have been expected; but, as Schiödte and others have remarked, this is not the case, and the cave-insects of the two continents are not more closely allied than might have been anticipated from the general resemblance of the other inhabitants of North America and Europe.[29]

Darwin's point is that, when we come to investigate the details of the similarities and differences among groups of organisms, his own proposal will offer answers to questions that rival approaches have to dismiss as unanswerable.

The third example that I shall consider is historically crucial, in that it represents the most promising field for the tradition of natural theology. Darwin confesses that his theory could not be admitted as satisfactory "until it could be shown how the innumerable species inhabiting this world have been modified, so as to acquire that perfection of structure and coadaptation which most justly excites our admiration."[30] However, he proposes that questions of adaptation, like questions of biogeography and organismic relationships, can be answered by rehearsing the historical process through which the adaptation emerged. The general form of question to be addressed is:

Why do organisms belonging to group G living in environment E have property P?

where the property P is a characteristic which appears to assist its bearers in environment E. A complete answer to this question would trace the Darwinian history of G from the time just prior to the first occurrence of P, showing how the variation producing P first arose, how it was advantageous to its bearer in the original environment, and how that advantage enabled P to become progressively more prevalent in subsequent generations of the lineage. (Here I am deliberately overdrawing the adaptationist commitments of Darwin's theory. I shall consider later whether Darwin allows a more pluralistic approach to the evolutionary explanation of apparently beneficial characteristics.)

As before, our understanding of the presence of properties in current groups of organisms is not dependent on our ability to recognize all the details of the historical processes through which those properties were selected. It is enough to understand the general character of the ancestral form, the way in which a variant might have arisen from that form, and the kinds of advantages that the variant could have been expected to serve. In different contexts, different features will require emphasis. So, for example, with "organs of extreme perfection" the trouble is to recognize the advantages that such structures might serve before they are fully developed.

Since the problem of adaptation is the stronghold of approaches that emphasize the design of nature, the *Origin* contains numerous passages in which Darwin contrasts the explanatory power of his own theory with the deficiencies of its main rival. In some places, he stresses the difficulty of finding any coherent account of the creative design that will do justice to the aspects of nature which are "abhorrent to our ideas of fitness."

> We need not marvel at the sting of the bee causing the bee's own death; at drones being produced in such large numbers for one single act and then being slaughtered by their sterile sisters; at the astonishing waste of pollen by our fir-trees; at the instinctive hatred of the queen bee for her own fertile daughters; at ichneumonidae feeding within the live bodies of caterpillars.[31]

Other passages descant on the "Panda's thumb theme,"[32] the existence of many cases in which it is evident that natural contrivances fall far short of the standards of good design we would expect from a competent engineer, and in which it is more plausible to suppose that the available materials dictated a clumsy solution to a design problem.

> He who believes in the struggle for existence and in the principle of natural selection, will acknowledge that every organic being is constantly endeavouring to increase in numbers; and that if any one being vary ever so little, either in habits or structure, and thus gain an advantage over some inhabitant of the country, it will seize on the place of that inhabitant, however different it may be from its own place. Hence it will cause him no surprise that there should be geese and

frigate-birds with webbed feet, either living on the dry land or most rarely alight-
ing on the water; that there should be long-toed corncrakes living in meadows
instead of in swamps; that there should be woodpeckers where not a tree grows;
that there should be diving thrushes, and petrels with the habits of auks.[33]

This theme receives its most detailed treatment in Darwin's book on orchids—
characterized as "a 'flank movement' on the enemy."[34] Again, it points toward the
same moral: questions that rival approaches must dismiss as unanswerable can be
tackled by adopting the Darwinian perspective.

IV

Darwin's theory is a collection of problem-solving patterns, aimed at answering
families of questions about organisms, by describing the histories of those organisms. The complete histories will always take a particular form in that they will trace
the modification of lineages of organisms in response to various factors—"Natural
Selection has been the main but not exclusive means of modification."[35] The time
has come to take a closer look at the notion of a Darwinian history and to distinguish "grades of Darwinian involvement."

There is a notion of Darwinian history that is minimal in the sense of embodying the fewest assumptions about the tempo and mode of evolutionary change. This
conception can be characterized as follows:

> A Darwinian history for a group G of organisms between t_1 and t_2 with respect to
> a family of properties F consists of a specification of the frequencies of the prop-
> erties belonging to F in each generation between t_1 and t_2.[36]

This minimal conception allows for evolutionary change, for the property frequencies may vary from generation to generation—indeed, properties initially absent
may ultimately be found in every member of the group—but it does not offer any
account of why this change occurs. At times, it appears that Darwin saw his primary
achievement in the *Origin* in terms of the introduction of the minimal conception
of a Darwinian history. Perhaps believing that half a loaf might be better than none,
Darwin responded to a criticism in the *Athenaeum* by claiming that the commitment to a particular view about how evolution has occurred "signifies extremely
little in comparison with the admission that species have descended from other
species and not been created immutable; for he who admits this as a great truth has
a wide field open to him for further inquiry."[37] Darwin was right to suggest that *some*
of the questions he proposed to answer could be undertaken by constructing
minimal Darwinian histories. Faced with a question of biogeography—for example,
the question of why the Galapagos contains endemic species of finches which are
similar to mainland South American forms—one might respond by describing a
history of descent with modification that offered no account of the modifying
factors.[38] Similarly, some questions about the relationships among groups of organisms can be addressed by rehearsing histories of descent with modification that do
not explore the causes of the alterations which have occurred in the relevant lineages. However, Darwin's own voyage to the advocacy of evolution makes it clear

that a minimal evolutionary theory which proposed to answer biological questions by offering minimal Darwinian histories would be vulnerable to serious challenges if it failed to specify any possible mechanisms for the modification of organisms.

More satisfactory is a suggestion that Darwin sometimes seems to favor in his most cautious moments—those moments at which he contends that the important point is to accept the existence of evolution, whatever one's views about the actual mechanisms of evolutionary change. Minimal Darwinian histories are to be used to answer biological questions, but, while we remain agnostic about the causes of modification in any particular case, we do regard ourselves as understanding the general ways in which evolutionary change is to be explained. Thus natural selection is identified as a *possible* agent of evolutionary change, in conjunction with such other agents as use and disuse, correlation and balance, direct action of the environment, stochastic factors, and so forth.[39] On this approach, we would not pretend to explain the modifications that have taken place along a particular lineage, and we would answer only those biological questions that can be addressed through the construction of minimal Darwinian histories. Quite evidently, we would have to forego attempts to tackle questions about organic adaptations.[40]

Numerous passages in Darwin's writings indicate that he preferred to be more ambitious.[41] A stronger conception of Darwinian history involves not only a speci-fication of the changes that take place from generation to generation in a group of organisms, but also a sequence of derivations that will infer the distribution of prop-erties in descendant generations from those in ancestral generations. These deriva-tions will exemplify certain patterns, patterns that reflect views about the agents of evolutionary change. The *selectionist* pattern proposes to derive increased frequen-cies of properties in descendant generations by identifying the advantages which those properties conferred on their bearers in ancestral generations. Ideally, one would show precisely how the possession of a property P gave to ancestral organ-isms an identifiable increase in the propensity for survival and reproduction, and how this exact enhancement of fitness led to the subsequent increase in the fre-quency of P. Other patterns involve use and disuse, and correlation and balance of characters. The former traces decreasing frequencies of structures in descendant generations to the fact that the structures were unused by the ancestors who pos-sessed them. The latter explains the increased frequency of a characteristic by con-tending that it is correlated with a property whose increased frequency can be explained in other ways, perhaps by invoking the selectionist pattern.[42] In each of these patterns one can give more or less scope to stochastic factors by allowing for greater or less disagreement between the expected outcome of a derivation of fre-quencies of properties and the actual distribution.[43]

Quite evidently, a commitment to a stronger conception of Darwinian history makes it possible to answer questions, such as those that involve "perfections of structure," which lie beyond the scope of minimal Darwinian histories. This com-mitment may be undertaken more or less pluralistically. That is, one may allow as equally appropriate a number of different patterns for deriving changes in property frequencies, or one may insist that a particular style of explanation should pre-dominate. So, for example, Darwin's suggestion that natural selection is the major agent of evolutionary change can be interpreted as a commitment to preferring to understand the distribution of characteristics in a group of organisms by invoking the selectionist pattern. However, there are several different possibilities of interpretation even here. One (implausibly strong) construal is to suppose that, for virtually any characteristic of virtually any organismic group, the prevalence of that characteristic is to be understood in terms of the advantages which the characteristic conferred on those ancestors who bore it. A more moderate interpretation is to suggest that prevalent characteristics are to be explained either directly, by citing their advantages, or indirectly, by pointing to correlations with characteristics which brought advantages. Or one may decrease the scope of selectionist explanations, proposing that they are appropriate for most instances of *major* evolutionary change, that is, for those cases in which we endeavor to understand the prevalence of some property which distinguishes an organismic group.

The point I have been making is that the *Origin* not only allows for the use of more or less ambitious notions of Darwinian history, but also covers a range of positions on the priority of selectionist explanations. My conclusion underscores a claim made by Stephen Gould and Richard Lewontin, who note that "the master's voice" is often more tolerant of alternatives than is usually thought.[44] In further support of this shared judgment, it is noteworthy that the correspondence between Darwin and Wallace on the origin of sterility shows that Darwin needed evidence in *favor* of a selectionist explanation, rather than holding that selectionist explanations were preferable until proved impossible.[45] Moreover, the following passage from the first edition[46] of the *Origin* shows Darwin's anticipation of the possibility that biologists might want to impose selectionist explanations as widely as possible.

> If green woodpeckers alone had existed, and we did not know that there were many black and pied kinds, I dare say that we should have thought that the green colour was a beautiful adaptation to hide this tree-frequenting bird from its enemies; and consequently that it was a character of importance and might have been acquired through natural selection.[47]

Darwin is sensitive to an important point. The presence of properties in contemporary organisms, even properties that suggest to us some benefit which they confer, is not necessarily to be explained by applying the selectionist pattern.

So far I have indicated a number of different theories which might be reconstructed from the *Origin*. These theories differ first in whether they attempt to explain changes in property frequencies along a lineage and, second, about the forms of explanation which they admit or to which they give emphasis. Unfortunately, this does not exhaust the variety of versions of Darwinism. Nothing I have said recognizes Darwin's commitment to evolutionary gradualism, nor have I allowed for a possible Darwinian flirtation with selection of groups rather than individuals. Both of these further variants can be accommodated within the framework I have proposed.

Huxley complained that "Mr. Darwin has unnecessarily hampered himself by adhering so strictly to his favourite '*Natura non facit saltum*,'"[48] Darwin's gradualism is not easy to characterize in general,[49] but I view it as the imposition of a constraint on Darwinian histories. An admissible specification of the successive frequencies of properties in a family F along a lineage must reveal distributions that are (in some sense) continuous and which are modified in "small" steps. At the very least, Darwin would ban histories in which a property absent in one generation is fixed in the next or in which a magnitude which admits of degrees shows an increase in degree without taking on intermediate values. Darwin's writings are full of passages that suggest a more stringent constraint.[50] But not all Darwinians agreed. Huxley preferred to allow for Darwinian histories which are liberated from any such requirement.

Finally, there are some passages in the *Origin* that may be read as indicating a different strategy of biological explanation than any so far considered. In his discussion of social insects, Darwin suggests that certain properties of communities of organisms, to wit, the existence within those communities of organisms with particular characteristics (for example, sterile workers), are present because those properties have proved advantageous in the past to the communities which possessed them.[51] These remarks point toward an alternative conception of a Darwinian history. All the notions discussed so far are *individual-oriented*. An individual-oriented history assigns frequency distributions of properties to successive generations of a lineage, and, if it is not minimal, supplies derivations of those frequencies, derivations that accord with particular preferred patterns. By contrast, a *group-oriented* history specifies the distributions of groups of organisms with particular properties at particular times, and attempts to derive these distributions using preferred patterns of reasoning. So, for example, it may be argued that the current dominance of groups of organisms in which reproduction is, at least occasionally, sexual, is the result of a historical process in which sexually reproducing groups of organisms have been able to produce more varied descendant groups and thus "to seize on many and widely diversified places in the polity of nature."[52]

Darwin's own preferred examples of "selection applied to the family"[53] are not developed in any great detail, so that it is hard for a contemporary champion of group selection to derive much support from them. Nevertheless, the ambiguous remarks about advantages that enable a species to give rise to descendants capable of occupying more niches do suggest another variant of Darwinism. Hence, among our versions of Darwinian evolutionary theory, minimal, pluralistic, selectionist, gradualistic, and so forth, I have included one which allows for group-oriented Darwinian histories.

V

We have begun to understand how the *Origin* might make a novel, controversial, and nontrivial contribution to biological theory—indeed how it might contain suggestions of a number of different theories with greater or lesser degrees of daring. Yet the identification of Darwin's theory with a collection of schemata for answering questions (or his theories with collections of schemata for answering questions) is only a beginning. To recognize the extent of Darwin's achievement we must give substance to the idea that Darwin reconstructed the field of biology. I shall now try to embed my account in a more general discussion of scientific change.

Any adequate conception of scientific change must contain a view of how the state of science at any given time is to be represented. For many philosophers, it is tacitly assumed that the representation will identify the language which is in use and the statements of the language which the scientists of that time accept. Thus changes within a field of science will be charted by looking for the ways in which the language of the field develops, how new statements come to be accepted, old statements rejected.[54] I believe that so simple an account of the state of science at a time will not do. If we are to understand the transition from the state of science at one time to its state at a subsequent time (or at subsequent times), we need a more complex and refined characterization of these states. To this end, I shall introduce the concept of a *scientific practice*.[55]

A *scientific practice* consists in a language, a set of statements in that language accepted by the scientists whose practice it is, a set of questions which are accepted as the important unanswered questions by those scientists, a set of schemata which specify the forms which answers to those questions are to take, a set of experimental techniques, and a set of methodological directives designed to aid scientists in assessing the credentials of rival proposals for answering open questions. For present purposes, I intend to concentrate solely on the first four components of the practice (language, accepted statements, important unanswered questions, and schemata for specifying the forms of answers to those questions). Darwin's achievement can best be understood, I think, by recognizing the ways in which he modified these four components of scientific practice.[56]

Let me preface my reconstruction of Darwin's transformation of scientific practice by noting explicitly that I doubt very much that all the episodes that are typically identified as major cases of "theoretical change" form a homogeneous class. I suspect that there are examples of theoretical change in physics—perhaps the case of the transition from classical Newtonian dynamics to the special theory of relativity is one—in which we can take the primary focus of the change to be the language of the practice and the set of accepted statements of the practice.[57] In instances like these, the traditional approach of concentrating on the introduction of new concepts and new general principles will prove adequate. Indeed, we may be able to identify the newly introduced theory by writing down some small set of theoretical postulates, we may be able to understand the efforts of the innovators as directed at confirming these postulates, and we may regard the newly accepted statements as the deductive consequences of the new postulates. But not all cases of major theoretical change in science are like this. Specifically, the case of Darwin is not.

Darwin's modification of the language of biology was relatively minor.[58] Certainly, after the acceptance of the *Origin*, it was necessary to abandon some criteria that had traditionally been used to identify the referent of "species," and Darwin introduced a new method of fixing the referent of "homology," but there is nothing comparable to the massive conceptual shifts we find in other cases during the course of the history of science.[59] Darwin did effect large changes in the set of accepted statements. His work introduced a large variety of new claims about particular organisms, their histories, relationships, distribution, and so forth. It would be counterintuitive to identify Darwin's theory with this motley of information, and as we have already seen, it is not easy to find a small set of general claims from which the descriptions of specific organisms are to flow. The problem with approaching the Darwinian revolution by asking what new statements Darwin advanced is not that he puts forth no novel assertions but that the *Origin* is a hodge-podge of specific original claims about barnacles, pigeons, South American mammals, social insects, arctic flowers, Scotch fir, and so forth. As Huxley noted, the *Origin* is a hard book, and the reader may all too easily find it "a sort of intellectual pemmican—a mass of facts crushed and pounded into shape, rather than held together by the ordinary medium of an obvious logical bond."[60] William Hopkins offered a more negative perspective on the same difficulty, commenting that "many details are apt to perplex the mind and to draw it off from general principles and real arguments."[61]

The trouble is that Darwin's primary achievement is the introduction of schemata for answering certain families of biological questions, and the identification of the questions that biologists should set for themselves. The mass of details is a cornucopia of illustrations. Darwin's initial claim is that certain questions—the questions of why organisms have the properties, distributions, and interrelationships that they do—should be taken as the central questions of biology. Because the principal reasons for *not* viewing these questions as the major unsolved problems of biology depended on the apparent impossibility of answering them, Darwin's principal task in introducing them lay in showing that it is indeed possible to provide informative answers to them.[62] Questions which had inevitably seemed to belong to the province of theological speculation were claimed for scientific discussion.[63] Quite evidently, Darwin's specification of the general forms that answers to questions about adaptations, homologies, distribution, and so forth, would take required him to defend the claim that his preferred schemata are applicable on a broad scale. As we shall see, much of the argumentative work of the *Origin* consists in attempting to demonstrate that the schemata advanced by Darwin can be broadly instantiated.

Finally, the introduction of the new schemata sets new questions for biology, in that, after Darwin, naturalists are given the tasks of (i) finding instantiations of the Darwinian schemata (i.e., developing Darwinian explanations of particular biological phenomena), (ii) finding ways of testing the hypotheses that are put forward in instantiating Darwinian schemata, (iii) developing theoretical accounts of the processes which are presupposed in Darwinian histories (specifically such processes as hereditary transmission, and the origination and maintenance of variation). These tasks arise in different ways. The first, (i), is simply the result of Darwin's claiming of the questions about distribution, adaptation, and relationships as legitimate questions for scientific investigation. (ii) is generated by the fact that, in attempting to instantiate Darwinian schemata, biologists are compelled to advance hypotheses about the historical development of life, and it is incumbent on them to specify ways of testing these hypotheses, so as to avoid the charge that evolutionary theory is simply an exercise in fantasizing.[64] Finally, (iii) stems from the fact that Darwin's theory is not only open-ended in provoking many specific inquiries into the properties, relationships, and distribution of particular organisms, but also in raising very general questions about the historical processes through which organisms have become modified.[65]

Thus, a summary of Darwin's modification of the state of science should take the following form. By including major families of questions that had traditionally been assigned to speculative theology, Darwin changed the set of biological questions accepted as important. He amended the orthodox views about how questions of these kinds should be answered (insofar as such questions were taken to be susceptible of treatment at all) by proposing schemata which answers should exemplify, schemata which invoke the general idea of understanding the current features of organisms by relating a history of descent with modification. Darwin's own efforts at instantiating these schemata led him to put forward hypotheses about the histories of particular organisms. As a result, the *Origin* contains a motley of new theses about individual types of organisms—thus taking on the character of an "intellectual pemmican" (in Huxley's phrase). Moreover, Darwin's schemata and his own instantiations of those schemata introduced new questions concerning the testing of hypotheses about the history of organisms and the general character of the processes presupposed in Darwinian histories. Finally, because of the general presuppositions of the notion of a Darwinian history—in particular the view that it is possible for descendants of one species to belong to a different species—it was necessary to modify the language of biology in certain respects.

What Darwin constantly emphasized, and what his contemporaries recognized, was that the *Origin* was not only a confession of ignorance but also a structuring of our ignorance.[66] As my summary indicates, its primary accomplishment lay in identifying the questions that biologists ought to ask. It is because of this primary accomplishment that Darwin may truly be said to have revolutionized the field. The nature of that revolution is captured in one of Hooker's letters to Darwin:

> But, oh Lord!, how little do we know and have known to be so advanced in knowledge by one theory. If we thought ourselves knowing dogs before you revealed Natural Selection, what d—d ignorant ones we must surely be now we do know that law.[67]

VI

I claim that if Darwin's achievement is construed in the way I have just suggested then we can give an illuminating reconstruction of his "long argument."[68] I divide the reasoning of the *Origin* into three main parts:

1. An attempt to show that it is possible to modify organisms extensively through a natural process (natural selection).
2. An attempt to show that, given the possibility of hypothesizing that organisms now classed in separate species (or higher taxa) are related by descent from a common ancestor, the introduction of such hypotheses would enable us to answer many questions about these organisms.
3. An attempt to respond to difficulties that threaten the introduction of hypotheses about common descent.

The early chapters are directed at (1), and it is in these chapters that the celebrated argument by analogy with artificial selection plays its crucial role.[69] Darwin adduces a number of examples, most prominently examples of different kinds of pigeons, to show that the conscious selection employed by plant and animal breeders has been able to produce striking modifications of organisms. Claiming that the struggle for existence imposes a selective process which is analogous to the deliberate selection of the breeder, he concludes that it is possible to suppose that large modifications can also be produced in nature. Hence it is unwarranted to maintain that hypotheses asserting the modification of an ancestral species to produce a quite different descendant are, in principle, inevitably false.

The role of the analogy with artificial selection is thus to clear the way for subsequent claims about the genealogical relationships of organisms. Darwin believes that he can support such claims by showing how they enable us to answer large numbers of questions about the characteristics, relationships, and distribution of organisms. This indirect support would be of little help if opponents could always charge that it is impossible that the attributions of descent with modification could be true.[70] The study of "variation under domestication" together with the recognition of variation and competition in nature blocks the charge by explaining how some natural modification of organisms is possible. Darwin's adversaries are thus compelled to meet his explanatory attributions of genealogical relationships with the claim that there are limits to the power of selection to modify a group of organisms.

Darwin's critics rose to the challenge. Several reviews of the *Origin* protested that Darwin had no direct evidence of large-scale modifications by natural or artificial selection. Typical were the comments of Thomas Vernon Wollaston:

> There is no reason why *varieties*, strictly so called, . . . and also geographical "subspecies," may not be brought about, even *as a general rule*, by this process of "natural selection": but this, unfortunately, expresses the limits between which we can imagine the law to operate, and which any evidence, fairly deduced from facts, would seem to justify: it is Mr. Darwin's fault that he presses his theory too far.[71]

Because Darwin could only suggest the *possibility* of unlimited variation, he was roundly chided by his critics for deserting the true path of science. Drawing an invidious contrast, William Hopkins descanted on the accomplishments of the physicists:

> They are not content to say that it *may* be so, and thus to build up theories based on bare possibilities. They *prove*, on the contrary, by modes of investigation that cannot be wrong, that phenomena exactly such as are observed would *necessarily*, not by some vague possibility, result from the causes hypothetically assigned, thus demonstrating those causes to be the true causes.[72]

In a letter to Asa Gray, Darwin explained clearly how Hopkins had failed to appreciate the force of his argument:

> I believe that Hopkins is so much opposed because his course of study has never led him to reflect much on such subjects as geographical distribution, homologies, &c., so that he does not feel it a relief to have some kind of explanation.[73]

Although Darwin took some trouble in the opening chapters of the *Origin* (and in the *Variation of Animals and Plants under Domestication*) to show that artificial selection is capable of producing quite dramatic modifications of organisms, his principal response to the charge that variation is only limited is that it is beside the point. The analogy with artificial selection is not intended to demonstrate—nor does it need to demonstrate—that variation is unlimited. Unless some reason can be given for supposing that there are limits to variation, then the explanatory power of the hypotheses that attribute descent with modification justifies us in accepting them, even though modifications as extensive as those which are hypothesized have not been directly observed. Only someone insensitive to the explanatory power of the novel theory—a nonbiologist like Hopkins, for example—will fail to realize that there is evidence for supposing that selection has quite extensive powers whose action cannot be directly demonstrated. The opening chapters of the *Origin* thus clear some space within which Darwin can defend his schemata for tackling biological questions by appealing to their power to unify the phenomena.[74]

If I am right, then the principal burden of argumentation should fall on the concluding chapters of the *Origin* in which the explanatory power of the theory is most extensively elaborated. Darwin himself seems to have seen his book in this way: he begs Lyell to keep his mind open until reading the "latter chapters, which are the most important of all on the favourable side."[75] Darwin's approach is to marshall an impressive array of puzzling cases of geographical distribution, affinity of organisms, adaptation, and so forth, aiming to convince his reader that there are numerous questions to which answers fitting his schemata would bring welcome relief. Consider, for example, Darwin's partial agenda for biogeography. After describing the "American type of structure" found in the birds and rodents of South America, Darwin suggests that biologists ought to ask what has produced this common structure. The similarities are too numerous just to be dismissed as beyond the province of scientific explanation.

> We see in these facts some deep organic bond, prevailing throughout space and time, over the same areas of land and water, and independent of their physical conditions. The naturalist must feel very little curiosity, who is not led to inquire what this bond is.
>   This bond, on my theory, is simply inheritance, that cause which alone, as far as we know, produces organisms quite like, or, as we see in the case of varieties, nearly like each other. The dissimilarity of the inhabitants of different regions may be attributed to modification through natural selection, and in a quite subordinate degree to the direct influence of physical conditions.[76]

The message of this passage—and of numerous similar passages that occur in the last four chapters of the *Origin*[77]—is clear. There are many details about particular organisms that cry out for explanation. Darwin's proposal to answer questions about distribution (and so forth) by instantiating a particular schema (particular schemata)[78] explains—or at least promises to explain—these otherwise inexplicable details.

But is the theory to be praised for its explanatory promise, or does it actually deliver explanations? Darwin was sometimes inclined to make the stronger claim:

> Thus, on the theory of descent with modification, the main facts with respect to the mutual affinities of the extinct forms of life to each other and to living forms,

seem to me explained in a satisfactory manner. And they are wholly inexplicable on any other view.[79]

Some reviewers were unconvinced. Hopkins protested:

A phenomenon is properly said to be *explained*, more or less perfectly, when it can be proved to be the necessary consequent of preceding phenomena, or more especially, when it can be clearly referred to some recognised cause; and any theory which enables us to do this may be said in a precise and logical sense, to explain the phenomenon in question. But Mr. Darwin's theory can explain nothing in this sense, because it cannot possibly assign any necessary relation between the phenomena and the causes to which it refers them.[80]

Hopkins's remarks make it clear that he regards Darwin's "explanations" as falling short in two main respects: the hypotheses about descent with modification which are invoked in answering biological questions are not independently confirmed, nor are those hypotheses linked by a gapless sequence of inferences to a description of the phenomena to be explained. The first demand is easily resisted. Darwin was fond of remarking that his proposal was no different from that of the physicists who introduced "the undulatory theory of light," without any direct demonstration of the passage of waves through the luminiferous ether, on the basis of its ability to explain the phenomena of diffraction, interference, polarization, and so forth.[81] The second point is more tricky. I suggest that Darwin appreciated the fact that claims that a theory explains the phenomena are ambiguous. Explanations are responses to questions, actual or anticipated, and what is enough to answer one question may not suffice to answer another, even a question posed in the same form of words. To ask why a group of organisms shares a common feature may simply be to wonder about the nature of the bond that unites them, or it may be already to presuppose the character of that bond and to inquire how the feature in question has been preserved through a course of modifications. The latter question will require a different—and more detailed—answer than the former.

Consider the case of the South American fauna. Darwin envisages a naturalist struck with the similar morphology of the South American rodents and the differences between these rodents and the European forms. The *first* question that arises is why the South American organisms are so similar to one another and distinct from the European rodents, and this question can be answered by pointing out that there is a history of descent with modification which traces all of the American organisms to a common ancestor more recent than any ancestor that they share with any of the European forms. The naive naturalist's puzzlement is completely answered when we know that *there is* a Darwinian history of this general form; we do not need to know the exact details of that history. Hence Darwin is entitled to claim that his theory does deliver some explanations, for it adequately answers some explanation-seeking questions. Equally, it is correct to note that Darwin only *promises* other explanations, for there are questions which can only be completely answered by recognizing much more of the detail of the Darwinian history of the organisms concerned. A more sophisticated naturalist, one who already presupposes that the common bond among the South American rodents is "simply inheritance," may inquire why the coypu and capybara are so similar, and in this context, what is needed is to relate enough of the Darwinian histories of these organisms to show how the common features have been preserved, while other characteristics have been modified.

Not only is there a distinction here, but it is (as I have already hinted) a distinction of which Darwin was aware. Although his published and unpublished writings are full of passages which claim that the theory of evolution provides explanations, Darwin explicitly notes that many questions require more detail than he is able to give:

> Very many difficulties remain to be solved. I do not pretend to indicate the exact lines and means of migration, or the reason why certain species and not others have migrated; why certain species have been modified and have given rise to new groups of forms, and others have remained unaltered.[82]

The right response to Hopkins is to maintain that his conditions on explanation are too restrictive, that the *Origin* already offers some explanations and that it indicates the lines along which further explanations are to be sought.

The final chapters of the *Origin* contain the most extensive discussions in which Darwin parades the power and promise of his theory. But there are earlier passages in which he tries to show how the rehearsing of Darwinian histories could enable us to understand some general features of the organic world. Thus Darwin is at pains to make clear how his theory accounts for the existence of discrete taxa and for the "great fact" that these taxa form nested sets.[83] Similarly, he attempts to explain why specific characteristics should be more variable than generic characteristics, why "a part developed in any species in extraordinary degree or manner, in comparison with the same part in allied species, tends to be highly variable," and why we find cases of "reversion," in which organisms of one species show characteristics found in allied or ancestral species.[84] Darwin's discussion of this last topic is especially interesting, in that it serves as the occasion for one of his most aggressive comparisons of the explanatory merits of his own theory with the deficiencies of previous views:

> He who believes that each equine species was independently created, will, I presume, assert that each species has been created with a tendency to vary, both under nature and under domestication, in this particular manner, so as often to become striped like other species of the genus; . . . To admit this view is, as it seems to me, to reject a real for an unreal, or at least for an unknown, cause. It makes the works of God a mere mockery and deception; I would almost as soon believe with the old and ignorant cosmogonists, that fossil shells had never lived, but had been created in stone so as to mock the shells now living on the sea-shore.[85]

Let us now turn to the third part of the argument of the *Origin*, which focuses on apparent difficulties that might be held to stand in the way of constructing Darwinian histories. To construct a Darwinian history will typically involve the scientist in advancing a hypothesis about the existence of certain ancestral organisms with particular properties. In many cases, the fossil record will contain no remnants of such organisms. How is this embarrassing lack of evidence to be understood?[86] Moreover, there are some properties of organisms which, in their final form, obviously assist their bearers, but which would appear to be at best useless if they were present in an incomplete state. If a Darwinian history is to show us how natural selection favored the emergence of a property of this kind—the presence of an eye, to cite the most hackneyed case—then it seems that we must show how, contrary to appearances, the incipient characteristics were themselves useful.[87] Similarly, Darwin devotes attention to the problem of understanding how the emergence of a sterility barrier might be explained. In all these cases, by turning back a challenge which would initially appear to limit the scope of the strategy of answering biological questions through constructing Darwinian histories, Darwin defends the broad claim that the entire families of questions that he has made central to biology can be answered in the way that he suggests.

Darwin's most acute critic, Fleeming Jenkin, saw clearly that the argument of the *Origin* ultimately rested on Darwin's contention that he could explain a very broad range of biological phenomena, a contention that could be undercut by showing that the class of explanations was far more limited than had been claimed.[88] Jenkin put the point as follows:

> The general form of his argument is as follows:—All these things have been, therefore my theory is possible, and since it is a possible one, all those hypotheses that it requires are rendered probable. There is little direct evidence that any of these maybe's actually *have been*.
>
> In this essay [Jenkin's review] an attempt has been made to show that many of these assumed possibilities are actually impossibilities, or at the best have not occurred in this world.[89]

Behind Jenkin's caricature of Darwin's argument is a sound point. If it could indeed be shown that some of the questions that Darwin hoped to claim for science were unanswerable in the ways that he suggested, then he would be vulnerable to the charge that his proposals for answering other questions of the same types were nothing more than idle speculations. Hence, the *Origin* contains numerous passages in which Darwin labors to show that apparent impossibilities are only apparent.[90] The three enterprises I have described dovetail to provide good reasons for modifying the practice of biology. Consider the situation from which biologists began. Certain features of the organic world had to be dismissed as brute facts, because it was felt that there was no means of answering the question of why they are present which would not appeal (at best, quite quickly, at worst, immediately) to the unfathomable fiat of a creator. Darwin's primary task is to show that such questions are indeed answerable. To do this he must emphasize the puzzling character of the phenomena to be explained and show how his schemata for answering the questions provide immediate relief from some forms of ignorance and promise relief from other such forms. He must also rebut two types of skepticism. One doubts that it is possible to achieve any kind of modification of organisms in nature. This worry is addressed by using the results of plant and animal breeders, and by showing how the struggle for limited resources provides a way for nature to select. The second type of skepticism objects to the possibility of applying the Darwinian strategies to all of the examples that they are intended to address. Darwin tackles this issue head on, by arguing that the difficulties with absent transitional forms, instincts, and complex adaptations dissolve under closer scrutiny.

Given that the argument for changing scientific practice so as to include questions about biogeography, adaptation, relationships, and so forth is cogent, then further modifications of that practice follow easily. The Darwinian schemata are introduced to specify the forms of admissible answers to the newly introduced questions. New statements are accepted because they form part of Darwinian answers to biological questions.[91] Linguistic usage is altered because it is no longer possible to maintain the theoretical presuppositions of certain terms. Any method of fixing the reference of the names of species taxa that presupposes that species are closed under reproduction will fail.[92] Finally, and most importantly, by introducing the Darwinian schemata, one recognizes further phenomena about which questions must subsequently be raised. Any Darwinian history presupposes variation, competition, and inheritance, and the restructuring of biology around the provision of Darwinian histories focuses attention on new theoretical issues surrounding these phenomena. How much variation is there in a naturally occurring population? How does this variation arise? How is it maintained? In what ways do different organisms (and different taxa) compete with one another? How are characteristics transmitted between generations? How are properties of organisms correlated with one another? These are large questions which assume great importance in the post-Darwinian context, and as both Darwin and Huxley foresaw,[93] the subsequent history of evolutionary theory is, in large measure, an attempt to find answers to them.

VII

Darwin's "long argument" does not explicitly confront an objection that was put forward by his most astute critics and that has played an important part in subsequent discussion of the merits of evolutionary theory. The criticism centers on the idea that it will be all too easy to produce stories about the histories of groups of organisms that meet the conditions imposed by Darwin's schemata. Because these conditions are so loose, the critics charge, one can always make up an appropriate account for whatever relationships, distributions, or characteristics one finds in the organic world. Hence the idea that evolutionary theory provides unified answers to questions that are unanswerable on rival approaches is simply false advertising.

Early reviewers of the *Origin* sought instances in which the application of a Darwinian schema would yield some definite statement whose truth value could be determined by observation. Each of them had his own favorite case in which it appeared that this *ought* to occur, and in which it seemed that the *Origin* frustrated legitimate expectations. Because definite predictions from Darwin's theory were so elusive, the critics concluded that the theory was equipped with devices that would permit it to dodge any uncomfortable observational finding. Pictet adduced a popular example, the lack of significant change in animals whose properties have been documented over centuries:

> If the 4000 years which separate us from the mummies of Egypt have been insufficient to modify the crocodile and the ibis, then Mr. Darwin can always reply that this period of time is really trifling. I dare not argue with such weapons whose range I cannot appreciate.[94]

Hopkins perceived troubles for Darwin in the richness of the fauna revealed in the oldest fossil deposits:

> Our author is perplexed with the existence of trilobites, comparatively highly-organized animals, in almost the earliest fossiliferous strata, and to make the fact square with his theory, he at once creates a hypothetical world of indefinite duration for the due elaboration of the ancestral dignity of these intrusive crustaceans.[95]

The most vigorous objection was made by Fleeming Jenkin. After challenging Darwin's suggestions that female choice might suffice to account for "the wonderful minutiae of a peacock's tail," Jenkin offered an inventory of Darwinian strategies for dodging refutations:

> A true believer can always reply, "You do not know how closely Mrs. Peahen inspects her husband's toilet, or you cannot be absolutely certain that under some unknown circumstances that insignificant feather was really unimportant"; or finally, he may take refuge in the word correlation, and say, other parts were useful, which by the law of correlation could not exist without these parts; and although he may not have one single reason to allege in favour of any of these statements, he may safely defy us to prove the negative, that they are not true. The very same difficulty arises when a disbeliever tries to point out the difficulty of believing that some odd habit or complicated organ can have been useful before fully developed. The believer who is at liberty to invent any imaginary circumstances, will very generally be able to conceive some series of transmutations answering his wants.
>
> He can invent trains of ancestors of whose existence there is no evidence; he can marshal hosts of equally imaginary foes; he can call up continents, floods, and peculiar atmospheres, he can dry up oceans, split islands, and parcel out eternity at will; surely with these advantages he must be a dull fellow if he cannot scheme some series of animals and circumstances explaining our assumed difficulty quite naturally.[96]

The objections leveled by Pictet, Hopkins, and Jenkin are seemingly very powerful, and much of the continued suspicion about the scientific status of evolutionary theory reflects the fact that, a century and a quarter after the *Origin*, it is still hard to say exactly what is wrong with them. The criticism can be presented in two different forms. One version begins from the premise that genuine scientific theories ought to be testable, uses the examples discussed by Jenkin et al. to deny that evolutionary theory lends itself to any possible test, and draws the obvious unflattering conclusion about Darwin's theory. The second version explicitly addresses Darwin's "long argument." As we have seen, that argument rests on the claim that the same schemata can be applied again and again to answer a host of otherwise unanswerable biological questions. The critic who has been frustrated by reading Darwin's responses to the apparent difficulties of applying these schemata (for example, in the cases of "organs of extreme perfection" and of lineages where the fossil record does not furnish traces of the alleged ancestors), may protest that the schemata are only so broadly applicable because there are no real constraints on instantiating them. The alleged unification achieved by evolutionary theory is therefore spurious.[97]

Defenders of Darwin sometimes insist that these criticisms are already fore-stalled in the *Origin*, and that Darwin is at pains to specify conditions under which his theory would be demonstrably inadequate. The defense is based on a much-quoted passage:

> If it would be proved that any part of the structure of any one species had been formed for the exclusive good of another species, it would annihilate my theory, for such could not have been produced through natural selection.[98]

But the critics will rightly point out that this passage only appears to give hostages to fortune. For how is one supposed to show that some characteristic of a species was "formed for the exclusive good of another species" (or, perhaps more exactly, was formed through some process which is not covered by Darwin's inventory of agents of evolutionary change)? The worry that pervades the remarks I have quoted from Jenkin, Hopkins, and Pictet is that there are ready-made stratagems which defenders of evolutionary theory can use to brush off any suggestions that such characteristics exist: one can conjure trains of ancestors to whom the property in question might be supposed to have been beneficial. Nor do contemporary defenders of Darwin succeed in doing better when they insist that the theory of evolution precludes the possibility of finding certain fossils, for example, hominid remains in pre-Cambrian deposits.[99] Such defenses are inadequate because they do not address the real worry, namely, that defenders of evolutionary theory have resources which enable them to reinterpret uncomfortable fossil findings, either by questioning the alleged connection between fossil and organism or by assigning a different age to the pertinent strata.

To turn back the criticism leveled by Jenkin et al., we must start with a relatively obvious distinction. In the history of science there have been some theories which, at an early stage of their development, have been difficult, even practically impossible, to test. Such theories are not to be confused with those objectionable doctrines which are impossible *in principle* to test. Someone who proposes that all natural phenomena are to be understood as effects of God's will *and* who also refuses to admit any independent way of fathoming the divine will may legitimately be reprimanded in the way that Jenkin reproved Darwin. In this case, there really are no checks on the ability of the proposal to match itself to whatever phenomena are found. But Darwin's approach is importantly different.

Consider the kind of application of a Darwinian schema which evidently worried Jenkin. Imagine that we are attempting to answer a question about the distribution of a group of organisms, and that, in doing so, we advance claims about the existence of ancestral forms that inhabited particular regions, about the previous connections of land masses, and the abilities of the organisms in question to disperse. Suppose that there is no fossil record of the alleged ancestral organisms. A naive opponent might think that this suffices to demonstrate that the supposed ancestors never existed, and that the Darwinian history is therefore incorrect. The chapters of the *Origin* which discuss the incompleteness of the fossil record are designed to show how naive this evaluation is. Yet that discussion lends plausibility to Jenkin's charge that, in emphasizing the fragmentary character of the record, Darwin is paving the way for accommodating *any* case in which there are no signs of the existence of hypothetical ancestors. But the charge is an overreaction. Jenkin and his fellow critics overlook the possibility of fashioning independent ways of specifying just how fragmentary the fossil record is. Unlike the person who appeals to the divine will without honoring any independent criterion for fathoming that will, Darwin allows for the possibility of a theory of fossilization, a theory which will generate well-founded expectations about the likelihood that records of ancestral organisms will be preserved. Indeed, we should go further. Evolutionary theory was committed, from the beginning, to the development of ancillary theories, of which our envisaged theory of fossilization is one, which could be used to supply constraints on Darwinian histories. This commitment results not simply from the need to remedy the initial difficulty of testing Darwinian histories, and thus to rebut the accusation that Darwin created a game without rules, but also from the need to decide among alternative Darwinian histories that might be proposed for the same phenomena. In the case at hand, development of a theory of fossilization would be expected not to test Darwinian histories individually but to provide an evaluation of the class of Darwinian histories actually proposed. We are to compare the totality of hypothetical ancestors for which no fossil forms are found with our theoretical knowledge of the fossilization process, asking whether it is probable that so many fossils should be missing.

Analogous points can be made about the Darwinian claims of previous continental arrangements and of possibilities for dispersal. Those who assert that the situations of land masses were formerly different are committed to finding some geological account of the hypothetical process through which the alteration has been effected. Darwin's own practice reveals how the deliverances of geology led him to test and reject an otherwise attractive hypothesis about animal distribution on oceanic islands. Hooker had urged the merits of a doctrine (due to Edward Forbes) which allowed the former extension of existing continents to make them continuous with what are now islands. Darwin replied:

> There never was such a predicament as mine: here you continental extensionists would remove enormous difficulties opposed to me, and yet I cannot honestly admit the doctrine, and must therefore say so. I cannot get over the fact that not a fragment of secondary or palaeozoic rock has been found on any island above 500 or 600 miles from a mainland.[100]

Similarly, Darwin's own work demonstrates the possibilities for testing claims about the dispersal of organisms. In his attempt to understand the distribution of plants, Darwin was concerned to discover the extent to which seeds could survive very harsh conditions. The *Origin* describes his careful experiments on seed germination after soaking in sea water, his discovery that seeds extracted from earth which had been enclosed in wood for fifty years could nonetheless germinate, and his investigations on the possibility of seed transport by birds.

In all these cases, Darwin is testing what initially appeared to him and his contemporaries as ambitious hypotheses about the dispersal powers of organisms, and he uncovers, in the process, some facts about germination that would initially have seemed highly unlikely.[101] Plant dispersal was one topic on which Darwin seemed to be driven to hypotheses with surprising consequences. Perhaps even more vexing was the problem of accounting for the distribution of fresh water molluscs. A post-script from a letter to Hooker indicates Darwin's sense that his ideas about the dispersal of these organisms seemed quite at odds with what was known about them.

> The distribution of fresh-water molluscs has been a horrid incubus to me, but I think I know my way now; when first hatched they are very active, and I have had thirty or forty crawl on a dead duck's foot; and they cannot be jerked off, and will live fifteen and even twenty-four hours out of water.[102]

Existing sciences, such as geology and physiology, thus provide ways of testing some of the claims advanced in Darwinian histories. In addition, as we have already seen, Darwin's reform of the practice of biology points the way to the construction of new sciences around the unanswered questions concerning variation and heredity. The pursuit of these new sciences makes Darwin's theory vulnerable from new directions. If it should be discovered that the principles that govern heredity cannot be integrated with the idea of modification through natural selection, or if it can be shown that organisms are not variable in the ways that Darwin's accounts require, then evolutionary theory will be tested and found wanting. Indeed, as the subsequent history of Darwinism clearly shows, it is precisely in the unfolding of the facts of heredity and variation that Darwinian evolutionary theory has faced some of its most serious challenges.[103]

I hope that this discussion makes it clear how the initial difficulty of testing Darwinian claims should not be confused with the view that Darwin's theory is in principle untestable and therefore worthy of the kinds of objections that Pictet, Hopkins, and Jenkin leveled against it. Let us now return to those objections, and confront them directly. The worries expressed by Hopkins and Pictet can be soothed by noting that there are sciences independent of evolutionary theory (sciences such as geology) which can, either in practice, or at least in principle, be employed to check the claims made in the Darwinian histories which the critics find objectionable. Jenkin's concern is more subtle. For Jenkin does not simply raise the problem of testing the hypotheses advanced in particular Darwinian histories; he charges that Darwinism is so flexible that the failure of particular Darwinian accounts need not prove troublesome; others can always be found. Thus, in the spirit of Jenkin's original critique, one might respond to the observations about the probative power of geology and physiology as follows: even if we grant that these other sciences can force the true believer to abandon a particular Darwinian history, that does not directly affect the main Darwinian claim, to wit, that there is some such history to be found; the Darwinian may simply set out to construct some other imaginative story to dodge the known difficulties, or if problems multiply and imagination runs out, may resort, *in extremis*, to the claim that there is some (unknown) Darwinian history which will overcome all the difficulties, and that the task of finding this history is an interesting research project for the theory.

This line of criticism seems to me to be implicit in Jenkin's original review, and I take it to constitute the most powerful methodological objection to Darwin's theory. Three points need to be made in response to it. First, one should not overlook the possibility of global challenges to the presuppositions of any Darwinian history, for example, results from the investigations of heredity and variation which would call into question the possibility of the processes which Darwinian histories constantly invoke. Thus the new sciences whose domains are described in Darwin's reform of the practice of biology may furnish tests of the theory and not simply of particular instantiations of it. Second, there are grounds for believing that Jenkin has overrated the flexibility of *at least part* of Darwinian evolutionary theory. In the light of the constraints imposed by geology, physiology, and morphology, minimal Darwinian histories which will address questions about organismic distribution and relationships may turn out to be rather hard to find. Finally, the idea that admission of ignorance and heralding a new research problem is a universal strategy for promoting the survival of doctrines in distress is seriously flawed.

Consider the second point in the light of our imagined question about organismic distribution. Suppose that we have originally proposed to explain the distribution of a group of organisms, some of which occur on an island, others on the nearest continent, by hypothesizing an earlier continental connection, subsequently submerged, followed by a modest amount of evolutionary divergence between the mainland and insular forms. On testing the geological claim we discover that the alleged continental connection is highly suspect. We now suggest that recent common ancestors of the continental and island organisms were able to traverse the sea that separates continent and island. But, when we investigate the dispersal powers of the contemporary organisms, we find that they are unable to swim the distance, that they are too large to be carried by birds, that they are unable to cling to pieces of driftwood sufficiently tightly to survive the rough seas, and so forth. Finally, we explore the possibility that the organisms in question are not related in the ways that we had originally conjectured. Here we are foiled by our knowledge of morphology, and by our practice of explaining such morphological relationships in other cases by appeal to the existence of a recent common ancestor. Consistency with other Darwinian explanations requires that we understand the morphological similarities by appealing to common ancestry, and our commitment to rapid evolutionary change among related organisms requires us to hold that the common ancestor is recent. Thus we are constrained, forced to construct a particular type of Darwinian history, and all the instances of this type that we are able to produce encounter difficulties with the geological and physiological findings.[104]

Hence there is reason to believe that, sometimes, when Darwinian attempts at explanation go awry, no substitute Darwinian histories will be readily available, so that the evolutionary theorist will be driven to a confession of ignorance. Jenkin perceives this ultimate evasion as a free move. However, familiar points about the character of scientific testing make it apparent that the usual ways of showing scientific theories to be inadequate involve demonstrating that unsolved "research problems" are multiplying at a much faster rate than successful solutions.[105] Immediate trouble can be avoided by pleading that the correct Darwinian history has not yet been conceived, but, if Darwinians are forced to make this plea again and again, then their claims will sound ever more hollow. This sad, imaginary, destiny for Darwin's theory would depict it as collapsing in just the ways that Ptolemaic astronomy collapsed in the sixteenth and seventeenth centuries and the pancorpuscularianism of some Newtonians collapsed in the eighteenth century.[106] Thus the comparison with other sciences, which Darwin liked to use in addressing methodological challenges, will serve to meet Jenkin's objection. Ptolemaic astronomers were compelled to insist, again and again, that there was some combination of allowed motions that would account exactly for the planetary orbits, and that the discovery of this combination was an important research problem for their theory. Darwinian evolutionary theory was potentially vulnerable to a similar predicament—and a similar fate.

A simple comparison between a new theory whose credentials are questioned and previous scientific theories can quell objections, but it does not provide a satisfactory explanation of why the new theory is methodologically sound. The previous analysis of Darwin's theory prepares the way for us to go a little further. On the account I have proposed, Darwin's theory is a collection of problem-solving patterns aimed at answering major families of questions. So construed, the theory plainly makes no definite predictions which can be evaluated by relatively direct observation. Indeed, the relation between theory and observation is doubly loose. In the first place, the theory does not dictate the particular Darwinian histories which are to be constructed. In the second place, individual Darwinian histories will not always imply definite claims about expected observational findings. (In fact, it will be relatively rare for a Darwinian history to imply any statement whose truth value can be ascertained by observation.) As we have already seen, the assessment of individual Darwinian histories must be undertaken with the aid of ancillary theories. Thus, in understanding the relationship between Darwinian theory and observation, one must consider a number of possible cases.

A. In attempting to answer a question about some group of organisms, one finds that there is only one Darwinian history that one can think of which is compatible with the constraints that have already been discovered. (These constraints are imposed by previous observational findings, together with ancillary theories and the prior practice of constructing Darwinian histories.)[107] This unique history implies claims about the existence of certain ancestral organisms (or of contemporary organisms with certain definite properties), claims which would not antecedently have been accepted. Evidently, these claims can be more or less improbable in the light of prior information. On investigation, we discover that there is observational evidence for the truth of some of the claims.

B. As for A, except that there are several available Darwinian histories, one of which implies existence claims for which there is observational evidence.

C. As for B, except that there are several alternative Darwinian histories, all of which imply common existence claims that are supported by observation. There is no evidence for any distinctive claim of any of these histories.

D. As for A, except that investigation does not reveal any evidence of the presence of the hypothetical organisms.

E. As for D, except that there are several alternative Darwinian histories none of which receives positive observational evidence for its existence claims.

F. As in A, there is, initially, a unique available Darwinian history. Observational evidence together with ancillary theories implies a statement that is inconsistent with some consequence of the unique Darwinian history.

G. As in *F*, except that there are, initially, several alternative Darwinian histories available. With respect to each of these, there is some body of observational evidence which, in conjunction with some body of ancillary theory, yields a consequence incompatible with some claim of the Darwinian history.

I do not want to suggest that these cases exhaust all the relevant possibilities. Rather I propose that any thorough reply to Jenkin's methodological criticism will need to take into account at least this much variety of relationships between Darwin's theory and observational findings.[108]

Consider, first, the cases that are clearly positive. Examples of type A redound to the credit of Darwin's theory, because they involve tests of the theory where it is apparently weakest.[109] In such instances, the resources of the theory are relatively impoverished, and there is only one available way of providing a theoretical explanation of the relevant phenomenon. Confirmation will be more or less dramatic according to the prior degree of improbability of the existence claims that receive observational support.[110] Examples of types B and C are somewhat less forceful in providing support for Darwinian theory, precisely because they show the theory succeeding in an area where it had more room for maneuver. (Instances of type B, unlike instances of type C, have the ability to confirm *particular* Darwinian accounts. This confirmation should be seen as dependent on support for the *general* theory, that is, the collection of problem-solving patterns.) In each of these three types of case we may view Darwin's theory as "leading to the discovery of new facts." I suggest that the frequent claims that the theory "uncovered new facts" are based on appreciation of this kind of relationship between theory and observation, and should not be confused with assertions that the theory makes predictions (where "prediction" is interpreted in the usual way, namely in terms of the deduction of some independently checkable statement from a body of theoretical statements and items of background knowledge).

Cases D and E are neutral with respect to the general theory. In examples of type D, as in examples of type B, support for Darwin's theory would provide reason to accept the individual Darwinian history, despite the fact that that history has no positive evidence in its favor. Such cases, like the more thoroughly neutral cases of type E, will usually invoke the fragmentary character of the fossil record to account for the disappointing absence of organisms which are hypothesized. As I have already noted, the strategy of making such invocations is subject to evaluation in the light of independent knowledge about the likelihood of fossilization. The consequences of this point will be apparent as the reply to Jenkin's objection is developed in more detail.

The last two types of case reveal the theory as encountering troubles. We have already seen the possibility of using ancillary theories together with observational findings to test a particular Darwinian history. Examples of types F and G involve tests with negative outcomes for all available Darwinian histories (with respect to some group of organisms and some question concerning those organisms). Faced with such cases, the defender of Darwinian evolutionary theory must plead temporary ignorance and recognize another "unsolved research problem." The admission is more serious in examples of type G, just as the support is strongest in cases of type A. One does the greatest damage to the theory by showing that it is inadequate to handle a problem for which it initially seemed to have abundant resources.[111]

At this stage, we can meet Jenkin's critique head-on, recognizing what is salutary in it and simultaneously clearing Darwin's theory of the main charge. The theory would indeed be methodologically suspect if it precluded the possibility of cases of types F and G. However, we have seen that such cases might occur, and it is relatively simple to describe conditions under which the occurrence of these cases would lead to the rational rejection of Darwin's theory. The most obvious dismal scenario is for the class of cases of types F and G to increase at a much more rapid rate than the class of cases of types A–C, and for the latter to contain few, if any, instances in which observation supports some antecedently highly improbable existence claim. To specify exactly the conditions under which it would be reasonable to abandon Darwin's theory in favor of a search for some alternative, one would need to describe a complicated function of several arguments: the rationality of rejection seems to depend upon the distribution of cases among the types A–G, the prior improbabilities of the existence claims that are supported in the positive cases, the extent to which the lack of evidence in the neutral cases diverges from our expectations about fossilization, the extent to which the "research problems" generated by the negative cases have resisted sustained attempts at solution, and, perhaps, the degree to which alternative approaches have already been explored. Luckily, for our present purposes, it is simply necessary to note that there are indeed *some* conditions under which it would be reasonable to reject Darwin's theory. We have enough understanding of the shape of the complicated function to recognize that rejection would be dictated if the arguments were to assume certain extreme combinations of values.

It is worth noting explicitly that mere multiplication of neutral cases, even without examples of types F and G, might lead to the downfall of Darwin's theory. If one were constantly to hypothesize organisms whose existence was never to be confirmed through observational investigation, then Darwin's theory would face an obvious difficulty: our understanding of the process of fossilization might teach us that the record is fragmentary, but it would be necessary to plead extreme bad luck, if remains of hypothetical, extinct organisms were never found. The situation would be less clear cut, though still damaging, if the record only occasionally revealed signs of the hypothetical organisms. More precisely, given a class of hypothetical ancestral organisms and a theory of fossilization, one will be able to make estimates of the probability that fossils of only n percent of the hypothetical organisms are found (where n is the frequency of the hypothetical organisms for which fossils are actually found), and the smaller this probability the more reasonable it will be to reject Darwin's theory.

Darwin did not defend his theory by pointing to definite predictions which were observationally verified. Noting the absence of predictions, some of his critics charged that the new theory was not a genuine piece of science. What is correct about their objections is the demand that the theory should be testable in principle and that its proponents should develop it so as to make it testable in practice.

As we have seen, despite the initial difficulties of testing Darwin's theory, that theory was, in principle, susceptible to test. I have described some scenarios which would have led to its rational rejection. However, these scenarios do not represent the actual course of history. Shortly after the publication of the *Origin*, Darwin's theory began to receive support in the ways that my account has suggested. Naturalists started to discover remains of organisms whose existence had been hypothesized in Darwinian histories. One of the most striking findings, the discovery of *Archaeopteryx*, was viewed by some of Darwin's supporters as an important boost for the new theory.[112] Darwin himself placed greater stock in a discovery about living organisms, the discovery of organs in nonelectric fish which are homologous to the electrical organs.

This example is a beautiful illustration of type A. In the *Origin*, Darwin confessed to two problems with the electric organs of fish. The first concerns the steps which led to the production of these organs. Darwin continues

> The electric organs offer another and even more serious difficulty; for they occur in only about a dozen fishes, of which several are widely remote in their affinities. Generally when the same organ appears in several members of the same class, especially if in members having very different habits of life, we may attribute its presence to inheritance from a common ancestor; and its absence in some of the members to loss through disuse or natural selection. But if the electric organs had been inherited from one ancient progenitor thus provided, we might have expected that all electric fishes would have been specially related to each other. Nor does geology at all lead to the belief that formerly most fishes had electric organs, which most of their modified descendants have lost.[113]

If we attempt to construct a Darwinian history for the electric fishes, then the constraints seem to rule out the two types of history that might antecedently have been viewed as most likely. We cannot treat the electric fish as a taxonomic group which share a more recent (electric) ancestor with one another than any common ancestor that they have with nonelectric fish; as Darwin notes, the electric fish are a diverse group, and the practice of constructing *other* Darwinian histories for the fish will rank some electric fish as evolutionarily quite distinct from other electric fish. Moreover, the fossil record militates against the alternative claim that the possession of an electric organ was a primitive condition that has been lost in most recent fish. The only possible evolutionary solution, as Darwin goes on to confess, seems to be that "natural selection, working for the good of each being and taking advantage of analogous variations, has sometimes modified in very nearly the same manner two parts in two organic beings which owe but little of their structure in common to inheritance from the same ancestor."[114]

Here, then, we seem to have a unique type of Darwinian history, to which Darwin's theory is forced, which will adduce multiple instances of unrelated productions of electric organs in fish. Combining this apparently necessary consequence with his knowledge of the morphology of fish, one of Darwin's contemporaries developed what he took to be a crucial objection to evolutionary theory. Darwin describes the episode, and its unexpected outcome, in the postscript of a letter to Lyell:

I must tell you one little fact which has pleased me. You may remember that I adduce electrical organs of fish as one of the greatest difficulties which have occurred to me, ... Well, McDonnell, of Dublin (a first-rate man), writes to me that he felt the difficulty of the whole case as overwhelming against me. Not only are the fishes which have electric organs very remote in scale, but the organ is near the head in some, and near the tail in others, and supplied by wholly different nerves. It seems impossible that there could be any transition. Some friend, who is much opposed to me, seems to have crowed over McDonnell, who reports that he said to himself, that if Darwin is right, there must be homologous organs both near the head and the tail in other nonelectric fish. He set to work, and, by Jove, he has found them! so that some of the difficulty is removed; and is it not satisfactory that my hypothetical notions should have led to pretty discoveries?[115]

So it appears that Darwin's theory was quickly tested in some of the ways that I have indicated, and that the happy outcome of those tests served to buttress Darwin's "long argument."

VIII

I began by promising an account of the structure of evolutionary theory and of the early arguments which were given in its favor that would enable us to see that the Darwinian revolution was settled by appeal to reason and evidence. In giving my account, I have departed from the main philosophical traditions about the structure of scientific theories and about the confirmation of theories. I have claimed that the theory (or, more precisely, the theories) contained in the *Origin* should be seen as a collection (or, collections) of problem-solving patterns, and that the decisive change effected by Darwin was the incorporation within biology of questions which had previously seemed inaccessible to science, together with strategies for answering those questions. Moreover, in analyzing the evidence which led Darwin's contemporaries to accept his theory—however tentative their acceptance may have been—I have emphasized the "long argument" of the *Origin*, rather than any "predictive successes" that might be attributed to Darwin's theory. Indeed, the burden of the previous section is that the relationship between Darwin's theory and observation is very loose, so that the absence of reports of confirmed predictions from the first edition of the *Origin* is no accident.

Are these departures from orthodox views really neccessary? Is it possible to provide an account of Darwin's achievement that will be more consonant with orthodox ideas about theory structure and about the confirmation of scientific theories? These are natural questions about the analysis that I have offered. In conclusion, I shall try briefly to answer them.

If we begin from the amorphous thesis that theories are sets of statements, the residue of what was once a complex and ambitious doctrine about the nature of scientific theories, then it is natural to attempt to find in the *Origin* some principles about organisms and the general features of their histories. As we discovered in section 2, the early chapters of the *Origin* do indeed contain some candidates. The trouble with the resultant collection of principles is that it turns out to be trivial and uncontroversial, and we do not achieve distinctively Darwinian doctrine even if we conjoin the thesis that species can be extensively modified so as to give rise to new taxonomic groups (the thesis of evolutionary change). A natural suggestion, at this stage, is to propose that (in Darwin's phrase) "Natural Selection has been the main but not exclusive means of modification."[116] Here, by adding a rather vague thesis, we obtain at last a collection of principles whose conjunction was first held by Darwin.

Is the result Darwin's theory? If it is, then Darwin's theory is plainly inferior to the scientific achievements which philosophers have admired and which have furnished the traditional views of the structure of theories. Crucial elements of the collection of principles we have assembled are deplorably vague, and there is no warrant for ascribing more precise versions of them to Darwin. Moreover, it is legitimate to wonder how there could be any interesting theoretical articulation of the "theory" so identified. In such theories as classical mechanics, electromagnetic theory, and quantum mechanics, we are used to seeing significant theoretical work: interesting and often surprising theorems are derived from the deductive depth. There is little point to heralding the principles we have collected as the "axioms of Darwin's theory" because there is so little of interest that we can derive from them. Yet the *Origin* is a long book, full of subtle discussions that one might naively think of as articulating evolutionary theory. All these discussions are bypassed when Darwin's theory is condensed in the suggested way, and we are left, in effect, with a single argument from chapter 4 of the *Origin* together with two imprecise claims that are made at various places in the book. The result is an impoverishment of Darwin's achievement.

The differences between Darwin's theory and a theory like classical mechanics can be made clear by considering one of the main achievements in the history of mathematics. As it was originally developed by Newton and Leibniz, the calculus had two parts. One part was a collection of claims about differentiation and integration. In this part, proceeding from (not very satisfactory) definitions of the derivative and the integral, Newton, Leibniz, and their successors deduced results about the derivatives and integrals of particular functions (for example, that the derivative of x^n is nx^{n-1}, for positive integral values of n). The second part consisted of proposed methods for using the concepts of derivative and integral in answering questions in geometry, kinematics, and dynamics. Here, Newton and Leibniz showed how one could find subtangents, subnormals, areas, and volumes, by employing the central concepts of the calculus. In its original form, the calculus appears as a deductively organized set of theorems about functions coupled with a set of techniques for solving traditional problems in geometry and mechanics.[117]

Some scientific theories are like the calculus from the very first moment of their careers. They are introduced with axioms containing their distinctive terms (or, perhaps, with explicit definitions of those terms) from which theorems are deduced. Often, these theorems obtain their interest because they can be applied to answer an entire class of questions: the theory offers general strategies for answering families of questions, and the theorem supplies information that enables the method to be used in a large subclass of cases. Classical mechanics, for example, gives a problem-solving pattern for addressing the questions of the trajectories of systems of bodies; the two-body theorem enables the method to resolve, at a stroke, an entire class of cases. But some areas of science need not proceed in this way. It is possible for a new theory—a *good* new theory—to be weak in terms of deriving theorems from axioms (or from definitions) and yet strong in its provision of problem-solving patterns for addressing important questions. This could have occurred in the case of the calculus: we can imagine that Newton and Leibniz had seen how to use derivatives and integrals in answering geometric and kinematic questions, but had failed to find any systematic way of computing derivatives and integrals. Moreover, we can expect this to occur in any domain of inquiry in which the problem-solving patterns involve a concept which, by the theory's own lights, can be realized in vast numbers of ways in nature. The primary achievement of plate tectonics lies in showing how such phenomena as mountain building, earthquake zones, geomagnetic variations, and so forth, can be understood in terms of the central concept of an interaction among plates. But the possibilities (and actualities) of plate interaction are numerous enough to defy any easy systematization of them. Thus, when one asks for the axioms of the theory of plate tectonics, the results are disappointing. By assuming that all good scientific theories must be cast in the mold exhibited in *one part of* the calculus, one prepares the way for dismissing such theories as plate tectonics and Darwinian evolutionary theory as poor theories.[118]

Because I believe that traditional views of scientific theories generate impoverished reconstructions of Darwin's theory, reconstructions that make the idea that the *Origin* provides a detailed articulation of that theory quite baffling, I believe that the apparatus I have introduced is necessary to make sense of Darwin's reform of biology.[119] I now want to defend my departure from traditional ideas about theory confirmation. On the account I have offered, Darwin is committed to certain higher-level claims, statements that do not directly describe organisms but which hail certain questions about organisms as important and which identify certain patterns for answering them. The "long argument" of the *Origin* defends these claims by appealing to a conception of the goals of science and arguing that the proposed change of biological practice will lead to the attainment of those goals. For Darwin, a principal goal of science is to achieve understanding, and this goal is attained by providing unified answers to questions about nature. Again and again, in the *Origin* and in his letters, Darwin sounds the theme of unification and advertises the unifying power of his theory.[120] His task in the *Origin* is to defend the unifying power of his problem-solving patterns, showing that it is in principle possible to instantiate them (the analogy with artificial selection), that they are broadly applicable (the lengthy rehearsal of the phenomena to which they can be applied), and that objections to the applicability of the patterns can be turned back (the responses to difficulties with "organs of extreme perfection," the fossil record, and so forth). It is worth asking what kinds of arguments are involved here. As I understand them, Darwin is drawing on premises about the main goals of science and on statements furnished by prior biological practice to argue *deductively* for the conclusion that his proposed modification of biological practice is designed to promote the aims of science.

On the usual approaches to the confirmation of scientific theories, it is assumed that the problem is to show how statements whose truth values can be ascertained on the basis of observation provide support for more general theoretical claims. Most attempts to explain the early acceptance of Darwin's theory presuppose the simplest form of hypothetico-deductivism.[121] Darwin is credited with having used his theory to derive statements which were not antecedently accepted and whose truth values are determinable by observation. Yet, even if hypothetico-deductivism is not hopeless in general, it is certainly inadequate in this particular instance. As I have noted in section 7, the connection between Darwin's theory and observation is doubly loose, and when we understand the character of this connection it is clear why claims about observational predictions are so remarkably absent from the *Origin*. (It is also evident why, a year after the publication of the *Origin*, Darwin should have expressed satisfaction that his ideas should have led to the discovery of new facts; this was not so familiar an occurrence that it could be taken for granted.) The hypothetico-deductivist reading of Darwin can only be sustained if we do not ask too pointedly what the supposed observational predictions are.

The analysis which I have given presents a quite different view of the justification of scientific claims. Darwin did not begin *ab initio*. He inherited from his predecessors and contemporaries a scientific practice which was justified because of its rational emergence from prior scientific practices. Darwin's task was to discover the best way to modify this practice in the light of his own experience. Using a wealth of statements bequeathed to him by earlier scientists, together with his own empirical findings and his understanding of the goals of science, he constructed an argument for abandoning the creationist assumptions favored by most of his contemporaries in favor of a new approach to biological phenomena. Once that approach had been adopted, it supplied a framework within which biologists could begin confirming hypotheses about the details of the history of life, according to the usual canons of inductive support. The heart of the "long argument" is the claim that Darwin's proposals for reforming biology satisfy canons of scientific inquiry that are unsatisfiable on the available rival approaches.

Those who championed Darwin's cause most fervently sometimes praised him by declaring that he had, at last, given biologists a hypothesis by which they could work. With that hypothesis—or, more exactly, that practice—in place, the detailed testing of Darwinian claims (along the lines indicated in section 7) could begin. Many of the tests were carried out by scientists who had already been persuaded by Darwin's "long argument." Their accounts of their reasons reveal the power of Darwin's contention that his theory promoted the goals of science by bringing new questions within its domain. Here, for example, is Asa Gray's comparison of the views of Darwin and Louis Agassiz:

> The one naturalist [Agassiz], perhaps too largely assuming the scientifically unexplained to be inexplicable, views the phenomena only in their supposed relation to the Divine Mind. The other, naturally expecting many of these phenomena to be resolvable under investigation, views them in their relations to one another, and endeavors to explain them as far as he can (can perhaps farther) through natural causes.[122]

The general view of the aims of science on which Darwin's argument turns is even more apparent in Huxley's evaluation of the qualities of Darwin's theory:

> In ultimate analysis everything is incomprehensible, and the whole object of science is simply to reduce the fundamental incomprehensibilities to the smallest possible number.[123]

My departures from the traditional ideas about scientific theories and confirmation are motivated by the desire to construct a perspective from which remarks like these can be taken seriously.

Notes

Earlier versions of parts of this essay were read at the University of Pittsburgh, the University of Michigan, and MIT. I am grateful to those who offered me comments on these occasions. Many of the suggestions I received have helped me to shape the final version of the essay. I am especially grateful to Gerald Massey, Larry Sklar, Peter Railton, and Paul Horwich, all of whom issued independent challenges to my thesis that the Darwinian revolution cannot be understood without departing from orthodox philosophical views of theories and confirmation. Although the present version of the essay does not contain a complete reply to their questions, I hope that it does indicate the lines along which my response would go. My final version of the essay has also benefited from constructive comments from Malcolm Kottler and Elliott Sober. Finally, I would like to thank the American Council of Learned Societies and Harvard University's Museum of Comparative Zoology for their support and hospitality during the time that I was at work on the early stages of this project.

1. Francis Darwin, ed., *The Life and Letters of Charles Darwin* (3 vols., London: John Murray, 1888); reprinted New York: Johnson, 1969, 2:23. (I shall henceforth cite this three-volume work as *Letters*, with volume and pagination.) For Darwin's hesitancy, see also F. Darwin, ed., *More Letters of Charles Darwin*, vol. 1 (London: John Murray, 1903), 47. (I shall henceforth cite this as *More Letters*.)

2. T. H. Huxley, *Darwiniana* (New York: Appleton, 1896), 120. The passage is quoted in *Letters* 3:132. Three years earlier, Darwin was prepared to talk of "the almost universal belief in the evolution (somehow) of species" (*More Letters* 1:304).

3. See *More Letters* 1:202. The passage is quoted on p. 54. Michael Ghiselin has given a concise analysis of Darwin's tactics. See Ghiselin, *The Triumph of the Darwinian Method* (Berkeley: University of California Press, 1969), esp. chapter 6.

4. See *Letters* 2:165, 216 ff., 232, 273, 302–303, 308, 330; 3:11. Two of these passages provide an interesting comparison. In November 1859, Darwin wrote to Huxley that, in advance, he had "fixed in [his] mind three judges, on whose decision [he] determined mentally to abide" (*Letters* 2:232). By early 1860, the trio had expanded to a quartet: Huxley, Hooker, and Lyell had been joined by Asa Gray. Darwin concluded a letter to Gray by writing, "It is the highest possible gratification to me to think that you have found my book worth reading and reflection; for you and three others I put down in my own mind as the judges whose opinions I should value most of all" (*Letters* 2:272). There is no inconsistency here, but the juxtaposition of the passages does reveal Darwin's tact.

5. See, for example, Huxley's account of how he came to espouse Darwin's theory in *Darwiniana*, p. 246. Huxley's letters provide an interesting perspective on his attitude toward Darwin during the 1850s. Before he became a defender of Darwin's theory (and a personal

friend of Darwin's) he offered a ranking of the leading British biologists of the day. Darwin figures in the second tier: he is described as "one who might be anything if he had good health." (See L. Huxley, ed., *The Life and Letters of Thomas Henry Huxley*, vol. 1 [London: Macmillan, 1902], 94.)

6. Huxley attributes the rapidity of the Darwinian revolution to the impression made by the *Origin*. See Huxley, *Darwiniana*, 286.

7. *Letters* 2:308; *More Letters* 1:157.

8. Mivart accused Darwin of citing and overpraising the work of his friends and supporters: "We allude to the terms of panegyric with which he introduces the names or opinions of every disciple of evolutionism, while writers of equal eminence, who have not adopted Mr. Darwin's views, are quoted, for the most part, without any commendation." (David Hull, ed., *Darwin and His Critics* [Cambridge, Mass.: Harvard University Press, 1973], 380; I shall henceforth refer to Hull's fine anthology as *Critics*.) Louis Agassiz attributed some of Darwin's success to his "attractive style" (see *Critics*, 440). However, such suggestions to the effect that Darwin substitutes political or aesthetic skill for argument are very rare, even in the writings of his most dedicated opponents.

9. *Origin*, 459. My references to this work will be to Ernst Mayr's edition of it (Cambridge, Mass.: Harvard University Press, 1964). This edition is a facsimile of the first edition (London: John Murray, 1859).

10. To recognize this, one need only follow the correspondence between Darwin and each of Hooker, Huxley, and Lyell during the period around the publication of the *Origin*. (See *Letters* 2.171, 172–173, 175, 197, 206, 221, 225, 228, 230–231.) Darwin is initially concerned about what his "three judges" will think. In a postscript to a letter to A. R. Wallace in November 1859, Darwin announces that "Hooker is a complete convert." He adds: "If I can convert Huxley I shall be content" (*Letters* 2:221). A letter from Huxley, written ten days later, fulfilled his wish. Huxley wrote: "Since I read von Baer's essays, nine years ago, no work on Natural History Science I have met with has made so great an impression upon me, and I do most heartily thank you for the great store of new views you have given me." (*Letters* 2:230). The correspondence reveals a similar progression in the support of Hooker and of Lyell.

11. Specifically, the excellent and informative accounts offered by Ghiselin (*The Triumph of the Darwinian Method*) and Michael Ruse (*The Darwinian Revolution* [Chicago: University of Chicago Press, 1979]) seem to me to be flawed by suggestions about the structure of Darwin's theory and about the evidence for it that cannot be sustained under careful scrutiny. Ghiselin, following Mayr, celebrates Darwin's use of the "hypotheticodeductive method." As will become clear below, Darwin did not use any such method, and Ghiselin's discussions of his alleged employment of it (*Triumph*, 63, 145–146) rest on very generous assessments of observational consequences obtainable from Darwin's theory. (See section 7, especially notes 98 and 99). In similar fashion, Ruse hails Darwin's arguments as "much closer to the hypothetico-deductive ideal than to anything, say, in Lyell" (*Darwinian Revolution*, 190). Despite the fact that Ruse and Ghiselin have provided the most thorough available analyses of how the Darwinian revolution was won, their narratives ought to foster suspicion in anyone who is prepared to look soberly at Darwin's alleged instantiation of the philosophical views that they attribute to him. To put the point bluntly, if Darwin was a scientist practicing by the canons favored by Ghiselin and Ruse, then he was a poor practitioner. On my account, Darwin will emerge as a successful scientist who answers to rather different methodological ideals.

12. Of course, there are periodic attempts to claim that Darwin was anticipated by earlier thinkers. Some popular candidates are discussed in Loren Eiseley, *Darwin's Century* (New York: Doubleday, 1958). For a lucid assessment of the forerunners and of Darwin's

originality, see Ernst Mayr, *The Growth of Biological Thought* (Cambridge, Mass.: Harvard University Press, 1982), chapters 8, 9, and 11, especially pp. 498–500.

13. For a sustained account of the received view and its problems, there is no better source than Fred Suppe's introduction to *The Structure of Scientific Theories*, 2d ed. (Urbana: University of Illinois Press, 1977).

14. Huxley, *Darwiniana*, 287; Lewontin, "The Units of Selection," *Annual Review of Ecology and Systematics* 1 (1970): 1.

15. Malcolm Kottler has pointed out to me that there is a parallel with the debates about Lyell's geological views, in that what was primarily in dispute was the significance of Lyell's claims about the time-scale. See M. J. S. Rudwick, "The Stategy of Lyell's *Principles of Geology*," *Isis* 61 (1970): 5–33.

16. Owen's complicated views about evolution and natural selection can be reconstructed from his (anonymous) review of the *Origin*. (See *Critics*, 175–213.) The review frequently compares Darwin unfavorably with "Professor Owen."

17. *Origin*, 62; the point is repeated at *Origin*, 319.

18. These are the words of a fellow-traveler, Henry Fawcett, in a review of the *Origin*. See *Critics*, 282. Darwin's American lieutenant, Asa Gray, saw the issue in a similar fashion (Gray, *Darwiniana* [Cambridge, Mass.: Harvard University Press, 1876], 30–31) as did two leading opponents, Mivart and Fleeming Jenkin. Mivart explicitly claims that "the one distinguishing feature of [Darwin's] theory was the all-sufficiency of 'natural selection'" (*Critics*, 356), and Jenkin, with characteristic precision, focuses the dispute as follows:

> All must agree that the process termed natural selection is in universal operation. The followers of Darwin believe that by that process differences might be added even as they are added by man's selection, though more slowly, and that this addition might in time be carried to so great an extent as to produce every known species of animal from one or two pairs, perhaps from organisms of the lowest type. (*Critics*, 303–304)

19. Typical of Darwin's many confessions of ignorance is *Origin*, 13. In a forthcoming book (*The Nature of Selection* [Cambridge, Mass.: Bradford, 1984]), Elliott Sober argues cogently that contemporary evolutionary theory—or, more exactly, the part of evolutionary theory that deals with the genetics of the evolutionary process—can be conceived as a theory of forces. Sober begins from the insight that the Hardy-Weinberg equilibrium principle plays the same role as the law of inertia in Newtonian mechanics, and he goes on to show how evolutionary theory is concerned with the forces (mutation, migration, selection) that perturb equilibrium. In my view, Sober's lucid discussion shows that a part of evolutionary theory that was unavailable to Darwin, can be reconstructed along the lines of traditional philosophical thinking about theories. However, I think that there are large parts of contemporary evolutionary theory—the parts associated with the names of Mayr and Simpson—that extend the work of Darwin, which are not susceptible to the usual philosophical reconstructions, and which can best be understood from the perspective I shall develop below.

20. Mary B. Williams "Deducing the Consequences of Evolution: A Mathematical Model," *Journal of Theoretical Biology* 29 (1970): 343–385. Williams's approach is refined, simplified, and defended by Alex Rosenberg in *The Structure of Biological Science* (Cambridge: Cambridge University Press, 1985).

21. In discussing "descent with modification" in section 16 of her essay, Williams attempts to allow for accidents that might interfere with the workings of selection by weakening the Darwinian conclusion that she is attempting to derive. Because drift is a factor in evolution, one obviously cannot commit Darwinian theory to the inevitable triumph of the fitter. Williams's solution to this problem—if I understand her correctly—is to formulate her

theorems about differential perpetuation not as claims about the increasing frequency of the fitter in *every* generation of a lineage, but as increasing frequency in a subsequence of future generations. However, this move will prove of no avail if there are some lineages in which the workings of selection are obliterated by some freak of nature that wipes out all the organisms bearing some advantageous mutation. I see no way to resolve this problem within Williams's framework without explicitly introducing probabilistic considerations, and because I believe that these considerations turn on probabilites which are very difficult to estimate, I suspect that any successful resolution would forfeit much of the precision of Williams's analysis.

22. I should note that there is far more to biogeography than these local concerns. Biologists are also interested in such general issues as why island faunas contain a large number of endemic species. For a classic work of recent theory, see R. H. MacArthur and E. O. Wilson, *The Theory of Island Biogeography* (Princeton, N.J.: Princeton University Press, 1967).

23. *More Letters*, 1:173. Darwin means that there is no group for which he can provide a complete account of the modifications of any characteristic.

24. *Letters* 3:25. I shall have more to say about Darwin's claims of the explanatory unification provided by his theory in later sections of this paper.

25. I have emphasized the idea that the sciences provide bases for acts of explanation, and that those who give explanations adapt material provided by the sciences to the needs of their actual or intended audiences, in the early sections of Philip Kitcher, "Explanatory Unification," *Philosophy of Science* 48 (1981): 507–531. Peter Railton articulates a similar approach by using the notion of an "ideal text," and by suggesting that we often perform explanatory acts by giving incomplete information about ideal texts. See Railton, "Probability, Explanation, and Information," *Synthese* 48 (1981): 233–256. Either approach can be used to elaborate the claims about Darwinian explanation that I make here.

26. *Origin* 352–353. See also *Origin*, 394.

27. *Letters* 1:336; 2:34. *More Letters* 1:118–119. Darwin's elder brother, Erasmus, confessed, "To me the geographical distribution, I mean the relation of islands to continents is the most convincing of the proofs, and the relation of the oldest forms to the existing species" (*Letters* 2:233). Huxley describes the importance of biogeography in his own reception of the theory of evolution at *Darwiniana*, 276. Darwin's official statement of the role of biogeography in his own route to evolution occurs in the opening sentences of the *Origin*.

28. Darwin provides a very clear account of the homology/analogy distinction and is well aware that his theory enables him to refine the concept of homology; see *Origin*, 427, and *More Letters* 1:306.

29. *Origin*, 138.

30. *Origin*, 3. Richard Lewontin perceptively discusses the way in which prior emphasis on the problem of design made the discussion of adaptation central to Darwin's evolutionary thinking, and how this fact, in turn, led Darwin to emphasize the selectionist commitments of his theory. See Lewontin's article "Adaptation," *Scientific American* 239 (1978): 212–231.

31. *Origin*, 472.

32. See the title essay in Stephen Jay Gould's collection *The Panda's Thumb* (New York: Norton, 1980). I think that Gould is correct to view Darwin's sounding of this theme as central to his case for evolution.

33. *Origin*, 186.

34. Darwin's book is entitled *On the Various Contrivances by Which British and Foreign Orchids Are Fertilized by Insects*. It was originally published in 1862. Ghiselin provides a penetrating analysis of the argumentative strategy and of its significance in Darwin's defense of evolution, and I cannot improve on his presentation of the crucial point:

[Darwin] attempted to show, in other words, that structures were not designed with the end in mind of engaging in their present biological role, but rather that they originated as parts adapted to quite different functions. The flower makes use of whatever parts happen to be available, and their availability and utility are purely accidental. (*Triumph*, 137)

Darwin's military characterization of the role of the orchid book is from *More Letters* 1:202.

35. *Origin*, 6. The difficulty of interpreting Darwin's sentence is obvious. Natural selection might be heralded as the force which produces most evolutionary changes, or, perhaps, as the force which produces the most important evolutionary changes. An alternative conception would be to suppose that the modifications produced by other forces are somehow impermanent, so that the large-scale course of evolution follows the trajectory laid down by selection. Finally, one might concentrate on those evolutionary changes which produce new taxa (for example, speciation events), understanding these as being effected by natural selection.

36. Here I mimic the approach of some of the most prominent models in population genetics, assuming discrete generations. An analogous conception is readily definable for the continuous case: the Darwinian history specifies a function that assigns to the properties in F their frequency values at each point of the interval between t_1 and t_2.

37. *Letters* 3:22.

38. I do not claim that this will be possible with respect to all questions of geographical distribution of organisms. There are obviously many instances in which understanding the range of a species will involve recognition of the competitive relations with other species, and in which resolution of the biogeographical questions will turn on issues of coadaptation. Nonetheless, it is sometimes possible to answer biogeographical questions without investigating issues of adaptation, and in such cases the history of descent with modification will suffice.

39. There is something peculiar about referring to all of these as *agents* of evolutionary change. Consider, for example, stochastic factors. These seem to be not so much a force of evolution as *filters* that modify the effects of other evolutionary forces. Similarly, correlation and balance—or, as contemporary evolutionary theory would put it, pleiotropy, linkage, and allometry—are constraints on the working of the force of selection.

40. A position akin to this seems to have been advanced by Niles Eldredge and Joel Cracraft, *Phylogenetic Patterns and the Evolutionary Process* (New York: Columbia University Press, 1980).

41. See, for example, *Origin*, 3, 84, 170.

42. I should note that, although Darwin sometimes invokes use and disuse as a separate agent of evolutionary change, it is quite common for him to reduce it to natural selection: useless structures disappear because organisms which squander resources on developing them are at a disadvantage in the struggle for existence (they will prove inferior to rivals that make better use of the resources in question). The most important nonselectionist strand in Darwin's thinking is the appeal to correlation and balance, an appeal that would now be understood in terms of linkage, allometry, pleiotrophy, and perhaps, some other forms of developmental constraint.

43. Darwin recognizes the importance of chance and rarity in questions about extinction. See *Origin*, 109.

44. Gould and Lewontin, "The Spandrels of San Marco and the Panglossian Paradigm: A Critique of the Adaptationist Programme," *Proceedings of the Royal Society of London* B. 205 (1979): 581–598. While I think that Gould and Lewontin are correct both

in their indictment of adaptationism and in their claim that Darwin is less selectionist than he is often read as being, the latter historical thesis seems overstated. As Ernst Mayr has noted ("How to Carry Out the Adaptationist Program," *American Naturalist*, 1983), many of the "alternative forces of evolution" discussed by Darwin are now widely discredited. What does remain—and this is sufficient to enable Gould and Lewontin to make their case—is the variety of factors that Darwin would have lumped together as "correlation and balance." The heart of the Gould and Lewontin critique is the idea that many characteristics of organisms may not be the direct target of selection, but may belong to ensembles that are selected as wholes. The answer to Mayr's charge that the assumption that a character has been shaped by selection is a necessary working hypothesis only to be discarded as a matter of last resort is that there are alternative ways to investigate the evolutionary history of such characters, namely by learning more about ontogeny and genetic connections. Hence, my judgment that the demise of some of Darwin's alternative agents of evolutionary change still allows for a cogent argument against adaptationism.

45. See *More Letters* 1:288ff. However, there are contrary suggestions at *More Letters* 1:306.

46. It is common for commentators to remark parenthetically that the first edition of the *Origin* is the least pluralistic, and I followed this practice in an earlier draft of the present essay. However, as Malcolm Kottler pointed out to me, this quick assessment is misleading. Line-by-line comparison of the editions shows Darwin inserting disjunctions of possible causes of evolutionary changes where he had previously appealed to natural selection alone, but there is no addition of a new mechanism in later editions. Hence, a more global comparison of the editions undermines the idea that Darwin became more pluralistic in response to critics who were skeptical about the efficacy of selection.

47. *Origin*, 197. The continuation explains the presence of the color by appealing to sexual selection, which Darwin often counts as an alternative mechanism to natural selection. To herald Darwin as a card-carrying antiadaptationist it would be more convincing to find him turning to "correlation and balance" at this stage, but the passage I have quoted does show that he is aware of some pitfalls of vulgar adaptationism.

48. Huxley, *Darwiniana*, 97, also quoted in *Letters* 2:231. See also Huxley, *Darwiniana*, 77.

49. This point is made by Peter Vorzimmer, in *Charles Darwin: The Years of Controversy* (Philadelphia: Temple University Press, 1970).

50. *Letters* 2:274–276; 3:33. *More Letters* 1:147–148. *Origin*, 32, 84, 95. For the relation to Darwin's selectionism, see *Origin*, 194, and, for a particularly strong statement of gradualism, *Origin*, 189.

51. *Origin*, 236ff. Darwin's remarks can be interpreted either as advocating group selection or as preliminary gropings towards the notion of inclusive fitness. For a concise and sensitive discussion of them, see Elliott Sober, *The Nature of Selection*.

52. *Origin*, 112. Those with a fine eye for anticipations may see glimmers here of the approach to the phenomenon of sex elaborated in Steven Stanley's *Macroevolution* (San Francisco: Freeman, 1979).

53. *Origin*, 237.

54. I take this to be the approach to scientific change which is tacit in the writings of the logical empiricists (for example, in the works of C. G. Hempel, Rudolf Carnap, and Ernest Nagel), and it is explicit in the discussions of scientific change within the Popper-Lakatos tradition. The approach to scientific theories and scientific change developed by Thomas Kuhn and the different ideas of Sylvain Bromberger mark a radical break with the idea of charting change by looking at changes in sets of statements. As I have argued elsewhere (Kitcher, *The Nature of Mathematical Knowledge* [New York: Oxford University Press,

1983], chapter 7), the concept of a paradigm, introduced by Kuhn in *The Structure of Scientific Revolutions* (Chicago: University of Chicago Press, 1970) serves two different functions: the emphasis on paradigm change is supposed to do justice both to the complexity of what it is that changes when a science evolves and to recognize important units of segmentation in the history of science. One can appreciate the insight behind the first function while remaining agnostic about preferred ways to divide up the history of the sciences. The account that follows owes considerable intellectual debts to Kuhn, although my reading of his seminal book is both idiosyncratic and selective. The influence of Bromberger's ideas about theories should also be apparent. I am especially indebted to his essays "A Theory about the Theory of Theories and about the Theory of Theory," in B. Baumrin, ed., *Philosophy of Science: The Delaware Seminar* (New York: Interscience, 1963); "Questions," *Journal of Philosophy* 63 (1966): 597–606; and "Science and the Forms of Ignorance," in M. Mandelbaum, ed., *Observation and Theory in Science* (Baltimore: Johns Hopkins University Press, 1971).

55. An analogous notion of a mathematical practice was characterized in chapters 7–9 of *The Nature of Mathematical Knowledge*. I have elaborated the idea of a scientific practice and attempted to use it to give an account of intertheoretic relations in genetics, in Kitcher, "1953 and All That: A Tale of Two Sciences," *Philosophical Review* (July 1984); see also chapter 1 of this book.

56. Darwin also deserves credit for introducing new techniques into experimental biology—see, for example, his careful analyses of seed dispersal. But this accomplishment pales in comparison with his more theoretical contributions. I shall not consider here whether he affected the methodological component of the practice of his day. Ruse claims that Darwin was greatly influenced by the influential Victorian philosophers of science, Herschel and Whewell. This evaluation only seems plausible to me if the philosophical standards are left in soft focus. Darwin's critics were vigorous in suggesting that he had deserted the true path of science, and they sometimes based their charges on the remarks of the contemporary philosophers. Many Victorian scientists paid lip service to the methodological dicta of the alleged experts. I think it would be interesting to explore in detail whether the work of men like Lyell and Owen really embodies the standards of the prevailing philosophy of science. In arguing below for an analysis of the testability of Darwin's theory, I think that I provide good reasons for believing that the methodology ascribed by Ruse (in *The Darwinian Revolution*) is not Darwin's.

By contrast, Michael Ghiselin emphasizes the originality of Darwin's methodological ideas, and in this he seems to me to be correct. (See *Triumph*, 4.) Unfortunately, Ghiselin develops his insight by interpreting Darwin as a proponent of the hypothetico-deductive method. The discussion of section 7 will show the difference between Darwin's work and the hypothetico-deductive ideal, and ironically, Ruse's account shows clearly that hypothetico-deductivism was hardly news to the Darwinian community.

57. In the discussion at Pittsburgh, Adolf Grunbaum pointed out to me that reconstructions in the history and philosophy of physics are often sensitive to the need to take into account changes in other components of the practice. This is correct, but it remains true that the most detailed philosophical models simplify the scientific changes by focusing on modification of the class of accepted statements. This is evident both in Bayesian accounts and in the quite different approach defended by Lakatos and his students.

58. In conversation, Kuhn has questioned this claim. His reason for skepticism is that it seems that Darwin affected a radical change in the meaning of the term *species*, and, because of the centrality of this term to biological discourse (at least after Darwin, if not before), this linguistic change can hardly be termed "minor." I reply that the refixing of the referent of "species" given by Darwin barely modified the division of organisms into species

taxa. As a result, there was virtually no breakdown of communication between Darwin and his opponents. As the references given in note 18 reveal, everybody could agree on a formulation of the issues. Thus the adjustment of the concept of species seems of far less import than Darwin's radical changes in the biological questions that were addressed and in the explanatory framework accepted as appropriate for biology.

59. Kuhn views such shifts as marking scientific revolutions. See Kuhn, "What Are Scientific Revolutions?" *The Structure of Scientific Revolutions* (MIT Center for Cognitive Science Working Paper), and Kuhn, "Commensurability, Comparability, Communicability," in Peter Asquith and Thomas Nickles, eds., *PSA 1982* [East Lansing, Mich.: Philosophy of Science Association, 1983], 669–688). I have tried to give a different interpretation of the phenomena to which Kuhn has drawn attention. See Kitcher, "Theories, Theorists, and Theoretical Change," *Philosophical Review* 87 (1978): 519–547; Kitcher, "Genes," *British Journal for the Philosophy of Science* 33 (1982): 337–359; Kitcher, "Implications of Incommensurability," *PSA 1982*, pp. 689–703. The construal of the case of phlogiston theory offered in the first and third of these papers reveals what I see as a much larger conceptual shift than anything that is found in the Darwinian revolution.

60. Gray, *Darwiniana*, 25. See also ibid., 286, where the *Origin* is described as "one of the hardest books to master."

61. *Critics*, 249.

62. The urgency of questions about the origins of organic forms is beautifully captured in a passage by Asa Gray. After emphasizing the importance of the search for unity in science, Gray remarks that we allow that "the inquiry transcends our powers, only when all endeavors have failed." See Gray, *Darwiniana*, 78–79.

63. In seeing this as Darwin's main contention, I hope to make it clear why Darwin argues hard against creationism. That the *Origin* is an onslaught on creationism has been clearly shown by Neal Gillespie in *Charles Darwin and the Problem of Creation* (Chicago: University of Chicago Press, 1979). But, as Elliott Sober has pointed out to me, the idea that Darwin's principal target is not a prevailing scientific tradition is something that requires explanation. In a nutshell, the explanation I offer is that there is no previous biological tradition, that Darwin defines biology as an area of inquiry, and that he does so by showing that it is possible to give scientific answers to questions that had previously been thought to lie outside science.

64. See section 7. As I shall argue, one also needs to show which areas of evolutionary theory are most open to test and confirmation. Those who are skeptical about unbridled selectionism (for example, Eldredge, Gould, and Lewontin) can be understood as giving *limited* endorsement to the traditional worry that evolutionary hypotheses are not readily tested and confirmed. This skepticism is quite compatible with acceptance of one of the more cautious theories outlined in section 4.

65. Darwin was very clear about the open-ended character of his theory, and about its potential to give rise to "new sciences" (see Bromberger, "A Theory about the Theory" for some suggestive ideas about the generation of new sciences). Not only is the last chapter of the *Origin* prophetic, but Darwin's letters also indicate his hopes for the future development of biology. See, for example, a letter of 1858 to Hooker:

> Whenever naturalists can look at species changing as certain, what a magnificent field will be open, —on all the laws of variation, —on the genealogy of all living beings, —on their lines of migration, &c &c. (*Letters* 2:128)

66. See *Origin*, 6, 13, 43, 73, 75, 78, 132, 167, 462, 486; Gray, *Darwiniana*, 207, 224–225; Huxley in *Letters* 2:197–198.

67. *More Letters* 1:135.

68. The reconstruction that I shall give is a philosophical elaboration of a scheme for interpreting the reasoning of the *Origin* presented by Huxley (*Darwiniana*, 72) and articulated in an illuminating review article by M. J. S. Hodge ("The Structure and Strategy of Darwin's 'Long Argument,'" *British Journal for the History of Science* 10, [1977]: 237–246).

69. For a capsule version of the argument, see *Origin*, 25.

70. Compare the fate of Wegener's theory of continental drift. Despite its apparent ability to answer certain outstanding questions in meteorology, biogeography, and geology, this theory was widely rejected by the geological community during the 1920s and 1930s precisely because it seemed impossible that there should be a mechanism for moving the continents. It is easy to imagine that, lacking a similar mechanism, Darwin's theory would have been equally vulnerable. Hence, even for the cautious Darwinian who remains agnostic about the causes of evolutionary changes in particular lineages (see above), it is still important to argue for the modifying power of natural selection. (For a concise account of Wegener's theory and its reception, see A. Hallam, *A Revolution in the Earth Sciences* [Oxford: Oxford University Press, 1972].)

71. *Critics*, 131. For similar remarks by Wollaston, Pictet, Haughton, Hopkins, and Jenkin, see *Critics*, 135, 145, 224, 253, 304ff.

72. *Critics*, 239.

73. *Letters* 2:237. Quoted in *Critics*, 229.

74. Darwin's main argument stresses the unifying power of his schemata (see the references in note 97). But he cannot resist giving subsidiary arguments. So, for example, the early chapters of the *Origin* campaign against the idea that there is a natural boundary around species. This subsidiary argument becomes very important to certain versions of Darwin's theory—for example, those which take a nominalistic approach to species and emphasize evolutionary gradualism. However, it is incidental to the more cautious versions of Darwinism.

75. *Letters* 2:166–167.

76. *Origin*, 349–350.

77. *Origin*, 318–319, 339–341, 394, 440–444, 452–453, 471–480.

78. There is obviously a tricky issue here. One might hold that the important unification is accomplished by a minimal version of Darwinian theory—for example, one which did not deploy the notion of natural selection in constructing particular Darwinian histories—and that the more ambitious claims about the power of selection and evolutionary gradualism are either otiose, or at best, only weakly supported. This was a position adopted by some scientists in Darwin's day, and it is accepted by some contemporary theorists. I take it to be a merit of my analysis of Darwin's theory that it focusses the disagreement on this traditionally vexed question.

79. *Origin*, 333. Note that a minimal version of Darwin's theory will achieve the explanatory dividends cited here.

80. *Critics*, 267. See also ibid., 268–269.

81. *Letters* 2:286 provides one example of this practice.

82. *Origin*, 380–381.

83. *Origin*, 111ff., especially 128.

84. *Origin*, 154–156; *Origin*, 150–154; *Origin*, 159–167.

85. *Origin*, 167.

86. Various forms of the problem of the poverty of the fossil record are posed forcefully at *Origin*, 280–281, 287–288, 292, 301–303. Wollaston saw the state of the fossil record as "the gravest of all objections" to Darwin's theory, but he noted Darwin's frankness in admitting the facts (*Critics*, 136).

87. Darwin's critics seized on the point. There are clear formulations of it by Wollaston, Pictet, and Jenkin (*Critics*, 133, 150, 314), and it became a major theme of Mivart's attack on the power of selection (see *On the Genesis of Species* [London: Macmillan, 1871]). Darwin anticipated the objection, and attempted to meet it, at *Origin*, 188–189, and he gives his strongest endorsement of selectionist gradualism in this context. For a lucid contemporary discussion of the objection and its resolution, see Ernst Mayr, "The Emergence of Evolutionary Novelties," in his *Evolution and the Diversity of Life* (Cambridge, Mass.: Harvard University Press, 1976).

88. This would then open the way for the use of different methods of explanation that might subsequently replace the evolutionary schemata. Many of Darwin's critics argued that Darwin was himself committed to two different "explanatory principles"

[Darwin] says that life has been breathed into the first primordial form. It is our creative force that has done it. Consequently, both theories acknowledge the existence of the two *forces* and differ only to the degree that each is employed. (*Critics*, 147–148)

Several of Darwin's critics can be viewed as limiting the power of selection to make room for creation, so that Darwin's softening of his naturalism about the origins of life may have been something of a tactical mistake.

89. *Critics*, 339.

90. Most prominent are chapters 6 and 7, which address the issues of "organs of extreme perfection" and the problems that arise in connection with instincts and social behavior, and chapter 9 "on the imperfection of the geological record."

91. Of course, this presupposes that there are ways of choosing among rival Darwinian histories. I shall consider this question in section 7.

92. The modification of linguistic usage can readily be understood from the perspective advanced in my papers cited in footnote 59. Huxley describes the conceptual shift in language which is very close to the terms of my analysis: he speaks of the criteria for species "falling apart" (Huxley, *Darwiniana*, 44).

93. Darwin, *Origin*, 484 ff.; Huxley in *Letters* 2:197–199.

94. *Critics*, 144.

95. *Critics*, 264.

96. *Critics*, 319; see also *Critics*, 342.

97. For a discussion of spurious unification, see the final section of my essay "Explanatory Unification." The emphasis on the power of Darwin's theory to explain biological phenomena by unifying them is notable both in his own writings and in those of some of his supporters. Again and again, Darwin admits his inability to "prove" his large claims about the history of life and describes himself as accepting those claims because they "explain large classes of facts." Moreover, he characterizes the explanations he has given of these classes (the affinities of organisms, the details of biogeography, and so forth) by suggesting that, on his theory, the "facts fall into groups." I take my analysis to make explicit the ideas that are tacit in numerous passages: *Letters* 2:13, 29, 78–79, 110, 121–122, 210–211, 240, 285, 327, 355, 362; 3:25, 44 (which advances similar arguments in favor of the abortive theory of pangenesis), 74; *More Letters* 1:139–140, 150, 156, 184; *Origin*, 188, 243–244, 482. Similar ideas are advanced by Huxley (in *Letters* 2:254, and in *The Life and Letters of Thomas Henry Huxley* 1:479), and by Asa Gray (*Darwiniana*, 19, 78–81, 88, 90, 195–196).

98. *Origin*, 201. Compare also *Origin*, 189, on the formation of complex organs. Ghiselin alludes to the former passage (*Triumph*, 63), using it to buttress his claim that Darwin obeyed the "falsification principle."

A theory is refutable, hence scientific, if it is possible to give *even one* conceivable state of affairs incompatible with its truth. Such conditions were specified by Darwin himself, who observed that the existence of an organ in one species, solely "for" the benefit of another species, would be totally destructive of his theory.

In paraphrasing Darwin, Ghiselin has dropped the crucial reference to *proving* the existence of the organ in question. Of course, it is trivial to state conditions that are incompatible with the truth of a theory T—the condition that T is false will do the trick. What is nontrivial is to find conditions that can be independently checked. As noted in the text, this is the trouble with Darwin's example, for it is far from obvious that there is any way to show that an organ was formed solely for the good of another species.

99. This suggestion about the possible falsification of evolutionary theory is made by Douglas Futuyma, *Science on Trial* (New York: Pantheon, 1983), 170. Besides the difficulty noted in the text, Futuyma's proposal faces the problem that what would have to be given up would be a particular claim about the history of life. It would remain logically possible to embrace Darwinian evolutionary theory and contend that humans are evolutionarily very old. Although this is hardly a plausible position, it does show that Futuyma's case does not directly falsify the theory whose falsifiability is at issue. For a discussion of the misleading use of a falsifiability criterion in debates about evolutionary theory, see chapter 2 of Kitcher, *Abusing Science* (Cambridge, Mass.: MIT Press, 1982).

100. *Letters* 2:80. See also *Letters* 2:72, 81. Ghiselin provides a very perceptive discussion of Darwin's use of independent evidence in advancing geological claims. See *Triumph*, 20, 40. The credentials of Forbes's theory are discussed in the *Origin*, 357–358.

101. *Origin*, 358ff.

102. *Letter* 2:93.

103. For example, in the early days of Mendelian genetics (that is, in the first decade of the century), many biologists believed that the new findings about heredity were incompatible with Darwin's theory of evolution by natural selection. The conflict was resolved by the development of theoretical population genetics. An excellent account of the difficulty and its resolution is given in William B. Provine, *The Origins of Theoretical Population Genetics* (Chicago: University of Chicago Press, 1971). Similarly, the investigation of the maintenance of variation in natural populations has led some biologists to advance claims about the importance of random factors in evolution. (From the perspective of the present article, neutralist proposals are minimal versions of Darwinism rather than accounts of "nonDarwinian evolution.") The controversies about variation are brilliantly analyzed in R. C. Lewontin, *The Genetic Basis of Evolutionary Change* (New York: Columbia University Press, 1974). One major challenge to classical Darwinian ideas that does not emerge from the development of new sciences of heredity and variation is the current proposal (due to Gould, Eldredge, Stanley, and others) that Darwinian gradualism should give way to punctuated equilibrium. The distinctions made in section iv of this paper offer a framework for seeing what is at stake in this dispute.

104. It is not so evident that we are similarly constrained when we attempt to construct selectionist histories for revealing certain characteristics as adaptations. There are some cases in which adaptationist hypotheses prove testable—the classic examples are industrial melanism in moths and cowbird parasitism of oropendulas. However, those who are skeptical of the adaptationist program can best be understood as arguing that, in many cases where selectionist stories are told, there are no ways of finding independent checks on the hypotheses that ascribe past advantages.

105. This approach to the question of analyzing the ways in which scientific theories (or programs of research) come to be rationally rejected absorbs a familiar Duhemian insight.

(See Pierre Duhem, *The Aim and Structure of Physical Theory* [New York: Atheneum, 1954]; I have discussed the implications of Duhem's point for naive falsificationism in chapter 2 of Kitcher, *Abusing Science*.) The approach is common to the work of thinkers as distinct as Kuhn and Lakatos.

106. See Thomas Kuhn, *The Copernican Revolution* (Cambridge, Mass.: Harvard University Press, 1957), and R. W. Schofield, *Mechanism and Materialism* (Princeton, N.J.: Princeton University Press, 1969). Although it is common to suppose that old theories are only rationally rejected when a new rival is available, these examples seem to me to show that the conventional wisdom is mistaken. It is perfectly reasonable to give up a decaying theory and to *look for* something better. I suspect that this is just what Copernicus did in the early decades of the sixteenth century, and what the first field theorists did in the mid-eighteenth century.

107. It is important to recognize that evolutionary theory itself supplies some constraints. Previously accepted problem solutions are not sacrosanct, but one cannot legitimately abandon a sizeable collection of past successes in the interests of fashioning one new solution.

108. The above cases are generated according to the following principle. There are two degrees of freedom: theory plus context (including work in ancillary sciences and past work on the theory itself) may allow a greater or lesser number of available solutions; the observational evidence may be positive, neutral, or negative. Quite evidently, the treatment is simplified by the fact that only one form of implication is considered (claims about the existence of particular organisms). It would surely be necessary to consider a broader class of implications from Darwinian histories if one were to assess the testability of selectionist histories.

109. Here I apply a methodological principle discussed in some detail by Richard Boyd in "Realism, Underdetermination, and a Casual Theory of Evidence," *Nous* 7 (1973): 1–12.

110. Of course, this is a classic result of Bayesian confirmation theory (which is not to say that it is unobtainable on rival approaches). The most famous example is the confirmation of the wave theory of light through observation of the Poisson bright spot. A similar example occurred in the early days of evolutionary theory (see below).

111. The brief analysis given in the text enables me to explain the excitement of some recent theoretical work in evolutionary theory. Great breakthroughs can be made if a scientist shows that problems for which no solution is available can be resolved by applying a new schema, one that was readily available within the theoretical framework but never antecedently recognized. The introduction of the notions of inclusive fitness and of evolutionary stable strategy seem to me to be breakthroughs of this type. W. D. Hamilton and John Maynard Smith demonstrated how fitness could be gained in subtle ways, so that characteristics which had previously seemed to be insusceptible of selectionist explanation could now be viewed as the products of natural selection. (See Hamilton, "The Genetical Theory of Social Behavior," *Journal of Theoretical Biology* 7 [1964]: 1–16, 17–51; Maynard Smith, *Evolution and the Theory of Games* [Cambridge, Mass.: Cambridge University Press, 1982].) Other subtle analyses of fitness that permit the broader application of selectionist schemata have been given by R. L. Trivers, E. O. Wilson, and George Oster. In all these cases, the initial situation reveals a characteristic of some organismic group for which there is no available selectionist Darwinian history. After certain unobvious ramifications of the concept of fitness have been exposed, one sees that it is possible to instantiate a selectionist schema. *It does not follow that the correct explanation of the presence of the characteristic is by appealing to natural selection.* For there may be a number of rival selectionist and nonselectionist explanations which cannot be discriminated by the evidence so far collected (or even by the evidence that one is in a position to collect). One may welcome the extension of the class of Darwinian problem-solving techniques while remaining agnostic about the application of

the new techniques to particular cases. In the terms of the analysis of the text, the break-through takes us from a position in which there was no available selectionist solution to a problem (although nonselectionist solutions may have been available) to a position of type C or type E.

112. See Huxley, *Darwiniana*, 234. Darwin is much more restrained (compare *Letters* 3:6). Perhaps he needed no further convincing.

113. *Origin*, 193.

114. *Origin*, 194. Note that Darwin's selectionism here is purely gratuitous. The issue can be treated without making any reference to natural selection at all. All that is needed is convergence *by some force or other*.

115. *Letters* 3:352–353.

116. *Origin*, 6.

117. See chapter 10 of *The Nature of Mathematical Knowledge* and the introduction to Judith V. Grabiner, *The Origins of Cauchy's Rigorous Calculus* (Cambridge, Mass.: MIT Press, 1981).

118. Of course, it would be possible to abandon the word *theory* in application to prob-lematic areas of science like biology and geology. We could retain the classic idea of a theory as a deductively organized set of statements whose axioms include general laws. There would be no harm in this so long as we recognized the existence of scientific disciplines with impor-tant, articulated accomplishments in which there are no theories, and so long as we freed ourselves from any prejudice to the effect that sciences which have theories are somehow superior.

Although I have developed my account primarily by opposing the residue of the "received view" of scientific theories, I think it right to note that the so-called semantic con-ception of theories seems no more adequate in characterizing Darwin's evolutionary theory. On the "semantic conception," a theory is given by specifying a type of system, and the the-orist then derives from the specifications conclusions about all systems of the type or about interesting subtypes. (This is a simplification of views presented with considerable sophisti-cation by Joseph Sneed, Bas van Fraassen, Fred Suppe, and others. See Sneed, *The Logical Structure of Mathematical Physics* [Dordrecht: Reidel, 1979]; van Fraassen, *The Scientific Image* [Oxford: Oxford University Press, 1980]; and pp. 221–230 of Suppe's introduction to *The Structure of Scientific Theories*.) If we now try to apply this approach to the case of Darwin, we encounter problems that are exactly parallel to those that beset the "received view." Darwin's specification of a type of system is as elusive as the set of axioms of Darwinian evolutionary theory. Moreover, seeing him as deriving results about "types of evolutionary systems" seems to me to have no more connection with the project of the *Origin* than an interpretation which supposes that the *Origin* contains derivations from axioms. Perhaps a more refined version of the "semantic conception" has the resources to overcome these problems, but it has appeared to me to be more promising to begin anew, and to develop an account of Darwin's theory which has some direct relevance to his text.

119. An additional advantage of my approach is that it makes sense of the varying com-mitments that we find in Darwin, and the varying commitments that are available for his successors. Many previous approaches to the *Origin* seem to err by failing to recognize the differences among the theories I have distinguished in section iv and the variations in argu-mentative strength that can be assembled for each of them. See, for example, Elisabeth A. Lloyd, "The Nature of Darwin's Support for the Theory of Natural Selection," *Philosophy of Science* 50 (1983): 112–129.

120. See note 97 and the reference cited therein.

121. This, of course, has been my complaint about the excellent and illuminating accounts offered by Ghiselin and Ruse. I have chosen their works for criticism not because

they are alone in applying a simple hypothetico-deductive methodology but because their application is made in the context of insightful historical analyses.

122. Gray, *Darwiniana*, 16.

123. Huxley, *Darwiniana*, 165. It is interesting to compare Michael Friedman's account of why unification produces understanding: "Our total picture of nature is simplified via a reduction in the number of independent phenomena that we have to accept as ultimate" ("Explanation and Scientific Understanding," *Journal of Philosophy* 71 [1974], pp. 5–19, 18).

4

The Return of the Gene (1988)

COAUTHORED WITH KIM STERELNY

We have two images of natural selection. The orthodox story is told in terms of individuals. More organisms of any given kind are produced than can survive and reproduce to their full potential. Although these organisms are of a kind, they are not identical. Some of the differences among them make a difference to their prospects for survival or reproduction, and hence, on the average, to their actual reproduction. Some of the differences which are relevant to survival and reproduction are (at least partly) heritable. The result is evolution under natural selection, a process in which, barring complications, the average fitness of the organisms within a kind can be expected to increase with time.

There is an alternative story. Richard Dawkins[1] claims that the "unit of selection" is the gene. By this he means not just that the result of selection is (almost always) an increase in frequency of some gene in the gene pool. That is uncontroversial. On Dawkins's conception, we should think of genes as differing with respect to properties that affect their abilities to leave copies of themselves. More genes appear in each generation than can copy themselves up to their full potential. Some of the differences among them make a difference to their prospects for successful copying and hence to the number of actual copies that appear in the next generation. Evolution under natural selection is thus a process in which, barring complication, the average ability of the genes in the gene pool to leave copies of themselves increases with time.

Dawkins's story can be formulated succinctly by introducing some of his terminology. Genes are *replicators* and selection is the struggle among *active germline replicators*. Replicators are entities that can be copied. Active replicators are those whose properties influence their chances of being copied. Germline replicators are those which have the potential to leave infinitely many descendants. Early in the history of life, coalitions of replicators began to construct *vehicles* through which they spread copies of themselves. Better replicators build better vehicles, and hence are copied more often. Derivatively, the vehicles associated with them become more common, too. The orthodox story focuses on the successes of prominent vehicles—individual organisms. Dawkins claims to expose an underlying struggle among the replicators.

We believe that a lot of unnecessary dust has been kicked up in discussing the merits of the two stories. Philosophers have suggested that there are important connections to certain issues in the philosophy of science: reductionism, views on causation and natural kinds, the role of appeals to parsimony. We are unconvinced. Nor do we think that a willingness to talk about selection in Dawkinspeak brings any commitment to the adaptationist claims which Dawkins also holds. After all, adopting a particular perspective on selection is logically independent from claiming that selection is omnipresent in evolution.

In our judgment, the relative worth of the two images turns on two theoretical claims in evolutionary biology.

1. Candidate units of selection must have systematic causal consequences. If Xs are selected for, then X must have a systematic effect on its expected representation in future generations.
2. Dawkins's gene selectionism offers a *more general theory* of evolution. It can also handle those phenomena which are grist to the mill of individual selection, but there are evolutionary phenomena which fit the picture of individual selection ill or not at all, yet which can be accommodated naturally by the gene selection model.

Those skeptical of Dawkins's picture—in particular, Elliott Sober, Richard Lewontin, and Stephen Jay Gould—doubt whether genes can meet the condition demanded in (1). In their view, the phenomena of epigenesis and the extreme sensitivity of the phenotype to gene combinations and environmental effects undercut genic selectionism. Although we believe that these critics have offered valuable insights into the character of sophisticated evolutionary modeling, we shall try to show that these insights do not conflict with Dawkins's story of the workings of natural selection. We shall endeavor to free the thesis of genic selectionism from some of the troublesome excrescences which have attached themselves to an interesting story.

I. Gene Selection and Bean-Bag Genetics

Sober and Lewontin[2] argue against the thesis that all selection is genic selection by contending that many instances of selection do not involve selection for properties of individual alleles. Stated rather loosely, the claim is that, in some populations, properties of individual alleles are not positive causal factors in the survival and reproductive success of the relevant organisms. Instead of simply resting this claim on an appeal to our intuitive ideas about causality, Sober has recently provided an account of causal discourse which is intended to yield the conclusion he favors, thus rebutting the proposals of those (like Dawkins) who think that properties of individual alleles can be causally efficacious.[3]

The general problem arises because replicators (genes) combine to build vehicles (organisms) and the effect of a gene is critically dependent on the company it keeps. However, recognizing the general problem, Dawkins seeks to disentangle the various contributions of the members of the coalition of replicators (the genome). To this end, he offers an analogy with a process of competition among rowers for seats in a boat. The coach may scrutinize the relative times of different teams but the competition can be analyzed by investigating the contributions of individual rowers in different contexts.[4]

Sober's Case

At the general level, we are left trading general intuitions and persuasive analogies. But Sober (and, earlier, Sober and Lewontin) attempted to clarify the case through a particular example. Sober argues that *heterozygote superiority* is a phenomenon that cannot be understood from Dawkins's standpoint. We shall discuss Sober's example in detail; our strategy is as follows. We first set out Sober's case: heterozygote superiority cannot be understood as a gene-level phenomenon, because only pairs of genes can be, or fail to be, heterozygous. Yet being heterozygous can be causally salient in the selective process. Against Sober, we first offer an analogy to show that there must be something wrong with his line of thought: from the gene's eye view, heterozygote superiority is an instance of a standard selective phenomenon, namely *frequency-dependent* selection. The advantage (or disadvantage) of a trait can depend on the frequency of that trait in other members of the relevant population.

Having claimed that there is something wrong with Sober's argument, we then try to say what is wrong. We identify two principles on which the reasoning depends. First is a general claim about causal uniformity. Sober thinks that there can be selection for a property only if that property has a positive uniform effect on reproductive success. Second, and more specifically, in cases where the heterozygote is fitter, the individuals have no uniform causal effect. We shall try to undermine both principles, but the bulk of our criticism will be directed against the first.

Heterozygote superiority occurs when a heterozygote (with genotype Aa, say) is fitter than either homozygote (AA or aa). The classic example is human sickle-cell anemia: homozygotes for the normal allele in African populations produce functional hemoglobin but are vulnerable to malaria, homozygotes for the mutant ("sickling") allele suffer anemia (usually fatal), and heterozygotes avoid anemia while also having resistance to malaria. The effect of each allele varies with context, and the contexts across which variation occurs are causally relevant. Sober writes:

> In this case, the a allele does not have a unique causal role. Whether the gene a will be a positive or a negative causal factor in the survival and reproductive success of an organism depends on the genetic context. If it is placed next to a copy of A, a will mean an increase in fitness. If it is placed next to a copy of itself, the gene will mean a decrement in fitness.[5]

The argument against Dawkins expressed here seems to come in two parts. Sober relies on the principle

(A)   There is selection for property P only if in all causally relevant background conditions P has a positive effect on survival and reproduction.

He also adduces a claim about the particular case of heterozygote superiority.

(B)   Although we can understand the situation by noting that the heterozygote has a uniform effect on survival and reproduction, the property of having the A allele and the property of having the *a* allele cannot be seen as having uniform effects on survival and reproduction.

We shall argue that both (A) and (B) are problematic.

Let us start with the obvious reply to Sober's argument. It seems that the heterozygote superiority case is akin to a familiar type of frequency-dependent selection. If the population consists just of AAs and a mutation arises, the *a*-allele, then initially *a* is favored by selection. Even though it is very bad to be *aa*, *a* alleles are initially likely to turn up in the company of A alleles. So they are likely to spread, and, as they spread, they find themselves alongside other *a* alleles, with the consequence that selection tells against them. The scenario is very similar to a story we might tell about interactions among individual organisms. If some animals resolve conflicts by playing hawk and others play dove, then, if a population is initially composed of hawks (and if the costs of bloody battle outweigh the benefits of gaining a single resource), doves will initially be favored by selection.[6] For they will typically interact with hawks, and, despite the fact that their expected gains from these interactions are zero, they will still fare better than their rivals whose expected gains from interactions are negative. But, as doves spread in the population, hawks will meet them more frequently, with the result that the expected payoffs to hawks from interactions will increase. Because they increase more rapidly than the expected payoffs to the doves, there will be a point at which hawks become favored by selection, so that the incursion of doves into the population is halted.

We believe that the analogy between the case of heterozygote superiority and the hawk-dove case reveals that there is something troublesome about Sober's argument. The challenge is to say exactly what has gone wrong.

Causal Uniformity

Start with principle (A). Sober conceives of selection as a *force*, and he is concerned to make plain the effects of component forces in situations where different forces combine. Thus, he invites us to think of the heterozygote superiority case by analogy with situations in which a physical object remains at rest because equal and opposite forces are exerted on it. Considering the situation only in terms of net forces will conceal the causal structure of the situation. Hence, Sober concludes, our ideas about units of selection should penetrate beyond what occurs on the average, and we should attempt to isolate those properties which positively affect survival and reproduction in every causally relevant context.

Although Sober rejects determinism, principle (A) seems to hanker after something like the uniform association of effects with causes that deterministic accounts of causality provide. We believe that the principle cannot be satisfied without doing violence to ordinary ways of thinking about natural selection, and once the violence has been exposed, it is not obvious that there is any way to reconstruct ideas about selection that will fit Sober's requirement.

Consider *the* example of natural selection, the case of industrial melanism.[7] We are inclined to say that the moths in a Cheshire wood, where lichens on many trees have been destroyed by industrial pollutants, have been subjected to selection pressure and that there has been selection for the property of being melanic. But a moment's reflection should reveal that this description is at odds with Sober's principle. For the wood is divisible into patches, among which are clumps of trees that have been shielded from the effects of industrialization. Moths who spend most of their lives in these areas are at a disadvantage if they are melanic. Hence, in the population comprising all the moths in the wood, there is no uniform effect on survival and reproduction: in some causally relevant contexts (for moths who have the property of living in regions where most of the trees are contaminated), the trait of being melanic has a positive effect on survival and reproduction, but there are other contexts in which the effect of the trait is negative.

The obvious way to defend principle (A) is to split the population into subpopulations and identify different selection processes as operative in different subgroups. This is a revisionary proposal, for our usual approach to examples of industrial melanism is to take a coarse-grained perspective on the environments, regarding the existence of isolated clumps of uncontaminated trees as a perturbation of the overall selective process. Nonetheless, we might be led to make the revision, not in the interest of honoring a philosophical prejudice, but simply because our general views about selection are consonant with principle (A), so that the reform would bring our treatment of examples into line with our most fundamental beliefs about selection.

In our judgment, a defense of this kind fails for two connected reasons. First, the process of splitting populations may have to continue much further—perhaps even to the extent that we ultimately conceive of individual organisms as making up populations in which a particular type of selection occurs. For, even in contaminated patches, there may be variations in the camouflaging properties of the tree trunks and these variations may combine with propensities of the moths to cause local disadvantages for melanic moths. Second, as many writers have emphasized, evolutionary theory is a statistical theory, not only in its recognition of drift as a factor in evolution but also in its use of fitness coefficients to represent the expected survivorship and reproductive success of organisms. The envisaged splitting of populations to discover some partition in which principle (A) can be maintained is at odds with the strategy of abstracting from the thousand natural shocks that organisms in natural populations are heir to. In principle, we could relate the biography of each organism in the population, explaining in full detail how it developed, reproduced, and survived, just as we could track the motion of each molecule of a sample of gas. But evolutionary theory, like statistical mechanics, has no use for such a fine grain of description: the aim is to make clear the central tendencies in the history of evolving populations, and to this end, the strategy of averaging, which Sober decries, is entirely appropriate. We conclude that there is no basis for any revision that would eliminate those descriptions which run counter to principle (A).

At this point, we can respond to the complaints about the gene's-eye view representation of cases of heterozygote superiority. Just as we can give sense to the idea that the trait of being melanic has a unique environment-dependent effect on survival and reproduction, so too we can explicate the view that a property of alleles, to wit, the property of directing the formation of a particular kind of hemoglobin, has a unique environment-dependent effect on survival and reproduction. The alleles form parts of one another's environments, and in an environment in which a copy of the A allele is present, the typical trait of the S allele (namely, directing the formation of deviant hemoglobin) will usually have a positive effect on the chances that copies of that allele will be left in the next generation. (Notice that the effect will not be invariable, for there are other parts of the genomic environment which could wreak havoc with it.) If someone protests that the incorporation of alleles as themselves part of the environment is suspect, then the immediate rejoinder is that, in cases of behavioral interactions, we are compelled to treat organisms as parts of one another's environments.[8] The effects of playing hawk depend on the nature of the environment, specifically on the frequency of doves in the vicinity.[9]

The Causal Powers of Alleles

We have tried to develop our complaints about principle (A) into a positive account of how cases of heterozygote superiority might look from the gene's-eye view. We now want to focus more briefly on (B). Is it impossible to reinterpret the examples of heterozygote superiority so as to ascribe uniform effects on survival and reproduction to allelic properties? The first point to note is that Sober's approach formulates the Dawkinsian point of view in the wrong way: the emphasis should be on the effects of properties of alleles, not on allelic properties of organisms (like the property of having an A allele) and the accounting ought to be done in terms of allele copies. Second, although we argued above that the strategy of splitting populations was at odds with the character of evolutionary theory, it is worth noting that the same strategy will be available in the heterozygote superiority case.

Consider the following division of the original population: let P_1 be the collection of all those allele copies which occur next to an S allele, and let P_2 consist of all those allele copies which occur next to an A allele. Then the property of being A (or of directing the production of normal hemoglobin) has a positive effect on the production of copies in the next generation in P_1, and conversely in P_2. In this way, we are able to partition the population and to achieve a Dawkinsian redescription that meets Sober's principle (A) — just in the way that we might try to do so if we wanted to satisfy (A) in understanding the operation of selection on melanism in a Cheshire wood or on fighting strategies in a population containing a mixture of hawks and doves.

Objection: the "populations" just defined are highly unnatural, and this can be seen once we recognize that, in some cases, allele copies in the same organisms (the heterozygotes) belong to different "populations." Reply: so what? From the allele's point of view, the copy next door is just a critical part of the environment. The populations P_1 and P_2 simply pick out the alleles that share the same environment. There would be an analogous partition of a population of competing organisms which occurred locally in pairs such that some organisms played dove and some hawk. (Here, mixed pairs would correspond to heterozygotes.)

So the genic picture survives an important initial challenge. The moral of our story so far is that the picture must be applied consistently. Just as paradoxical conclusions will result if one offers a partial translation of geometry into arithmetic, it is possible to generate perplexities by failing to recognize that the Dawkinsian *Weltanschauung* leads to new conceptions of environment and of population. We now turn to a different worry, the objection that genes are not "visible" to selection.

2. Epigenesis and Visibility

In a lucid discussion of Dawkins's early views, Gould claims to find a "fatal flaw" in the genic approach to selection. According to Gould, Dawkins is unable to give genes "direct visibility to natural selection."[10] Bodies must play intermediary roles in the process of selection, and since the properties of genes do not map in one-one fashion onto the properties of bodies, we cannot attribute selective advantages to individual alleles. We believe that Gould's concerns raise two important kinds of issues for the genic picture: (i) Can Dawkins sensibly talk of the effect of an individual allele on its expected copying frequency? (ii) Can Dawkins meet the charge that it is the phenotype that makes the difference to the copying of the underlying alleles, so that, whatever the causal basis of an advantageous trait, the associated allele copies will have enhanced chances of being replicated? We shall take up these questions in order.

Do Alleles Have Effects?

Dawkins and Gould agree on the facts of embryology that subvert the simple Mendelian association of one gene with one character. But the salience of these facts to the debate is up for grabs. Dawkins regards Gould as conflating the demands of embryology with the demands of the theory of evolution. While genes' effects blend in embryological development, and while they have phenotypic effects only in concert with their gene-mates, genes "do not blend as they replicate and recombine down the generations. It is this that matters for the geneticist, and it is also this that matters for the student of units of selection."[11]

Is Dawkins right? Chapter 2 of *The Extended Phenotype* is an explicit defense of the meaningfulness of talk of "genes for" indefinitely complex morphological and behavioral traits. In this, we believe, Dawkins is faithful to the practice of classical geneticists. Consider the vast number of loci in *Drosophila melanogaster* which are labeled for eye-color traits—white, eosin, vermilion, raspberry, and so forth. Nobody who subscribes to this practice of labeling believes that a pair of appropriately chosen stretches of DNA, cultured in splendid isolation, would produce a detached eye of the pertinent color. Rather, the intent is to indicate the effect that certain changes at a locus would make against the background of the rest of the genome.

Dawkins's project here is important not just in conforming to traditions of nomenclature. Remember: Dawkins needs to show that we can sensibly speak of alleles having (environment-sensitive) effects, effects in virtue of which they are selected for or selected against. If we can talk of a gene for X, where X is a selectively important phenotypic characteristic, we can sensibly talk of the effect of an allele on its expected copying frequency, even if the effects are always indirect, via the characteristics of some vehicle.

What follows is a rather technical reconstruction of the relevant notion. The precision is needed to allow for the extreme environmental sensitivity of allelic causation. But the intuitive idea is simple: we can speak of genes for X if substitutions on a chromosome would lead, in the relevant environments, to a difference in the X-ishness of the phenotype.

Consider a species S and an arbitrary locus L in the genome of members of S. We want to give sense to the locution 'L is a locus affecting P' and derivatively to the phrase 'G is a gene for $P*$' (where, typically, P will be a determinable and $P*$ a determinate form of P). Start by taking an *environment* for a locus to be an aggregate of DNA segments that would complement L to form the genome of a member of S together with a set of extraorganismic factors (those aspects of the world external to the organism which we would normally count as part of the organism's environment). Let a set of variants for L be any collection of DNA segments, none of which is debarred, on physico-chemical grounds, from occupying L. (This is obviously a very weak constraint, intended only to rule out those segments which are too long or which have peculiar physico-chemical properties.) Now, we say that L is a locus affecting P in S relative to an environment E and a set of variants V just in case there are segments s, $s*$, and $s**$ in V such that the substitution of $s**$ for $s*$ in an organism having s and $s*$ at L would cause a difference in the form of P, against the background of E. In other words, given the environment E, organisms who are $ss*$ at L differ in the form of P from organisms who are $ss**$ at L and the cause of the difference is the presence of $s*$ rather than $s**$. (A minor clarification: while $s*$ and $s**$ are distinct, we do not assume that they are both different from s.)

L is a locus affecting P in S just in case L is a locus affecting P in S relative to any standard environment and a feasible set of variants. Intuitively, the geneticist's practice of labeling loci focuses on the "typical" character of the complementary part of the genome in the species, the "usual" extraorganismic environment, and the variant DNA segments which have arisen in the past by mutation or which "are likely to arise" by mutation. Can these vague ideas about standard conditions be made more precise? We think so. Consider first the genomic part of the environment. There will be numerous alternative combinations of genes at the loci other than L present in the species S. Given most of these gene combinations, we expect modifications at L to produce modifications in the form of P. But there are likely to be some exceptions, cases in which the presence of a rare allele at another locus or a rare combination of alleles produces a phenotypic effect that dominates any effect on P. We can either dismiss the exceptional cases as nonstandard because they are infrequent or we can give a more refined analysis, proposing that each of the nonstandard cases involves either (a) a rare allele at a locus L' or (b) a rare combination of alleles at loci L', L''. . . such that that locus (a) or those loci jointly (b) affect some phenotypic trait Q that dominates P in the sense that there are modifications of Q which prevent the expression of any modifications of P. As a concrete example, consider the fact that there are modifications at some loci in *Drosophila* that produce embryos that fail to develop heads; given such modifications elsewhere in the genome, alleles affecting eye color do not produce their standard effects!

We can approach standard extragenomic environments in the same way. If L affects the form of P in organisms with a typical gene complement, except for those organisms which encounter certain rare combinations of external factors, then we may count those combinations as nonstandard simply because of their infrequency. Alternatively, we may allow rare combinations of external factors to count provided that they do not produce some gross interference with the organism's development, and we can render the last notion more precise by taking nonstandard environments to be those in which the population mean fitness of organisms in S would be reduced by some arbitrarily chosen factor (say, $\frac{1}{2}$).

Finally, the feasible variants are those which actually occur at L in members of S, together with those which have occurred at L in past members of S and those which are easily attainable from segments that actually occur at L in members of S by means of insertion, deletion, substitution, or transposition. Here the criteria for ease of attainment are given by the details of molecular biology. If an allele is prevalent at L in S, then modifications at sites where the molecular structure favors insertions, deletions, substitutions, or transpositions (so-called "hot spots") should count as easily attainable even if some of these modifications do not actually occur.

Obviously, these concepts of "standard conditions" could be articulated in more detail, and we believe that it is possible to generate a variety of explications, agreeing on the core of central cases but adjusting the boundaries of the concepts in different ways. If we now assess the labeling practices of geneticists, we expect to find that virtually all of their claims about loci affecting a phenotypic trait are sanctioned by all of the explications. Thus, the challenge that there is no way to honor the facts of epigenesis while speaking of loci that affect certain traits would be turned back.

Once we have come this far, it is easy to take the final step. An allele A at a locus L in a species S is for the trait P^* (assumed to be a determinate form of the determinable characteristic P) relative to a local allele B and an environment E just in case (a) L affects the form of P in S, (b) E is a standard environment, and (c) in E organisms that are AB have phenotype P^*. The relativization to a local allele is necessary, of course, because, when we focus on a target allele rather than a locus, we have to extend the notion of the environment—as we saw in the last section, corresponding alleles are potentially important parts of one another's environments. If we say that A is for P^* (period), we are claiming that A is for P^* relative to standard environments and common local alleles or that A is for P^* relative to standard environments and itself.

Now, let us return to Dawkins and to the apparently outré claim that we can talk about genes for reading. Reading is an extraordinarily complex behavior pattern and surely no adaptation. Further, many genes must be present and the extraorganismic environment must be right for a human being to be able to acquire the ability to read. Dyslexia might result from the substitution of an unusual mutant allele at one of the loci, however. Given our account, it will be correct to say that the mutant allele is a gene for dyslexia and also that the more typical alleles at the locus are alleles for reading. Moreover, if the locus also affects some other (determinable) trait, say, the capacity to factor numbers into primes, then it may turn out that the mutant allele is also an allele for rapid factorization skill and that the typical allele is an allele for factorization disability. To say that A is an allele for $P*$ does not preclude saying that A is an allele for $Q*$, nor does it commit us to supposing that the phenotypic properties in question are either both skills or both disabilities. Finally, because substitutions at many loci may produce (possibly different types of) dyslexia, there may be many genes for dyslexia and many genes for reading. Our reconstruction of the geneticists' idiom, the idiom which Dawkins wants to use, is innocent of any Mendelian theses about one-one mappings between genes and phenotypic traits.

Visibility

So we can defend Dawkins's thesis that alleles have properties that influence their chances of leaving copies in later generations by suggesting that, in concert with their environments (including their genetic environments), those alleles cause the presence of certain properties in vehicles (such as organisms) and that the properties of the vehicles are causally relevant to the spreading of copies of the alleles. But our answer to question (i) leads naturally to concerns about question (ii). Granting that an allele is for a phenotypic trait $P*$ and that the presence of $P*$ rather than alternative forms of the determinable trait P enhances the chances that an organism will survive and reproduce and thus transmit copies of the underlying allele, is it not $P*$ and its competition which are directly involved in the selection process? What selection "sees" are the phenotypic properties. When this vague, but suggestive, line of thought has been made precise, we think that there is an adequate Dawkinsian reply to it.

The idea that selection acts directly on phenotypes, expressed in metaphorical terms by Gould (and earlier by Ernst Mayr), has been explored in an interesting essay by Robert Brandon [12] Brandon proposes that phenotypic traits screen off genotypic traits (in the sense of Wesley Salmon [13]):

$$\Pr(O_n/G\&P) = \Pr(O_n/P) \neq \Pr(O_n/G)$$

where $\Pr(O_n/G\&P)$ is the probability that an organism will produce n offspring given that it has both a phenotypic trait and the usual genetic basis for that trait, $\Pr(O_n/P)$ is the probability that an organism will produce n offspring given that it has the phenotypic trait, and $\Pr(O_n/G)$ is the probability that it will produce n offspring given that it has the usual genetic basis. So fitness seems to vary more directly with the phenotype and less directly with the underlying genotype.

Why is this? The root idea is that the successful phenotype may occur in the presence of the wrong allele as a result of judicious tampering, and conversely, the typical effect of a "good" allele may be subverted. If we treat moth larvae with appropriate injections, we can produce pseudomelanics that have the allele which normally gives rise to the speckled form and we can produce moths, foiled melanics, that carry the allele for melanin in which the developmental pathway to the emergence of black wings is blocked. The pseudomelanics will enjoy enhanced reproductive success in polluted woods and the foiled melanics will be at a disadvantage. Recognizing this type of possibility, Brandon concludes that selection acts at the level of the phenotype.[14]

Once again, there is no dispute about the facts. But our earlier discussion of epigenesis should reveal how genic selectionists will want to tell a different story. The interfering conditions that affect the phenotype of the vehicle are understood as parts of the allelic environment. In effect, Brandon, Gould, and Mayr contend that, in a polluted wood, there is selection for being dark colored rather than for the allelic property of directing the production of melanin, because it would be possible to have the reproductive advantage associated with the phenotype without having the allele (and conversely it would be possible to lack the advantage while possessing the allele). Champions of the gene's-eye view will maintain that tampering with the phenotype reverses the typical effect of an allele by changing the environment. For these cases involve modification of the allelic environment and give rise to new selection processes in which allelic properties currently in favor prove detrimental. The fact that selection goes differently in the two environments is no more relevant than the fact that selection for melanic coloration may go differently in Cheshire and in Dorset.

If we do not relativize to a fixed environment, then Brandon's claims about screening off will not generally be true.[15] We suppose that Brandon intends to relativize to a fixed environment. But now he has effectively begged the question against the genic selectionist by deploying the orthodox conception of environment. Genic selectionists will also want to relativize to the environment, but they should resist the orthodox conception of it. On their view, the probability relations derived by Brandon involve an illicit averaging over environments (see note 15). Instead, genic selectionists should propose that the probability of an allele's leaving n copies of itself should be understood relative to the total allelic environment, and that the specification of the total environment ensures that there is no screening off of allelic properties by phenotypic properties. The probability of producing n copies of the allele for melanin in a total allelic environment is invariant under conditionalization on phenotype.

Here too the moral of our story is that Dawkinspeak must be undertaken consistently. Mixing orthodox concepts of the environment with ideas about genic selection is a recipe for trouble, but we have tried to show how the genic approach can be thoroughly articulated so as to meet major objections. But what is the point of doing so? We shall close with a survey of some advantages and potential drawbacks.

3. Genes and Generality

Relatively little fossicking is needed to uncover an extended defense of the view that gene selectionism offers a more general and unified picture of selective processes than can be had from its alternatives. Phenomena anomalous for the orthodox story of evolution by individual selection fall naturally into place from Dawkins' viewpoint. He offers a revision of the "central theorem" of Darwinism. Instead of expecting individuals to act in their best interests, we should expect an animal's behavior "to maximize the survival of genes 'for' that behavior, whether or not those genes happen to be in the body of that particular animal performing it."[16]

The cases that Dawkins uses to illustrate the superiority of his own approach are a somewhat motley collection. They seem to fall into two general categories. First are outlaw and quasi-outlaw examples. Here there is competition among genes which cannot be translated into talk of vehicle fitness because the competition is among cobuilders of a single vehicle. The second group comprises "extended phenotype" cases, instances in which a gene (or combination of genes) has selectively relevant phenotypic consequences which are not traits of the vehicle that it has helped build. Again the replication potential of the gene cannot be translated into talk of the adaptedness of its vehicle.

We shall begin with outlaws and quasi-outlaws. From the perspective of the orthodox story of individual selection, "replicators at different loci within the same body can be expected to 'cooperate.'" The allele surviving at any given locus tends to be one best (subject to all the constraints) for the whole genome. By and large this is a reasonable assumption. Whereas individual outlaw organisms are perfectly possible in groups and subvert the chances for groups to act as vehicles, outlaw genes seem problematic. Replication of any gene in the genome requires the organism to survive and reproduce, so genes share a substantial common interest. This is true of asexual reproduction, and, granting the fairness of meiosis, of sexual reproduction too.

But there's the rub. Outlaw genes are genes that subvert meiosis to give them a better than even chance of making it to the gamete, typically by sabotaging their corresponding allele (EP 136). Such genes are *segregation distorters* or *meiotic drive* genes. Usually, they are enemies not only of their alleles but of other parts of the genome, because they reduce the individual fitness of the organism they inhabit. Segregation distorters thrive, when they do, because they exercise their phenotypic power to beat the meiotic lottery. Selection for such genes cannot be selection for traits that make organisms more likely to survive and reproduce. They provide uncontroversial cases of selective processes in which the individualistic story cannot be told.

There are also related examples. Altruistic genes can be outlawlike, discriminating against their genome mates in favor of the inhabitants of other vehicles, vehicles that contain copies of themselves. Start with a hypothetical case, the so-called "green beard" effect. Consider a gene Q with two phenotypic effects. Q causes its vehicle to grow a green beard and to behave altruistically toward green-bearded conspecifics. Q's replication prospects thus improve, but the particular vehicle that Q helped build does not have its prospects for survival and reproduction enhanced. Is Q an outlaw not just with respect to the vehicle but with respect to the vehicle builders? Will there be selection for alleles that suppress Q's effect? How the selection process goes will depend on the probability that Q's cobuilders are beneficiaries as well. If Q is reliably associated with other gene kinds, those kinds will reap a net benefit from Q's outlawry.

So altruistic genes are sometimes outlaws. Whether coalitions of other genes act to suppress them depends on the degree to which they benefit only themselves. Let us now move from a hypothetical example to the parade case.

Classical fitness, an organism's propensity to leave descendants in the next generation, seems a relatively straightforward notion. Once it was recognized that Darwinian processes do not necessarily favor organisms with high classical fitness, because classical fitness ignores indirect effects of costs and benefits to relatives, a variety of alternative measures entered the literature. The simplest of these would be to add to the classical fitness of an organism contributions from the classical fitness of relatives (weighted in each case by the coefficient of relatedness). Although accounting of this sort is prevalent, Dawkins (rightly) regards it as just wrong, for it involves double bookkeeping and, in consequence, there is no guarantee that populations will move to local maxima of the defined quantity. This measure and measures akin to it, however, are prompted by Hamilton's rigorous development of the theory of inclusive fitness (in which it is shown that populations will tend toward local maxima of inclusive fitness).[17] In the misunderstanding and misformulation of Hamilton's ideas, Dawkins sees an important moral.

Hamilton, he suggests, appreciated the gene selectionist insight that natural selection will favor "organs and behavior that cause the individual's genes to be passed on, whether or not the individual is an ancestor."[18] But Hamilton's own complex (and much misunderstood) notion of inclusive fitness was, for all its theoretical importance, a dodge, a "brilliant last-ditch rescue attempt to save the individual organism as the level at which we think about natural selection."[19] More concretely, Dawkins is urging two claims: first, that the uses of the concept of inclusive fitness in practice are difficult, so that scientists often make mistakes; second, that such uses are conceptually misleading. The first point is defended by identifying examples from the literature in which good researchers have made errors, errors which become obvious once we adopt the gene selectionist perspective. Moreover, even when the inclusive fitness calculations make the right predictions, they often seem to mystify the selective process involved (thus buttressing Dawkins's second thesis). Even those who are not convinced of the virtues of gene selectionism should admit that it is very hard to see the reproductive output of an organism's relatives as a property of that organism.

Let us now turn to the other family of examples, the "extended phenotype" cases. Dawkins gives three sorts of "extended" phenotypic effets: effects of genes—indeed key weapons in the competitive struggle to replicate—which are not traits of the vehicle the genes inhabit. The examples are of artifacts, of parasitic effects on host bodies and behaviors, and of "manipulation" (the subversion of an organism's normal patterns of behavior by the genes of another organism via the manipulated organism's nervous system).

Among many vivid, even haunting, examples of parasitic behavior, Dawkins describes cases in which parasites synthesize special hormones with the consequence that their hosts take on phenotypic traits that decrease their own prospects for reproduction but enhance those of the parasites.[20] There are equally forceful cases of manipulation: cuckoo fledglings subverting their host's parental program, parasitic queens taking over a hive and having its members work for her. Dawkins suggests that the traits in question should be viewed as adaptations—properties for which selection has occurred—even though they cannot be seen as adaptations of the individuals whose reproductive success they promote, for those individuals do not possess the relevant traits. Instead, we are to think in terms of selectively advantageous characteristics of alleles which orchestrate the behavior of several different vehicles, some of which do not include them.

At this point there is an obvious objection. Can we not understand the selective processes that are at work by focusing not on the traits that are external to the vehicle that carries the genes, but on the behavior that the vehicle performs which brings those traits about? Consider a spider's web. Dawkins wants to talk of a gene for a web. A web, of course, is not a characteristic of a spider. Apparently, however, we could talk of a gene for web building. Web building is a trait of spiders, and, if we choose to redescribe the phenomena in these terms, the extended phenotype is brought closer to home. We now have a trait of the vehicle in which the genes reside, and we can tell an orthodox story about natural selection for this trait.

It would be tempting to reply to this objection by stressing that the selective force acts through the artifact. The causal chain from the gene to the web is complex and indirect; the behavior is only a part of it. Only one element of the chain is distinguished, the endpoint, the web itself, and that is because, independently of what has gone on earlier, provided that the web is in place, the enhancement of the replication chances of the underlying allele will ensue. But this reply is exactly parallel to the Mayr-Gould-Brandon argument discussed in the last section, and it should be rejected for exactly parallel reasons.

The correct response, we believe, is to take Dawkins at his word when he insists on the possibility of a number of different ways of looking at the same selective processes. Dawkins's two main treatments of natural selection, *The Selfish Gene* and *The Extended Phenotype*, offer distinct versions of the thesis of genic selectionism. In the earlier discussion (and occasionally in the later) the thesis is that, for any selection process, there is a uniquely correct representation of that process, a representation which captures the causal structure of the process, and this representation attributes causal efficacy to genic properties. In *Extended Phenotype*, especially in chapters 1 and 13, Dawkins proposes a weaker version of the thesis, to the effect that there are often alternative, equally adequate representations of selection processes and that, for any selection process, there is a maximally adequate representation which attributes causal efficacy to genic properties. We shall call the strong (early) version *monist genic selectionism* and the weak (later) version *pluralist genic selectionism*. We believe that the monist version is faulty but that the pluralist thesis is defensible.

In presenting the "extended phenotype" cases, Dawkins is offering an alternative representation of processes that individualists can redescribe in their own preferred terms by adopting the strategy illustrated in our discussion of spider webs. Instead of talking of genes for webs and their selective advantages, it is possible to discuss the case in terms of the benefits that accrue to spiders who have a disposition to engage in web building. There is no privileged way to segment the causal chain and isolate the (really) real causal story. As we noted two paragraphs back, the analog of the Mayr-Gould-Brandon argument for the priority of those properties which are most directly connected with survival and reproduction—here the webs themselves—is fallacious. Equally, it is fallacious to insist that the causal story must be told by focusing on traits of individuals which contribute to the reproductive success of those individuals. We are left with the general thesis of pluralism: there are alternative, maximally adequate representations of the causal structure of the selection process. Add to this Dawkins's claim that one can always find a way to achieve a representation in terms of the causal efficacy of genic properties, and we have pluralist genic selectionism.

Pluralism of the kind we espouse has affinities with some traditional views in the philosophy of science. Specifically, our approach is instrumentalist, not of course in denying the existence of entities like genes, but in opposing the idea that natural selection is a force that acts on some determinate target, such as the genotype or the phenotype. Monists err, we believe, in claiming that selection processes must be described in a particular way, and their error involves them in positing entities, "targets of selection," that do not exist.

Another way to understand our pluralism is to connect it with conventionalist approaches to space-time theories. Just as conventionalists have insisted that there are alternative accounts of the phenomena that meet all our methodological desiderata, so too we maintain that selection processes can usually be treated, equally adequately, from more than one point of view. The virtue of the genic point of view, on the pluralist account, is not that it alone gets the causal structure right but that it is always available.

What is the rival position? Well, it cannot be the thesis that the only adequate representations are those in terms of individual traits which promote the reproductive success of their bearers, because there are instances in which no such representation is available (outlaws) and instances in which the representation is (at best) heuristically misleading (quasi-outlaws, altruism). The sensible rival position is that there is a hierarchy of selection processes: some cases are aptly represented in terms of genic selection, some in terms of individual selection, some in terms of group selection, and some (maybe) in terms of species selection. Hierarchical monism claims that, for any selection process, there is a unique level of the hierarchy such that only representations that depict selection as acting at that level are maximally adequate. (Intuitively, representations that see selection as acting at other levels get the causal structure wrong.) Hierarchical monism differs from pluralist genic selectionism in an interesting way: whereas the pluralist insists that, for any process, there are many adequate representations, one of which will always be a genic representation, the hierarchical monist maintains that for each process there is just one kind of adequate representation, but that processes are diverse in the kinds of representation they demand.[21]

Just as the simple orthodoxy of individualism is ambushed by outlaws and their kin, so too hierarchical monism is entangled in spider webs. In the "extended phenotype" cases, Dawkins shows that there are genic representations of selection processes that can be no more adequately illuminated from alternative perspectives. Since we believe that there is no compelling reason to deny the legitimacy of the individualist redescription in terms of web-building behavior (or dispositions to such behavior), we conclude that Dawkins should be taken at face value: just as we can adopt different perspectives on a Necker cube, so too we can look at the workings of selection in different ways.[22]

In previous sections, we have tried to show how genic representations are available in cases that have previously been viewed as troublesome. To complete the defense of genic selectionism, we would need to extend our survey of problematic examples. But the general strategy should be evident. Faced with processes that others see in terms of group selection or species selection, genic selectionists will first try to achieve an individualist representation and then apply the ideas we have developed from Dawkins to make the translation to genic terms.

Pluralist genic selectionists recommend that practicing biologists take advantage of the full range of strategies for representing the workings of selection. The chief merit of Dawkinspeak is its generality. Whereas the individualist perspective may sometimes break down, the gene's-eye view is apparently always available. Moreover, as illustrated by the treatment of inclusive fitness, adopting it may sometimes help us to avoid errors and confusions. Thinking of selection in terms of the devices, sometimes highly indirect, through which genes lever themselves into future generations may also suggest new approaches to familiar problems.

But are there drawbacks? Yes. The principal purpose of the early sections of this paper was to extend some of the ideas of genic selectionism to respond to concerns that are deep and important. Without an adequate rethinking of the concepts of population and of environment, genic representations will fail to capture processes that involve genic interactions or epigenetic constraints. Genic selectionism can easily slide into naive adaptationism as one comes to credit the individual alleles with powers that enable them to operate independently of one another. The move from the "genes for P" locution to the claim that selection can fashion P independently of other traits of the organism is perennially tempting.[23] But, in our version, genic representations must be constructed in full recognition of the possibilities for constraints in gene-environment coevolution. The dangers of genic selectionism, illustrated in some of Dawkins's own writings, are that the commitment to the complexity of the allelic environment is forgotten in practice. In defending the genic approach against important objections, we have been trying to make this commitment explicit, and thus to exhibit both the potential and the demands of correct Dawkinspeak. The return of the gene should not mean the exile of the organism.[24]

Notes

Kim Sterelny and I coauthored this essay, which was written when we discovered that we were writing it independently. We would like to thank those who have offered helpful suggestions to one or both of us, particularly Patrick Bateson, Robert Brandon, Peter Godfrey-Smith, David Hull, Richard Lewontin, Elisabeth Lloyd, Philip Pettit, David Scheel, and Elliott Sober. Kenneth Waters has independently developed a view similar to ours.

1. The claim is made in Dawkins, *The Selfish Gene* (New York: Oxford University Press, 1976); and, in a somewhat modified form, in Dawkins, *The Extended Phenotype* (San Francisco: Freeman, 1982). We shall discuss the difference between the two versions in the final section of this essay, and our reconstruction will be primarily concerned with the later version

of Dawkins's thesis. To forestall any possible confusion, our reconstruction of Dawkins's position does not commit us to the provocative claims about altruism and selfishness on which many early critics of *The Selfish Gene* fastened.

2. Elliott Sober and Richard Lewontin, "Artifact, Cause and Genic Selection," *Philosophy of Science* 49 (1982): 157–180.

3. See Elliott Sober, *The Nature of Selection* (Cambridge, Mass.: MIT Press, 1984), chapters 7–9, esp. 302–314.

4. Dawkins, *Selfish Gene*, 40–41, 91–92; Dawkins, *Extended Phenotype*, 239.

5. Sober, *Nature of Selection*, 303.

6. For details, see John Maynard Smith, *Evolution and the Theory of Games* (New York: Cambridge, 1982); for a capsule presentation, see Philip Kitcher, *Vaulting Ambition: Sociobiology and the Quest for Human Nature* (Cambridge, Mass.: MIT Press, 1985), 88–97.

7. The locus classicus for discussion of this example is H. B. D. Kettlewell, *The Evolution of Melanism* (New York: Oxford University Press, 1973).

8. In the spirit of Sober's original argument, one might press further. Genic selectionists contend that an A allele can find itself in two different environments, one in which the effect of directing the formation of a normal globin chain is positive and one in which that effect is negative. Should we not be alarmed by the fact that the distribution of environments in which alleles are selected is itself a function of the frequency of the alleles whose selection we are following? No. The phenomenon is thoroughly familiar from studies of behavioral interactions—in the hawk-dove case we treat the frequency of hawks both as the variable we are tracking and as a facet of the environment in which selection occurs. Maynard Smith makes the parallel fully explicit in his paper, "How to Model Evolution," in John Dupré, ed., *The Latest on the Best: Essays on Optimality and Evolution* (Cambridge, Mass.: MIT Press, 1987), 119–131, esp. 125–126.

9. Moreover, we can explicitly recognize the coevolution of alleles with allelic environments. A fully detailed general approach to population genetics from the Dawkinsian point of view will involve equations that represent the functional dependence of the distribution of environments on the frequency of alleles, and equations that represent the fitnesses of individual alleles in different environments. In fact, this is just another way of looking at the standard population genetics equations. Instead of thinking of W_{AA} as the expected contribution to survival and reproduction of (an organism with) an allelic pair, we think of it as the expected contribution of copies of itself of the allele A in environment A. We now see W_{AS} as the expected contribution of A in environment S and also as the expected contribution of S in environment A. The frequencies p, q are not only the frequencies of the alleles, but also the frequencies with which certain environments occur. The standard definitions of the overall (net) fitnesses of the alleles are obtained by weighting the fitnesses in the different environments by the frequencies with which the environments occur.

Lewontin has suggested to us that problems may arise with this scheme of interpretation if the population should suddenly start to reproduce asexually. But this hypothetical change could be handled from the genic point of view by recognizing an alteration of the coevolutionary process between alleles and their environments: whereas certain alleles used to have descendants that would encounter a variety of environments, their descendants are now found only in one allelic environment. Once the algebra has been formulated, it is relatively straightforward to extend the reinterpretation to this case.

10. Stephen Jay Gould, "Caring Groups and Selfish Genes," in Gould, *The Panda's Thumb* (New York: Norton, 1980), 90. There is a valuable discussion of Gould's claims in Sober, *Nature of Selection*, 227ff.

11. Dawkins, *Extended Phenotype*, 117.

12. Gould, "Caring Groups and Selfish Genes"; Ernst Mayr, *Animal Species and Evolution* (Cambridge, Mass.: Harvard University Press, 1963), 184; Robert Brandon, "The Levels of Selection," in Brandon and Richard Burian, eds., *Genes, Organisms, Populations* (Cambridge, Mass.: MIT Press, 1984), 133–141.

13. Brandon refers to Salmon's "Statistical Explanation," in Wesley Salmon, ed., *Statistical Explanation and Statistical Relevance* (Pittsburgh: University of Pittsburgh Press, 1971). It is now widely agreed that statistical relevance misses some distinctions that are important in explicating causal relevance. See, for example, Nancy Cartwright, "Causal Laws and Effective Strategies," *Noûs* 13 (1979): 419–437; Sober, *Nature of Selection*, chapter 8; Salmon, *Scientific Explanation and the Causal Structure of the World* (Princeton, N.J.: Princeton University Press, 1984).

14. Unless the treatments are repeated in each generation, the presence of the genetic basis for melanic coloration will be correlated with an increased frequency of grandoffspring, or of great-grandoffspring, or of descendants in some further generation. Thus, analogs of Brandon's probabilistic relations will hold only if the progeny of foiled melanics are treated so as to become foiled melanics, and the progeny of pseudomelanics are treated so as to become pseudomelanics. This point reinforces the claims about the relativization to the environment that we make below. Brandon has suggested to us in correspondence that now his preferred strategy for tackling issues of the units of selection would be to formulate a principle for identifying genuine environments.

15. Intuitively, this will be because Brandon's identities depend on there being no correlation between O_n and G in any environment, except through the property P. Thus, ironically, the screening-off relations only obtain under the assumptions of simple bean-bag genetics! Sober seems to appreciate this point in a cryptic footnote (*Natural Selection*, 229–230).

To see how it applies in detail, imagine that we have more than one environment and that the reproductive advantages of melanic coloration differ in the different environments. Specifically, suppose that E_1 contains m_1 organisms that have P (melanic coloration) and G (the normal genetic basis of melanic coloration), that E_2 contains m_2 organisms that have P and G, and that the probabilities $Pr(O_n/G\&P\&E_1)$ and $Pr(O_n/G\&P\&E_2)$ are different. Then, if we do not relativize to environments, we shall compute $Pr(O_n/G\&P)$ as a weighted average of the probabilities relative to the two environments.

$$Pr(O_n/G\&P) = Pr(E_1/G\&P) \cdot Pr(O_n/G\&P\&E_1) + Pr(E_2/G\&P) \cdot Pr(O_n/G\&P\&E_2)$$
$$= m_1/(m_1 + m_2) \cdot Pr(O_n/G\&P\&E_1) + m_2/(m_1 + m_2) \cdot Pr(O_n/G\&P\&E_2)$$

Now, suppose that tampering occurs in E_2 so that there are m_3 pseudomelanics in E_2. We can write $Pr(O_n/P)$ as a weighted average of the probabilities relative to the two environments.

$$Pr(O_n/P) = Pr(E_1/P) \cdot Pr(O_n/P\&E_1) + Pr(E_2/P) \cdot Pr(O_n/P\&E_2).$$

By the argument that Brandon uses to motivate his claims about screening off, we can take $Pr(O_n/G\&P\&E_i) = Pr(O_n/P\&E_i)$ for $i = 1, 2$. However, $Pr(E_1/P) = m_1/(m_1 + m_2 + m_3)$ and $Pr(E_2/P) = (m_2 + m_3)/(m_1 + m_2 + m_3)$, so that $Pr(E_i/P) \neq Pr(E_i/G\&P)$. Thus $Pr(O_n/G\&P) \neq Pr(O_n/P)$, and the claim about screening off fails.

Notice that, if environments are lumped in this way, then it will only be under fortuitous circumstances that the tampering makes the probabilistic relations come out as Brandon claims. Pseudomelanics would have to be added in both environments so that the weights remain exactly the same.

16. Dawkins, *Extended Phenotype*, 223.

17. For Hamilton's original demonstration, see "The Genetical Evolution of Social Behavior I," in G. C. Williams, ed., *Group Selection* (Chicago: Aldine, 1971), 23–43. For a brief presentation of Hamilton's ideas, see Kitcher, *Vaulting Ambition*, 77–87; for penetrating diagnoses of misunderstandings, see A. Grafen, "How Not to Measure Inclusive Fitness," *Nature* 298 (1982): 425–426; and R. Michod, "The Theory of Kin Selection," in Brandon and Burian, *Genes, Organisms, Populations*, 203–237.

18. Dawkins, *Extended Phenotype*, 185.

19. Ibid., 187.

20. For a striking instance, see ibid., 215.

21. In defending pluralism, we are very close to the views expressed by Maynard Smith in "How to Model Evolution." Indeed, we would like to think that Maynard Smith's article and the present essay complement one another in a number of respects. In particular, as Maynard Smith explicitly notes, "recommending a plurality of models of the same process" contrasts with the view (defended by Gould and by Sober) of "emphasizing a plurality of processes." Gould's views are clearly expressed in "Is a New and General Theory of Evolution Emerging?" *Paleobiology* 6 (1980): 119–130; Sober's ideas are presented in *Nature of Selection*, chapter 9.

22. Dawkins, *Extended Phenotype*, chapter 1.

23. At least one of us believes that the claims of the present paper are perfectly compatible with the critique of adaptationism developed in Gould and Lewontin, "The Spandrels of San Marco and the Panglossian Paradigm: A Critique of the Adaptationist Programme," in Elliott Sober, ed., *Conceptual Problems in Evolutionary Biology* (Cambridge, Mass.: MIT Press, 1984). For discussion of the difficulties with adaptationism, see Kitcher, *Vaulting Ambition*, chapter 7, and Kitcher, "Why Not the Best?" in Dupré, *The Latest on the Best*.

24. As, we believe, Dawkins himself appreciates. See the last chapter of *The Extended Phenotype*, especially his reaction to the claim that "Richard Dawkins has rediscovered the organism" (251).

5

Species (1984)

1. Pluralistic Realism

The most accurate definition of "species" is the cynic's. Species are those groups of organisms that are recognized as species by competent taxonomists. Competent taxonomists, of course, are those who can recognize the true species. Cynicism is attractive for the weary systematist who despairs of doing better. But I think that philosophers and biologists need not despair. Despite the apparently endless squabbles about how species are to be characterized, it is possible to defend an account of the species category that will do justice to the insights of several divergent approaches.[1]

I shall try to explain a position about species that I shall call *pluralistic realism*, and to indicate in a general way why I think that this position is true. In particular, I want to defend four theses.

1. Species can be considered to be sets of organisms, so that the relation between organism and species can be construed as the familiar relation of set-membership.
2. Species are sets of organisms related to one another by complicated, biologically interesting relations. There are many such relations which could be used to delimit species taxa. However, there is no unique relation which is privileged in that the species taxa it generates will answer to the needs of all biologists and will be applicable to all groups of organisms. In short, the species category is heterogeneous.
3. The species category is heterogeneous because there are two main approaches to the demarcation of species taxa and within each of these approaches there are several legitimate variations. One approach is to group organisms by structural similarities. The taxa thus generated are useful in certain kinds of biological investigations and explanations. However, there are different levels at which structural similarities can be sought. The other approach is to group organisms by their phylogenetic relationships. Taxa resulting from this approach are appropriately used in answering different kinds of biological questions. But there are alternative ways to divide phylogeny into evolutionary units. A pluralistic view of species taxa can be defended because the structural relations among organisms and the phylogenetic relations among organisms provide common ground on which the advocates of different taxonomic units can meet.

4. Pluralism about species taxa is not only compatible with realism about species. It also offers a way to disentangle various claims that can be made in maintaining that "species are real entities existing in nature, whose origin, persistence, and extinction require explanation." [2]

I do not intend to provide a complete defense of all these claims. I shall concentrate primarily on the first three theses, saying little about the issue of realism about species, although I hope that my explanations of theses (1)–(3), together with the discussion in section 5, will make it possible to see how to avoid the charge that species are merely fictions of the systematist's imagination.[3]

2. Sets versus Individuals

My first thesis seems banal. After all, who would think of denying that species are sets of organisms? However, a number of philosophers and biologists—most prominently, David Hull and Michael Ghiselin—have recently campaigned against the notion that species are (what they call) "spatio-temporally unrestricted classes" and they have urged that species should be viewed as individuals.[4] Strange though this proposal may initially appear, it cannot be lightly dismissed. Hull and Ghiselin argue that their account of species is far more consonant with our current understanding of the evolutionary process than the view that they seek to replace.[5]

Let me begin by explaining what I take to be the commitments of the traditional idea that species are sets. First, there is no inconsistency in claiming that species are sets and denying that the members of these sets share a common property. Unless 'property' is used in an attenuated sense, so that all sets are sets whose members share one trivial property—namely, the property of belonging to that set—then there are sets whose members are not distinguished by any common property. In particular, believing that species are sets does not entail believing that there is some homogeneous collection of morphological properties such that each species taxon is the set of organisms possessing one of the morphological properties in the collection. So we can accept (1) while endorsing Mayr's celebrated critique of the morphological concept of species.[6]

Let me now turn to the main arguments that have been offered for thinking that the view of species as sets is at odds with our best biological theorizing. One of these arguments claims that construing species as sets is incompatible with the doctrine that species evolve.[7] Here is the starkest version:[8] "Species evolve. Sets are atemporal entities. Hence sets cannot evolve. Therefore species are not sets." Quite evidently, there is a fallacy here, the fallacy of incomplete translation. It would be futile to think that mathematicians need to revise their standard ontology because of the following argument: "Curves have tangents. Sets of triples of numbers are nonspatial entities. Hence sets of triples of real numbers cannot have tangents. Therefore curves are not sets of triples of real numbers." The correct response to the latter argument is to insist that, in the reduction of geometry to real arithmetic, the property of being a tangent is itself identified in arithmetical terms. Once the property has been so identified, it is possible to see how sets of triples of real numbers can have it. Only incomplete translation deludes us into thinking that sets of triples of real numbers cannot have tangents. An exactly parallel response is available in the case of species.

Assume, for the sake of the present argument, that a species is a set of organisms consisting of a founder population and some (but not necessarily all) of the descendants of that population. I make this assumption in order to show that there is a set-theoretic equivalent of the approach to species that Hull favors. For any given time, let the *stage* of the species at that time be the set of organisms belonging to the species that are alive at that time. To say that the species evolves is to say that the frequency distribution of properties (genetic or genetic plus phenotypic) changes from stage to stage.[9] To say that the species gives rise to a number of descendant species is to claim that the founding populations of those descendant species consist of organisms descending from the founding population of the original species. By proceeding in this way it is relatively easy to reconstruct the standard claims about the evolutionary behavior of species.

A second major theme in Hull's attack on the tradition is his suggestion that recognizing species as individuals will enable us to understand why there are no biological laws about particular species.

> If species are actually spatio-temporally unrestricted classes, then they are the sorts of things which can function in laws. "All swans are white," if true, might be a law of nature and generations of philosophers have treated it as such. If statements of the form "Species X has property Y" were actually laws of nature, one might rightly expect biologists to be disturbed when they are proven false. To the contrary, biologists expect exceptions to exist. At any one time, a particular percentage of a species of crows will be non-back. No one expects this percentage to be universal or to remain fixed. Species may be classes, but they are not very important classes because their names function in no scientific laws. Given the traditional analyses of scientific laws, statements which refer to particular species do not count as scientific laws, as they should not if species are spatio-temporally localized individuals.[10]

Ignoring all sorts of interesting issues, I shall concentrate on two central points. First, it seems to me that Hull is correct to dismiss statements like "All swans are white" as candidates for being laws of nature. But I think that he offers an incorrect explanation of why such statements are not laws. Second, I claim that he is far too quick to conclude that there are no laws about individual species. When we understand why "All swans are white" isn't a candidate for a law of nature—since it is neither lawlike nor true—we shall be able to recognize the possibility of laws about particular species.

Why isn't "All swans are white" a law? The answer is relatively obvious, given our understanding of the process of evolution: even if it had been true that all members of some swan species—*Cygnus olor*, for example—were white, then this would have been an evolutionary fluke. Organisms flouting the generalization could easily have been produced without any large-scale disruption of the course of nature. A small mutation or chromosomal change could easily modify biosynthetic pathways, and thus result in differently colored plumage. Thus I suggest that "All swans are white" is what it appears to be, a generalization, but a generalization which fails to be lawlike. Biologists are unsurprised when generalizations like this prove to be false, because, given their understanding of the workings of evolution, they would be flabbergasted if there were no exceptions.

In the light of this explanation, we can see what conditions would have to be met for a statement of form "All *S* are *P*," where *S* is a species and *P* a property, to count as a law. Mutations or chromosomal novelties producing the absence of *P* in progeny of members of *S* would have to be so radical that they fell into one of two categories: (a) changes giving rise to inviable zygotes, (b) changes with effects large enough to count as events of instantaneous speciation. In other words, the property *P* would have to be so deeply connected with the genetic constitution of members of the species that alterations of the genome sufficient to lead to the absence of *P* would disrupt the genetic organization, leading to inviable offspring or to offspring of a new species.[11] So, *if* there are developmental systems whose modification in certain respects would generate either "hopeful" or "hopeless" monsters, then statements ascribing to members of a species appropriately chosen properties would be candidates for laws about the species. These laws, I suggest, would have the same status as low-level laws of chemistry, generalizations like "DNA molecules contain adenine and thymine molecules in (almost) equal numbers." While they are more particular than the grand equations of physics, these generalizations are scientifically significant, and are featured in numerous explanations.

So Hull is far too quick to foreclose the possibility of biological laws about particular species. Let me now consider the third main strand in his argument for the idea that species are individuals. What moves Hull is a sense of disanalogy between the set of atoms of an element and a typical biological species. Apparently, atoms of gold might occur anywhere in the universe, while members of *Rattus rattus* are bound to be much more localized. Now, despite the fact that Hull typically formulates the issue by claiming that species are spatio-temporally localized,[12] the root of his observation is the connectedness of species rather than their boundedness in space-time. The following passage contains the main idea:

> If a species evolved which was identical to a species of extinct pterodactyl save origin, it would still be a new distinct species. Darwin himself notes, "When a species has once disappeared from the face of the earth, we have reason to believe that the same identical form never reappears. . . ." Darwin presents this point as if it were a contingent state of affairs, when actually it is conceptual. Species are segments of the phylogenetic tree. Once a segment is terminated, it cannot reappear somewhere else in the phylogenetic tree. . . .
>
> If species were actually spatiotemporally unrestricted classes, this state of affairs would be strange. If all atoms with atomic number 79 ceased to exist, gold would cease to exist, although a slot would remain open in the periodic table. Later when atoms with the appropriate atomic number were generated, they would be atoms of gold regardless of their origins. But in the typical case, to *be* a horse one must be *born* of horse.[13]

Let us say that a set of organisms is *historically connected* just in case any organism belonging to the set is either a member of the initial population included in the set or else an immediate descendant of members of the set. Hull's argument can be reformulated as follows: if species were "spatiotemporally unrestricted classes" then species could be historically disconnected; since no species can be historically disconnected, species are not "spatiotemporally unrestricted classes."

One way to respond would be to concede that species are special kinds of sets (namely historically connected sets). To reply in this way would be to acquiesce in Hull's interpretation of biological practice, but to claim that a different ontological reconstruction of that practice is possible, a reconstruction whose chief merit is that it allows a perspicuous way of raising questions about the internal structure of species taxa. However, this reply grants too much. To be sure, one part of biological inquiry focuses on relations of descent in the phylogenetic nexus. But this is by no means the only type of inquiry with which biologists are concerned, nor should one develop one's approach to the ontology of species in such a way as to foreclose possibilities which are useful in some biological contexts.

More concretely, there are cases in which it would be proper to admit a historically disconnected set as a species. Let me offer an example that is based on an actual event of species formation through hybridization. In the lizard genus *Cnemidophorus*, several unisexual species have arisen through hybridization. In particular, the lizard *Cnemidophorus tesselatus* has resulted from a cross between *C. tigris* and *C. septemvittatus*.[14] Although there are important differences between bisexual and unisexual species, the practice of naturalists and theoretical biologists has been to count *C. tesselatus* as a distinct species, whose status is not impugned by its unisexual character. In fact, *C. tesselatus* has served as a test case for comparing genetic diversity in bisexual and unisexual species.

C. tesselatus is probably not historically disconnected. But it might all too easily have been. The actual species probably originated when peripheral populations of the ancestral species came into contact. Clones could even have been established on many different occasions from parental individuals belonging to different breeding populations. A more radical type of discontinuity is also possible. Imagine that the entire initial population of *C. tesselatus* was wiped out and that the species was rederived after a second incident of hybridization between the two parental species. I claim that this would have been the correct description to give of a sequence of events in which first hybridization was followed by extinction and later by second hybridization. For, supposing that the clones founded in the first hybridization fall within the same range of genetic (morphological, behavioral, ecological) variation present in the population that has persisted to the present, what biological purpose would be served by distinguishing two species? To hypothesize "sibling species" in this case (and in like cases) seems to me not only to multiply species beyond necessity but also to obfuscate all the biological similarities that matter. Hence I conclude that Hull is wrong to chide Darwin for confusing contingent state of affairs with a conceptual point. In most groups of organisms, historically disconnected species are unlikely—and conceding the logical possibility that *Homo sapiens* might reevolve after a holocaust does not offer us any genuine comfort. But it is not necessary, and it may not even be true, that all species are historically connected.[15]

3. The Troubles of Monism

The traditional thesis that species are sets provides us with a framework within which we can investigate the species category, and this framework is not at odds with insights drawn from evolutionary theory. But if species are sets, what kind of sets are they?[16] The twentieth-century literature in biology is strewn with answers to this question. Most popular has been the so-called biological species concept, developed with great care by Ernst Mayr. According to Mayr's definition, species are "groups of interbreeding natural populations that are reproductively isolated from other such groups."[17] A somewhat different approach, developed in different ways by G. G. Sampson, Willi Hennig, E. O. Wiley, and others,[18] is to regard the notion of a speciation event as the basic notion and to take a species to be the set of organisms in a lineage (a sequence of ancestral-descendant populations) bounded by successive speciation events.[19] Speciation events themselves can be understood either as events in which a descendant population becomes reproductively isolated form its ancestors (Simpson) or as events in which an ancestral population gives rise to two descendant populations that are reproductively isolated from one another (Wiley and Hennig).[20] A more radical departure from traditional concepts of species is effected by viewing speciation as a process in which descendant populations are ecologically differentiated from their ancestors.[21] And there are still other approaches. In the early 1960s there arose an influential school of taxonomy that proclaimed the virtues of dividing organisms into species by constructing a measure of overall similarity and taking species to be sets of organisms which are clustered by this measure.[22] Finally, in the last decade, another taxonomic school, the so-called pattern cladists, have proposed that a species is a set of organisms distinguished by their common possession of a "minimal evolutionary novelty.[23]

I do not have space here to explain in detail what these various proposals are, much less to examine their merits. So I shall simply give a brief, dogmatic statement of my main claim and then offer a quick illustration of it. Most of the suggestions that I have mentioned can be motivated by their utility for pursuing a particular type of biological inquiry. But, in each case, the champions of the proposal contend that their species concept can serve the purposes of all biologists. In this I think that they err.

Consider Mayr's biological species concept. There is no doubting the importance of reproductive isolation as a criterion for demarcating certain groups of organisms. To cite a classic example, it was a major achievement to separate six sibling species within the *Anopheles* complex of mosquitoes, and thus to understand the distribution of malarial infection in Europe.[24] This example shows the biological species concept in its native habitat: reproductive isolation is important to recognize when we have organisms with overlapping ranges that are morphologically similar but which do not interbreed.

But it is all too familiar that there are difficult cases. Consider the plight of the paleontologist concerned to understand the rates of evolution in different lineages. Quite evidently, there is no way to evaluate directly some hypothesis about whether two forms, long extinct, were or were not reproductively isolated from one another. Thus conclusions about the succession of species in an evolving lineage must be based upon morphological data. Only the most enthusiastic operationalist would conclude directly from this that the paleontological species concept ought to be morphological. As has been repeatedly pointed out,[25] one can search for correlations between morphological changes and the changes that lead to reproductive isolation, using such correlations to reconstruct the division of the lineage into biological species. However, this response to the operationalist's recommendation misses one important feature of the continued insistence by some paleontologists that the biological species concept will not serve all their purposes. There is a perfectly legitimate paleontological question which focuses on the rates and patterns of morphological diversification within evolving lineages, and paleontologists pursue this question by dividing lineages into species according to morphological changes. To insist that they should always formulate their inquiries by using the biological species concept is to make them take a risky trip around Robin Hood's barn. (For further discussion of this point, see section 5.)

But paleontology is not the only place in which there are shortcomings of the biological species concept. That concept also fails in application to organisms that do not reproduce themselves sexually. The typical response to that failure reveals a mistake that pervades much traditional thinking about the concept of species.

In an early explanation and defense of the biological species concept, Mayr acknowledged that there is a problem with asexual organisms, but this problem was not to be taken to be particularly threatening.

> There is, however, some question as to whether this species definition can also be applied to aberrant cases, such as the mating types of protozoa, the self-fertilizing hermaphrodites, animals with obligatory parthenogenesis, and certain groups of parasites and host specialists. . . . The known number of cases in which the above species definition may be inapplicable is very small, and there seems to be no reason at the present time for "watering" our species concept to include these exceptions.[26]

Two interesting features of this passage set the tone for most subsequent defenses of the biological species concept. First, the problem is seen as one of *application*. How do we apply the criterion of reproductive isolation to organisms that do not mate? Second, Mayr attempts to minimize the scope of the problem. Only a few difficult cases are known, and it is suggested hopefully that these may disappear if we learn more about the organisms concerned. The joint effect of these two claims is to portray the biological species concept as a valuable *instrument*. It is recommended to us on the grounds that it will almost always pick out the right groups — as if it were a diagnostic machine that could reveal the patient's malady in 999 cases out of 1000.

This way of looking at the situation is curious. For it seemed originally that the biological species concept was intended as an *analysis* of previous discourse. For centuries, botanists, zoologists, field naturalists, and ordinary people have responded to the diversity of the living world by dividing organisms into species. The biological species concept appeared to offer a reconstruction of their remarks—we were to be given a description of what the species are which would parallel the chemist's account of what the elements are. But, in Mayr's response to the problem of asexuality, the goals of the enterprise seem to shift. The biological species concept is no longer seen as identifying the fundamental feature on which organismic diversity rests; it is viewed as a handy device for leading us to the right groups.

Theoretical systematics often seems to presuppose that there is a fundamental feature of organismic diversity, common to all groups of organisms, that taxonomists try to capture by making judgments of the form "A and B are distinct species." Accounts of the species category propose explanations of what these judgments mean, by offering hypotheses about what the fundamental fact of organismic diversity is. The biological species concept claims that what constitutes the ground of diversity is the reproductive isolation of groups of populations. Asexual organisms teach us that this cannot be the ground of diversity in all groups of organisms. We can react to this lesson in one of a number of ways. One is to deny that there is any fundamental phenomenon of diversity among asexual organisms, abandoning judgments of form "A and B are distinct species," in cases where A and B are sets of asexual organisms. But those who work with asexual organisms contend that there are theoretically significant distinctions among such organisms which defy any such radical revision of taxonomic practice. A second response, developed by Mayr, is to count morphological differences as indicators of species distinctness, treating sexual and asexual organisms alike. But this does not touch the real question which theoretical systematics seemed to address. For what we want to know is *what morphological difference is an indicator of*, what we are after when we attend to morphological distinctness.[27] If it is suggested that, in the case of asexual organisms there is nothing more fundamental than morphological difference, that here clustering in morphological space is not evidence of species distinctness but *constitutive* of species distinctness, then we should ask why we fail to attend to this patterning of organismic diversity in the case of sexual organisms as well. Why isn't morphological distinctness *always* constitutive of species distinctness?

It is here that the difficulties of the biological species concept expose an important moral. Although the biological species concept brings out an important pattern in the diversity of nature—the division of organisms into groups that are reproductively isolated from one another is theoretically significant—this is not the only important pattern of organisms diversity. Champions of the biological species concept—and defenders of alternative approaches to the species category—are too quick to assume that problematic groups of organisms can be dismissed as irritating exceptions, or that they can be handled by adding disjuncts to a definition of "species." By contrast, I suggest that the problem cases should be taken seriously, in that they point to distinctions among organisms which can be used to generate alternative legitimate conceptions of species. I shall now try to explain why it is to be expected that biology needs a number of different approaches to the division of organisms, a number of different sets of "species."

4. The Possibility of Pluralism

In the writings of great systematists, there are occasional passages in which the author recognizes the needs of different groups of biologists. Typically, these passages precede the moment at which monism takes over and the writer becomes an advocate for a single conception of species which is to answer to the interests of every one. An excellent example occurs at the beginning of Hennig's classic work on systematics,[28] where he emphasizes the multiplicity of admissible approaches to classification. Yet, within a few pages,[29] Hennig reformulates the question in a way that makes it clear that some one of the systems is to be regarded as privileged, that biology must have a single general reference system.

I shall try to show why it is both desirable and possible to resist the Hennigian move. I begin with an important distinction due to Mayr. Pointing out that biology covers "two largely separate fields," Mayr claims that practitioners in one field ("functional biology") are primarily interested in questions of "proximate causation," while those in the other field ("evolutionary biology") are primarily concerned with issues of "ultimate causation."[30] Mayr's choice of terms suggests his own predilections and threatens his own fundamental insight. There are indeed two kinds of biological investigation that can be carried out relatively independently of one another, neither of which has priority over the other. These kinds of investigation demand different concepts of species. In fact, as I shall suggest, each main type of biological investigation subdivides further into inquiries that are best conducted by taking alternative views of the species category.

The main Mayrian division is easily explained by example. One interesting biological project is to explain the properties of organisms by means of underlying structures and mechanisms. A biologist may be concerned to understand how, in a particular group of bivalve molluscs, the hinge always comes to a particular form. The explanation that is sought will describe the developmental process of hinge formation, tracing the final morphology to a sequence of tissue or cellular interactions, perhaps even identifying the stages in ontogeny at which different genes are expressed. Explanations of this type abound in biology: think of the mechanical accounts of normal (and abnormal) meiosis, of respiration and digestion, of details of physiological functioning in all kinds of plants and animals. For obvious reasons, I shall call these explanations "structural explanations."[31] They contrast with *historical explanations,* accounts that seek to identify the evolutionary forces that have shaped the morphology, behavior, ecology, and distribution of past and present organisms. So, for example, our imagined biologist—or, more likely, a colleague— may be concerned to understand why the bivalves evolved the form of hinge that they did. Here, what is sought is an evolutionary history that will disclose why the genes regulating the particular hinge morphology became fixed in the group of bivalves.

Neither mode of explanation is more fundamental than the other. If I want to relieve my ignorance about the structures and mechanisms underlying a morphological trait, then I cannot receive enlightenment from an account which tells me (for example) how natural selection favored the emergence of the trait. Equally, I can be well acquainted with the developmental details underlying the presence of a feature and still legitimately wonder why the structures and mechanisms concerned have come to be in place. This is not to deny that structural and historical investigations can prompt further historical and structural inquiries. As we understand more about the structures that underlie facets of morphology or pieces of behavior, new questions arise about the historical processes through which those structures emerged. In similar fashion, deeper understanding of evolutionary history raises new questions about the structures instantiated in the organisms who participated in the historical process. A study of a particular organism can easily give rise to a sequence of questions, some structural and some historical, with structural answers raising new historical questions and historical answers raising new structural questions. We should not confuse ourselves into thinking that one type of answer is appropriate to both types of questions or that one type of question is more "ultimate" than the other. The latter mistake is akin to thinking of even numbers as more "advanced" on the grounds that each odd number is followed by an even number.

I claim that these two main types of biological inquiry generate different schemes for classifying organisms. Consider the enterprise of structural explanation as it might be developed in microbiological investigations. Our study of viruses initially reveals certain patterns of morphological and physiological similarity and difference: we discover that there are different shapes and constitutions of the viral protein sheaths and that there are differences in the abilities of viruses to replicate in various hosts. These initial discoveries prompt us to ask certain questions: Why does this virus have a protein sheath of this shape? Why is it able to replicate on this host but not on that? Viral genetics proves some answers. We learn that the features that originally interested us depend upon certain properties of the viral genome. At this point our inquiries are transformed. We now regard viruses as grouped not by the superficial patterns that first caught our attention, but by similarities in those properties of the genome to which we appeal in giving our explanations. Our reclassification may prompt us to differentiate viruses that we would formerly have lumped together, or to regard as mere "variants" organisms previously viewed as of radically different types. But, irrespective of any reforms it may induce, the achievement of an explanatory framework goes hand in hand with a scheme for delineating the "real kinds" in nature.[32]

This example mixes science with science fiction. We at present know an enormous amount about the genetics of some viruses, enough to discern minute details of the process of sheath synthesis and even of viral replication. Fiction enters in my suggestion that knowledge of this sort is available across the board, so that we can actually reclassify viruses on the basis of genetic discontinuities. To the best of my knowledge, microbiologists are not currently in a position to apply explicit genetic criteria to demarcate structural species of viruses. Nevertheless, it is not hard to envisage the possibility that future science may operate with a species concept in which microorganisms are divided by particular differences in their genetic material, and in which these differences are regarded as "real" whether or not they correspond to morphological or physiological distinctions, whether or not they coincide with the groupings produced by the evolutionary process.

Consider, by contrast, the enterprise of historical explanation. Again, our inquiries may begin with an unfocused question. We notice a pattern of similarities and differences among certain animals, carnivorous mammals, for example, and we ask how this diversity has arisen. Our project may initially be formulated in quite inadequate terms: we may begin by excluding giant pandas (because they are her-

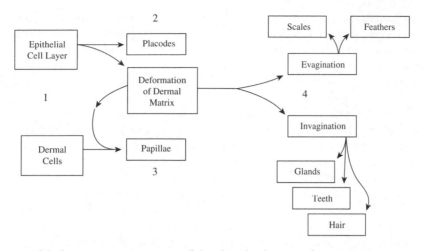

FIGURE 5.1 Diagrammatic summary of the skin developmental program. From G. Oster and P. Alberch, "Evolution and Bifurcation of Developmental Programs," *Evolution* 36 (1982): 444–459.

bivores), hyenas may be classified with cats, marsupials like the Tasmanian "wolf" may be included, and so forth. As we proceed to reconstruct the phylogeny of the carnivores our groupings change, reflecting the recency of common ancestry. We learn to see the "important" similarities (like skull morphology) and to ignore "plastic" traits (like body size). In this way a new classification is produced, which may override similarities in gross morphology, in behavior, in ecology, even, in principle, in genetic structure.

So far I have outlined two main approaches to the classification of organisms, but within each of these more general schemes there are particular variations. Some patterns of organismic diversity may be explained by reference to structural similarities at different levels. When thinking about structural explanation, there is a strong temptation to adopt a reductionist perspective, to hold that the fundamental distinctions among organisms must be made in genetic terms. My example about the viruses exploits the hold that reductionism exerts on our thinking. Yet we should acknowledge that there may be phenomena whose structural explanation will ultimately be given by appealing to discontinuities in the architecture of chromosomes.[33] Another possibility is that some biological phenomena—like those of phenotypic stability—may be explained by identifying developmental programs, conceived as flow charts that trace cell movements and tissue interactions (see figure 5.1). So we might arrive at a structural conception of species that identified a species as a set of organisms sharing a common program, without committing ourselves to the idea that there is any genetic similarity that covers exactly those organisms instantiating the program. The situation I envisage is easily understood by taking seriously the metaphor of a *program*. Organisms may be divided into species according to their possession of a common "software," and this division might cut across the distinctions drawn by attending to genetic "hard wiring."

At present, we can only speculate about the possibilities for structural concepts of species. A far more detailed case can be made for pluralism about historical species concepts. Let me begin with an obvious point. The enterprise of phylogenetic reconstruction brings home to us the importance of the principle of grouping organisms according to recency of common ancestor. But that principle, by itself, does not legislate a division into kinds. It must be supplemented with a principle of *phylogenetic division*, something that tells us what the important steps in evolution are, what changes are sufficiently large to disrupt phylogenetic connections and to give rise to a new evolutionary unit.

There are three main views about the kinds of evolutionary change that break lineages: the production of reproductively isolated branches,[34] the attainment of ecological distinctness, and the development of a new morphology. Each of these principles of division identifies a relationship among organisms that is intrinsically of biological interest. Each can be used to yield an account of the species category in which the units of evolution are taken to correspond to the major types of discontinuity. Alternatively, each can be used in subordination to the principle of grouping organisms according to recency of common ancestor, and this approach generates another three different accounts of species.

Historical species concepts arise from applying two principles. The principle of continuity demands that *a* and *b* be more closely related than *c* and *d* if and only if *a* and *b* have a more recent common ancestor than *c* and *d*. The principle of division, of which there are three versions, takes the general form of specifying the conditions under which *a* and *b* are evolutionarily distinct. The candidate conditions are: (i) *a* and *b* belong to populations that are reproductively isolated from one another, (ii) *a* and *b* belong to different ecological (or adaptive) zones, (iii) *a* and *b* are morphologically distinct. Some currently popular approaches to species give precedence to the principle of continuity, using some favored version of the principle of division to segment lineages. Other conceptions are generated by focusing first on the criterion of division, using common ancestry only as a means of assigning borderline cases (for example, deviant organisms or evolutionary intermediates).[35]

This taxonomy of species concepts (figure 5.2) already helps us to see how different views of species may be produced by different biological priorities. There are three important types of division among organisms, and each of these three types of division can rightly be viewed as the criterion for disrupting phylogenetic continuity or as a phenomenon of interest in its own right. I have already remarked on the way in which the biological species concept illuminated the issue of the distribution of mosquitoes in the *Anopheles maculipennis* complex. Yet it should be evident that distinction according to reproductive isolation is not always the important criterion. For the ecologist concerned with the interactions of obligatorily asexual organisms on a coral reef, the important groupings may be those that trace the ways in which ecological requirements can be met in the marine environment and which bring out clearly the patterns of symbiosis and competition. Similarly, paleontologists reconstructing the phylogenies of major classes of organisms will want to attend primarily to considerations of phylogenetic continuity, breaking their lineages into species according to the considerations that seem most pertinent to

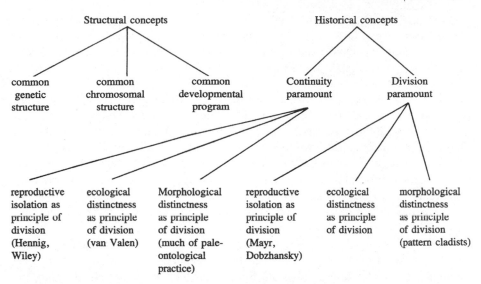

FIGURE 5.2

the organisms under study: reproductive isolation of descendant branches, perhaps, in the case of well-understood vertebrates; ecological or morphological discontinuities, perhaps, in the cases of asexual plants or marine invertebrates. I suggest that when we come to see each of these common biological practices as resulting from a different view about what is important in dividing up the process of evolution we may see all of them as important and legitimate.

Although he did not articulate the point as I have done, Hennig appreciated the diversity of biological interests. Why then did he feel it necessary to demand for biology a single general reference system? Perhaps the most obvious worry about the pluralism that I recommend is that it will engender a return to Babel, a situation in which biological discourse is plunged into confusion. But I think that biology has already been forced to cope with a different case of the same general problem, and that it has done so successfully. One of the lessons of molecular biology is that there is no single natural way to segment DNA into functional units. Present uses of *gene* sometimes refer to segments whose functional activity affects the phenotype at the level of protein formation, sometimes to segments whose functional activity affects more gross aspects of the phenotype. Even if we pretend that all genes function to produce porteins there is no privileged characterization of genes as functional units.[36] Yet geneticists (and other biologists) manage their investigations quite well, and the use of a plurality of gene concepts does not generate illusions of agreement and disagreement.

This happy state of affairs rests on the following features of the current practice of genetics. (1) For many general discussions about "genes," no particular principle of segmentation of DNA needs to be chosen. The questions that arise can be recast as questions about the *genetic material* without worrying about how that material divides up into natural units. For example, the issue of how genes replicate is reformulated as the question of the mode of replication of the genetic material. *Whatever* view one takes about the segments that constitute genes, the challenge is to understand how DNA makes copies of itself. (2) When general inquiries about genes do depend crucially on the segments of DNA identified as genes, it is important for investigators to note explicitly the principle of segmentation that is being used. So, for example, in introducing his thesis about genic selection Richard Dawkins takes pains to identify the units that he will count as genes.[37]

The case of the many genes shows how the multiplicity of overlapping natural kinds can be acknowledged without either arbitrary choice or inevitable confusion. Similar resources are available with respect to the species category. Just as there are many ways to divide DNA into "natural functional units," so there are many ways to identify sets of "structurally similar" organisms or to pick out "units of phylogeny." In some discussions of species, what is important to the issue is not dependent on any particular criterion for dividing an evolving lineage into species. When ecologists discuss reproductive strategies, distinguishing between *k*-selected and *r*-selected species, for example, their remarks can be understood independently of any particular proposal for lineage division. Species are conceived as sets of organisms forming part of a lineage, and the distinction at hand is drawn by considering the characteristics of their stages. But in other cases the principle of segmentation is crucial. Paleontologists concerned with comparing species turnover in a group of lineages are likely to misunderstand one another unless they make clear their principle of lineage division.

As Hempel remarked long ago in his celebrated critique of operationalism, the risk of equivocation is ever present in scientific discourse.[38] To guard against confusion it is futile to attempt to fashion some perfectly unambiguous language. Instead, responsible scientists should recognize where dangerous ambiguities are likely to occur and should be prepared to forestall misunderstandings. Biologists have already learned to be responsible in discussions of genes. The same responsibility can be attained in the case of species. To allow pluralism about species and to deny the need for a "general reference system" in biology is not to unlock the doors of Babel.[39]

5. Three Consequences

I have tried to outline and to motivate a general approach to the category of species. I want to conclude by drawing three morals, one for an area of current biological dispute, one for a question in the philosophy of science, and one which overlaps biology and philosophy. I shall begin with the biological issue.

Paleontologists are currently divided on a number of important issues about the tempo and mode of evolution. In an important and much discussed contribution to these debates, Peter Williamson[40] provides extensive documentation of the fossil record of several mollusc lineages from the Lake Turkana Basin. Williamson's data (see figure 5.3) reveal abrupt changes in phenotype punctuating periods of phenotypic stasis. Moreover, the episodes of phenotypic change are themselves associated with an increase in phenotypic variability. Williamson draws attention

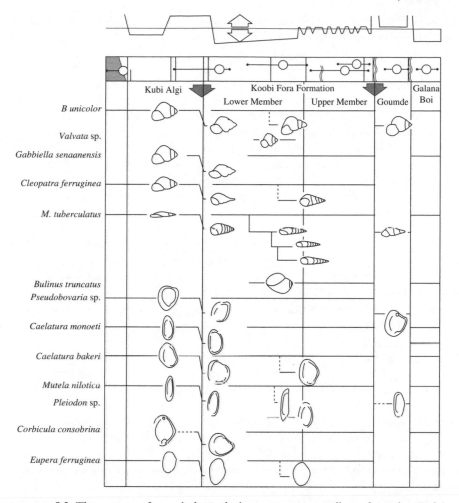

FIGURE 5.3 The pattern of morphological change in some molluscs from the Turkana basin. From P. Williamson, "Paleontological Documentation of Speciation in Cenozoic Molluscs from Turkana Basin," *Nature* 293 (1981): 437–443.

to this association, and goes on to make some speculations about the genetics of speciation.[41]

There are two important ways in which Williamson's data may be interpreted. The first is to suppose that Williamson is employing Mayr's biological species concept, and that he intends to study transitions between biological species. When we choose this reading certain questions about the data become relevant. In particular, we have to ask if the species boundaries identified on the basis of phenotypic considerations coincide with the attainment of reproductive isolation.[42] Thus one contribution that the essay makes is toward advancing our understanding of speciation, *conceived as a process in which descendant populations achieve*

reproductive isolation from a persisting ancestral population. If Williamson's findings are interpreted in this way, they bear on *one* issue of the tempo of evolutionary change and *one* issue of the genetics of speciation. Is the attainment of
reproductive isolation a process that occurs rapidly, punctuating long periods of
stasis? What mechanisms of population genetics underlie this process?

The second construal ignores any considerations about reproductive isolation.
Williamson's data reveal a pattern of phenotypic change, and we can concentrate
on this pattern without linking it to claims about reproductive isolation. If processes
of speciation are simply identified with the rapid morphological transitions that
Williamson describes, then we can inquire about the tempo of these processes and
about their underlying genetic basis. Nor are these uninteresting questions. It is no
less significant to ask after the tempo and mode of speciation, *conceived as a process
of morphological discontinuity*, than it is to inquire about the attainment of reproductive isolation. Williamson's suggestions about genetic mechanisms can be construed as hypotheses about the genetic changes that underlie those episodes of
phenotypic modification (with increased phenotypic variability) which are recorded
in his data. We do not need to introduce the idea that these episodes lead to reproductive isolation.

Williamson has sometimes been criticized on the grounds that his morphological findings do not rule out the possibility of cryptic "speciation events" during
periods of alleged stasis.[43] Whether or not these criticisms succeed against the first
interpretation, they are plainly irrelevant to the second. The pattern of phenotypic
change, a pattern which the fossil record wears on its face, can itself serve as the
basis for some important inquiries about the tempo and mode of evolution. By
separating different conceptions of species and of speciation, we can disentangle
different important issues that arise in biology, and recognize the significance of
investigating a number of different patterns in the diversity of life.

At this point let me take up the question that is common to philosophy and
biology, the question of the "reality" of species. It is important to understand that
realism about species is quite independent of the view that species are individuals.
Notice first that if realism about species is construed as the bare claim that species
exist independently of human cognizance of them, then anyone who accepts a
modest realism about sets can endorse realism about species. Organisms exist and
so do sets of those organisms. The particular sets of organisms that are species exist
independently of human cognition. So realism about species is trivially true.

To make realism come so cheap is obviously not to recognize what provokes
biologists and philosophers to wave banners for the objectivity of systematics.[44] What
is at issue here is whether the division of organisms into species corresponds to something in the *objective structure* of nature. Articulating this realist claim is difficult.
But I suggest that however it is developed, it will prove compatible with pluralism
about species. *Pluralistic* realism rests on the idea that our objective interests may
be diverse, that we may be objectively correct in pursuing biological inquiries which
demand different forms of explanation, so that the patterning of nature generated
in different areas of biology may cross-classify the constituents of nature.[45] Despite
the fact that realist theses about the objectivity of classification cry out for analysis,

we can recognize the plausibility of those theses when we reflect on Williamson's molluscs. Williamson's lineages should remind us that there are a number of objective patterns of evolutionary change. The pluralistic realist is someone who is concerned to understand all of them.

Finally, let me turn to the moral for philosophy of science. In thinking about the general problem of conceptual change in science, we are inclined to consider two main possibilities. With the advantage of hindsight, we see that our predecessors were referring to natural groups of things, about which they may have had radically false beliefs. Or, perhaps, we view them as referring to sets that cut across the natural kinds in bizarre ways. The example of the concept of species—and, I would suggest, that of the concept of gene[46]—reveals an intermediate situation. Here we find ourselves unable to provide some short description that will finally reveal the natural group that our predecessors struggled to characterize, but neither are we willing to dismiss them as simply producing an uninteresting heterogeneous collection. The set of species taxa is heterogeneous, but it is not wrongheaded in the way that some early attempts at chemical classification are.

If I am right, then there will be no simple description that will pick out exactly those sets of organisms that some biologists reasonably identify as species taxa. We shall not be able to reconstruct the language of biology and to trace its historical development in the way in which we have been able to cope with cases of conceptual change in chemistry. But this does not mean that we are swept into the cynic's view of species. For although it may be true that species are just those sets of organisms recognized as species by competent taxonomists, there is a way to understand why just those sets have been picked out. That way is not the familiar way of using current theory as an Archimedean point from which we can, at last, provide a single descriptive characterization of the groups to which our benighted predecessors have referred. Instead, we must recognize that there are many different contexts of investigation in which the concept of species is employed, and that the currently favored set of species taxa has emerged through a history in which different groups of organisms have been classified by biologists working on different biological problems. The species category can be partitioned into sets, each of which is a subset of some category of kinds. We can conceive of it as generated in the following way. A number of biologists, $B_1 \ldots, B_n$, each with a different focus of interest, investigate parts of the natural world. For each B_i there is a subset of the totality of organisms, O_i, which are investigated. B_i identifies a set of kinds, K_i, the kinds appropriate to her interest—that partition O_i. The set of species taxa bequeathed to us is the union of the K_i. In areas where the O_i overlap, of course, there may be fierce debate. My suggestion is that we recognize the legitimacy of all those natural partitions of the organic world of which at least one of the K_i is a part.

This schematic account of the set of species taxa we have inherited is intended to make clear the moral of my story. To appreciate the rationale for the species category we must reconstruct the history of our discourse about species, and there is no quick substitute for that reconstruction. The cynic's definition may be the beginning of wisdom about species. But it is only the beginning.

Notes

An earlier version of this essay was given at the Eastern Division meeting of the American Philosophical Association in December 1982. I am very grateful to my commentator, Elliott Sober, for some helpful criticisms and suggestions, and to Alex Rosenberg, who chaired the session and later supplied me with valuable written comments. I also thank David Hull for his detailed response to a much longer manuscript on this topic. Finally I acknowledge the enormous amount I have learned from correspondence and conversations with numerous biologists and philosophers, most notably: John Beatty, Jonathan Bennett, Bill Fink, Sara Fink, Steve Gould, Marjorie Grene, Kent Holsinger, Dick Lewontin, Gregory Mayer, Ernst Mayr, Brent Mishler, Michael Ruse, Husain Sarkar, Laurance Splitter, and Ernest Williams. Residual errors are probably my own.

1. Several—but not all of those—have figured in the recent taxonomic literature. In particular, I hold no brief for phenetics.

2. N. Eldredge and J. Cracraft, *Phylogenetic Patterns and the Evolutionary Process* (New York: Columbia University Press, 1980), 15.

3. The person who comes closest to advocating a realistic version of pluralism about species is John Dupré, who defends what he calls (borrowing a name from John Perry) "promiscuous realism" (Dupré, "Natural Kinds and Biological Taxa," *Philosophical Review* 90 [1981]: 66–90). Dupré's defense is brief (since the primary concern of his 1981 article is to address some important issues in philosophy of language) and, to my mind, unconvincing. Pointing out that biological taxa cut across the divisions of organisms introduced by gastronomes hardly shows that there are alternative sets of kinds internal to biology. Nor does it help to note (ibid., 83) that ecologists use the concept of population, for this does not indicate any commitment to alternative species taxa. Hence, although I find Dupré's short discussion of promiscuous realism provocative, I don't think he has made out a case for this view.

4. Loci classici are the following: M. Ghiselin, "A Radical Solution to the Species Problem," *Systematic Zoology* 23 (1974): 536–544; D. Hull, "Are Species Really Individuals?" *Systematic Zoology* 25 (1976): 174–191; Hull, "A Matter of Individuality," *Philosophy of Science* 45 (1978): 335–360; and Hull, "Individuality and Selection," *Annual Review of Ecology and Systematics* 11 (1980): 311–332. A helpful summary is that in A. Rosenberg, *Sociobiology and the Pre-Emption of Social Science* (Baltimore: Johns Hopkins University Press, 1981). My discussion will be directed primarily at the arguments advanced by Hull. To the best of my knowledge, Ghiselin deserves credit for his original presentation of the thesis that species are individuals, but Hull's articles are more systematic and detailed in arguing for the thesis.

5. An exhaustive evaluation of this claim would require discussion of the merits and shortcomings of main features of phylogenetic systematics (cladism). For reasons of space, I have concentrated here on the main philosophical arguments.

6. Ernst Mayr, *Systematics and the Origin of Species* (New York: Columbia University Press, 1942); Mayr, *Animal Species and Evolution* (Cambridge, Mass.: Harvard University Press, 1963); Mayr, *Principles of Systematic Zoology* (Cambridge, Mass.: Harvard University Press, 1969); Mayr, *Populations, Species, and Evolution* (Cambridge, Mass.: Harvard University Press, 1970); Mayr, *The Growth of Biological Thought* (Cambridge, Mass.: Harvard University Press, 1982).

7. Another general worry about construing species as sets was voiced by Elliott Sober. As Sober rightly points out, his own existence is not essential to the existence of *Homo sapiens*: there are worlds in which Sober does not exist but in which the species does exist. Hence,

he contends, the species is not to be identified with the set of humans. I reply that this conclusion does not follow. In different worlds, *Homo sapiens* is a different set. Just as the extension of 'car' varies form world to world, so does the referent of *Homo sapiens*.

8. In fairness to Hull, I should note that he does not advocate any argument that is as stark as the one presented here. However, he sometimes comes very close: see, for example, Hull, "Kitts and Kitts and Caplan on Species," *Philosophy of Science* 48 (1981): 141–152.

9. As Bill Fink pointed out to me, this allows for relatively trivial changes—such as chance fluctuations in frequency—to count as cases of evolution. Quite evidently, one can attempt to circumscribe the "genuine forces" of evolution and use the resultant characterization to generate a more stringent conception of evolutionary change. Any such conception can easily be applied to the present context.

10. Hull, "A Matter of Individuality," 353.

11. For those who are inclined to believe that the inviability of a zygote because of some genetic disruption does not signal a species boundary, let me suggest a slightly different criterion. One might propose that mutations or chromosomal novelties giving rise to the absence of *P* generate inviable gametes. In this way, the effect of the disruption of the genome is felt at the prezygotic stage. (I am grateful to Elliott Sober for bringing to my attention the possibility that an inviable zygote may not indicate a species boundary.)

12. The issue of the spatiotemporal localization is a tricky one. (For an illuminating discussion of localizability of the extensions of predicates and the character of natural laws, see John Earman, "The Universality of Laws," *Philosophy of Science* 45 [1978]: 173–181.) Given contemporary cosmology, it appears that the extension of "atom of gold," no less that that of "organism belonging to *Rattus rattus*," is spatiotemporally localized (as noted in D. B. Kitts and D. J. Kitts, "Biological Species as Natural Kinds," *Philosophy of Science* 46 [1979]: 613–622). Hull's most explicit discussion of this issue runs as follows: "Biological species are spatiotemporally localized in a way in which physical substances and elements are not. No spatio-temporal restrictions are built into the definitions of 'gold' and 'water'" (Hull, "Kitts and Kitts and Caplan," 148–149). It seems to me that this response confuses semantical and ontological issues. A defender of the view that species are sets (an ontological view) is free to adopt a number of different theses about how the names of species are defined (or how their referents are fixed). I do not see that remarks about the semantical features of "gold," "*Homo sapiens*," and so forth cut any ontological ice. We can use proper names (e.g., "2," "n") to refer to sets, and it's possible that our only way of referring to a person (a paradigm individual) should be via a description (e.g., "the first person to make fire"). Interestingly, Hull immediately proceeds from the passage I have cited to the point about the *connectedness* of species—the point that I regard as central to his case. I see this as reflecting the fact that the official notion of a "spatiotemporally unrestricted class" is unworkable for Hull's purposes: in one sense, far too many classes are spatiotemporally restricted; in another, the distinction only holds with respect to class *names*.

13. Hull, "A Matter of Individuality," 349.

14. E. D. Parker and R. Selander, "The Organization of Genetic Diversity in the Parthenogenetic Lizard *Cnemidophorus tesselatus*," *Genetics* 84 (1976): 791–805; Parker, "Phenotypic Consequences of Parthenogenesis in *Cnemidophorus* Lizards: 1. Variability in Parthenogenetic and Sexual Populations," *Evolution* 33 (1979): 1150–1166.

15. Let me briefly respond to an obvious objection. It may be held that the set-theoretic reformulation of discourse about species—specifically, the translation sketched in section 2, "Sets versus Individuals"—grants Hull everything he wants. At this stage, it ought to be clear that this is not so. At least two of the main consequences of the doctrine that species are individuals (the thesis that species are historically connected, and the explanation of the nonexistence of laws about particular species) do not follow from my set-theoretic

account. Indeed, I would contend that all of the apparently exciting results which Hull has wanted to establish are not honored by the set-theoretic version.

16. There is a short answer: species are natural kinds. I accept this answer, but I don't adopt all the implications some may want to draw from it. In particular, I want to remain agnostic on the issue of whether any species taxon has a nontrivial essence. But what then distinguishes a natural kind? I suggest that natural kinds are the sets that one picks out in giving explanations. They are the sets corresponding to predicates that figure in our explanatory schemes. Are kinds then the extensions of predicates that occur in laws? Possibly—but not necessarily. The account of explanation I favor (see P. S. Kitcher, "Explanatory Unification," *Philosophy of Science* 48 [1981]: 507–531) does not require that all explanation involve derivation from laws. One of the central features of that account is that the generality of a scientific explanation need not consist in its using some lawlike premise but in its instantiating a pattern exemplified in numerous other explanations. Hence, though I link natural kinds to the predicates that occur in scientific explanations, I do not require that there be laws about all kinds.

Subsequent discussion in this article will not rest on this all-too-brief elaboration of the idea that species are natural kinds. I am grateful to a number of people, most notably Alex Rosenberg, for helping me to see the relation between my own views and the traditional idea of species as natural kinds.

17. Mayr, *Populations, Species, and Evolution*, 12; Mayr, *Principles of Systematic Zoology*, 26.

18. G. G. Simpson, *Principles of Animal Taxonomy* (New York: Columbia University Press, 1961); W. Hennig, *Phylogenetic Systematics* (Urbana: University of Illinois Press, 1966); E. O. Wiley, *Phylogenetics* (New York: Wiley, 1981).

19. I should point out that this proposal for demarcating species taxa is the one most congenial to the Hull-Ghiselin thesis. The difficulties that arise for the Simpson-Hennig-Wiley approach provide more reasons to adopt the position defended in section 2.

20. Wiley and Hennig diverge from Simpson in disallowing speciation through anagenesis. Wiley, unlike Hennig, is prepared to grant that a species may persist through a speciation event.

21. L. van Valen, "Ecological Species, Multispecies, and Oaks," *Taxon* 25 (1976): 233–239.

22. R. Sokal and P. Sneath, *Principles of Numerical Taxonomy* (San Francisco: Freeman, 1961); Sneath and Sokal, *Numerical Taxonomy* (San Francisco: Freeman, 1973).

23. G. Nelson and N. Platnick, *Systematics and Biogeography: Cladistics and Vicariance* (New York: Columbia University Press, 1981), 12; also D. Rosen, "Fishes from the Upland Intermontane Basins of Guatemala: Revisionary Studies and Comparative Geography," *Bulletin of the American Museum of Natural History* 162 (1979): 269–375; perhaps also Eldredge and Cracraft, *Phylogenetic Patterns*, 92.

24. For a classic discussion, see Mayr, *Animal Species and Evolution*, 35–37; Mayr, *Populations, Species, and Evolution*, 24–25.

25. D. Hull, "The Operational Imperative: Sense and Nonsense in Operationism," *Systematic Zoology* 16 (1968): 438–457; Simpson, *Principles of Animal Taxonomy*.

26. Mayr, *Systematics and the Origin of Species*, 121–122.

27. There are some curious twists in recent versions of the biological species concept, including what appears to be a flirtation with essentialism. Consider the following recent statement by Ernst Mayr: "In spite of the variability caused by the genetic uniqueness of every individual, there is a species-specific unity to the genetic program (DNA) of nearly every species" (Mayr, *The Growth of Biological Thought*, 297). Similar suggestions have been voiced by others (e.g., N. Eldredge and S. J. Gould, "Punctuated Equilibria: An Alternative

to Phyletic Gradualism," in T. J. M. Schopf, ed., *Models in Paleobiology* [San Francisco: Freeman, 1972]), and they reinforce the idea that morphological difference and reproductive isolation are indicators of a more fundamentl cleavage among organisms.

28. Hennig, *Phylogenetic Systematics*, 5.

29. Ibid., 9.

30. E. Mayr, "Cause and Effect in Biology," in Mayr, *Evolution and the Diversity of Life* (Cambridge, Mass.: Harvard University Press, 1976), 360.

31. In choosing this label, I don't intend to downplay the role of physiological (as opposed to anatomical) considerations. The contrast is between appeals to structure and present function, on the one hand, and appeals to history, on the other. (I am grateful to Marjorie Grene for suggesting to me that my label might mislead.)

32. Evidently, this scenario recapitulates the views of H. Putnam (*Philosophical Papers*, Vol. 2 [Cambridge, Mass.: Cambridge University Press, 1975]) and S. Kripke (*Naming and Necessity* [Cambridge, Mass.: Harvard University Press, 1980]) about the conceptualization of natural kinds.

33. See, for example, M. J. D. White, *Modes of Speciation* (San Francisco: Freeman, 1978).

34. I should point out that the criterion of reproductive isolation can itself be applied in two different ways to divide lineages. One can count two stages of a lineage as parts of different species if they are reproductively isolated, or one can view speciation events as occurring only when one species gives rise to descendant populations that coexist and are reproductively isolated from one another. The first criterion is problematic unless certain theses about the geometry of evolution are true; the second represents the approach of Hennig, Wiley, and some other cladists.

35. This type of approach seems to be used by Nelson and Platnick (*Systematics and Biogeography*) and by Eldredge and Cracraft (*Phylogenetic Patterns*).

36. For amplification of these points, see Kitcher, "Genes," *British Journal for the Philosophy of Science* 33 (1982): 337–359. As Alex Rosenberg has pointed out to me, the increasing complexity of the systems revealed in molecular biology underscores the pluralism about genes defended in that article.

37. R. Dawkins, *The Selfish Gene* (Oxford: Oxford University Press, 1976), 34; there is a much more refined discussion of the same point in Dawkins, *The Extended Phenotype* (New York: Freeman, 1982).

38. C. G. Hempel, "A Logical Appraisal of Operationism," in Hempel, *Aspects of Scientific Explanation* (Glencoe, N.J.: Free Press, 1965), 126–127; Hempel, *Philosophy of Natural Science* (Englewood Cliffs, N.J.: Prentice Hall, 1966), 92–97.

39. Thus there is no univocal answer to the question of how to describe the type of hypothetical situation beloved of philosophers. Suppose we have a species S and discover the existence of a historically unrelated group of organisms that agree with the members of S in any respect we choose (reproductively compatible, genetically similar, and so forth). Does the group count as a subset of S? I claim that the answer must be relative to a *prior decision* on whether or not to employ a historical species concept. Use of such a concept is not forced on us, and it may prove helpful in seeing this to consider a range of organisms and a range of biological investigations. What we may be inclined to say when S is *Rattus rattus* may well be different from what we say when S is the bacteriophage T_4. (I am grateful to Jonathan Bennett for prodding me into making this pioint explicit.)

40. P. Williamson, "Paleontological Documentation of Speciation in Cenozoic Molluscs from Turkana Basin," *Nature* 293 (1981): 437–443.

41. Ibid., 442–443.

42. There are complications here. One of the lineages (*Melanoides tuberculata*) is asexual. Hence, Williamson's claim must be that the morphological discontinuities correspond to the lineage divisions marked out by reproductive isolation—*where demarcation by reproductive isolation is possible*. This example underscores the point made in section 3.

43. Schopf makes a similar point against claims of documentation of punctuated equilibrium. See T. J. M. Schopf, "Punctuated Equilibrium and Evolutionary Stasis," *Paleobiology* 7 (1981): 156–166.

44. For a clear explanation of this point, see E. Sober, "Evolution, Population Thinking, and Essentialism," *Philosophy of Science* 47 (1980): 350–383.

45. There are suggestions about how to articulate this point in R. Boyd, "Metaphor and Theory Change: What Is 'Metaphor' a Metaphor For?" in A. Ortony, ed., *Metaphor and Thought* (Cambridge, Mass.: Cambridge University Press), and in Kitcher, "Genes."

46. In Kitcher, "Genes."

6

Some Puzzles about Species (1989)

In the fall of 1974, my second year of teaching philosophy, I was giving a course in philosophy of science to a dozen or so bright undergraduates. After about three weeks, one of the students came to see me in my office. "We find this material interesting," he explained, "but most of us are pre-meds, and the science we know best is biology. It would really help if you could give us some examples from biology, and not talk about physics quite so much." The point was a good one. Like many philosophers of science of my generation, I offered standard examples from physics—when I needed an illustration, I pointed to Newtonian dynamics, optics, electromagnetic theory, thermodynamics, and only occasionally ventured as far afield as chemistry. However, it was clear that the course would be improved if I honored my student's reasonable request, so I set off for the library in search of a key to reform.

I was lucky. There on the shelves was David Hull's *Philosophy of Biological Science*, relatively newly published in the Prentice-Hall series I knew and loved. I took it out and began to read. Almost immediately it was clear that this would not simply be a Useful Source of Improving Examples (although it did fulfill that function for my grateful students). Reading David's lucid discussions of reductionism and of the character of evolutionary theory, I realized that there were deep and important issues of which I had previously been ignorant, and a body of science that I would find difficult to integrate with the philosophical ideas I had absorbed in graduate school. It was clear that I needed reeducation, and David's book pointed the way.

Other philosophers of biology of my generation probably have similar stories to tell. All of us owe David Hull an enormous debt. For, at a time when biology was almost invisible in the graduate education of philosophers of science, he showed how exciting and significant the philosophy of biology could be. Moreover, the high scientific standards set in David's work made it clear that there could be no room for mere dabbling: biology, like physics, is serious, difficult, and demanding, and those who philosophize about it had better do their homework. David's example led many of us to the ever-hospitable Museum of Comparative Zoology at Harvard and to regular interchanges with the local population of biologists.

As I have eradicated some of my initial innocence about biology, I have learned more and more from David's own work. On many topics, his discussions have influenced my own ways of thinking, probably beyond the extent to which I am aware.

But there is one issue on which we are in deep disagreement. Following a provocative article by Michael Ghiselin, David has argued at considerable length for a view
of species that seems to me to bypass the main questions that arise in this area of
the philosophy of biology.[1] The aim of the present essay is to continue the debate
between us. But it seemed to me wrong to launch into the arguments without some
prefatory acknowledgement of my intellectual debts. And perhaps those who champion David's view of species may draw the obvious moral from my story: the reeducation stopped too soon.

I. Individuality Again

According to Ghiselin and Hull,[2] biological species are not "spatiotemporally unrestricted classes" but "historical individuals." What does this claim mean? And why
does it matter?

I have argued[3] that there are conceptual difficulties in the position that
Ghiselin and Hull wish to oppose: they are stalking a broken-backed chimera. What
is a spatiotemporally unrestricted class? The obvious response is to say that a class
is spatiotemporally unrestricted just in case, for any finite region of space-time that
one chooses, there are members of the class that lie outside the region. But this will
not do, since no class of physical objects is spatiotemporally unrestricted in this
sense. Hull recognizes the point and proposes that a class is spatiotemporally unrestricted if its definition allows for the presence of instances that lie outside a spatiotemporal boundary. But this, I suggest, is a confused hybrid notion. Classes (or
sets) as I understand them are entities that have their properties independently of
the particular ways in which we choose to talk about them. Set-theoretic identity
is extensional: $a = b$ just in case a and b have the same members.

So what? Well, let $\{a_1, \ldots a_n\}$ be any finite set of physical objects. Let B be
some finite region of space-time that includes all the a_i. We can pick out the set in
two different ways: as the extension of the predicate '$x = a_1 \lor x = a_2 \lor \ldots \lor x = a_n$'
or as the extension of the predicate "$(x = a_1 \lor x = a_2 \lor \ldots \lor x = a_n)$ & x lies within
B." Here I assume that the names a_i do not pick out their referents in ways that
restrict those referents to particular regions of space-time. (If they do, choose different names.) Now we ask if the set $\{a_1, \ldots, a_n\}$ is spatiotemporally unrestricted.
Answer: yes, because the first way of specifying it sets no spatiotemporal boundary
within which its members must lie. Answer: no, because the second way of picking
it out does set a spatiotemporal boundary, *viz.* B, within which its members must
lie. Both definitions identify sets with exactly the same members, and therefore, by
the extensionality of set-theoretic identity, they pick out the same set. So, the set we
have identified is both spatiotemporally unrestricted and spatiotemporally
restricted. But that set was an arbitrary finite set of physical objects. Thus we can
conclude that any finite set of physical objects is both spatiotemporally unrestricted
and spatiotemporally restricted.

The contradiction arises because the notion of the spatiotemporally unrestricted class with which we have been working mixes properties of entities with
properties of their definitions. The *first* issue about the ontology of species is whether

species are sets (with organisms as members) or whether they are mereological wholes (with organisms as parts). In two papers, Hull[4] offers a number of arguments for thinking that species are individuals. He appeals to the character of evolutionary theory, the nature of natural selection, and the absence of laws about individual species. If these arguments are taken as directed at the conclusion that species are wholes rather than sets, then I think they fail to reach their target. As I have argued at length,[5] all of our discourse about evolution can be reconstructed equally well within set theory or within mereology. The moral that I draw from this—and that I shall develop in some detail below—is that the point Hull (and Ghiselin) really want to make has nothing to do with ontology. There is a *second* issue about the delineation of the species category on which Hull and Ghiselin offer a significant (though controversial) proposal, and this issue is orthogonal to the question whether species are individuals or sets.

Before presenting that issue, I want to consider a line of argument that Hull has recently offered.[6] The kind of reasoning that leads us to think of a species as a set of organisms, he suggests, should also induce us to think of an organism as a set of cells. Because of our size and perceptual abilities, we are able to see the gaps that separate the parts of species from one another, and thence arises the temptation to view the species as a set of organisms. But the accidents of epistemological access should not lead us to attribute an ontological difference where there is none.

I find this argument interesting, challenging, and ultimately unsuccessful. First, let us ask why we do not think that organisms are sets of cells. One important, and fairly obvious, point is that an organism consists of cells and extracellular matrix and the latter may play a crucial role in its development and physiology. Another is that the organism (conceived as existing over time) would be better viewed set-theoretically as a function mapping any time at which it exists onto the set of space points occupied at that time. Since Carnap and Reichenbach, this has been a standard way of thinking about physical objects in general, and organisms can be treated as special cases.

But there is a deeper point that can be appreciated by recognizing that there are some organisms that we can easily conceive as collections of cells (or, more accurately, there are some *stages* of organisms that we can view in this way). In such organisms as *Hydra* and *Dictyostelium* cells can function with a high degree of independence, and we can think of the organism as continuing to survive (albeit in a different form) even when the cells are dissociated. But this is not the rule with organisms. The distinction between an organism and a set of cells is vividly brought home to us when we recognize that it is in principle possible for the organization of the cells that make up a complex organism to be destroyed while each cell persists. The set of cells remains but it is no longer an organism.[7]

Let us ask the analogous question about species. Does a species continue to exist when we disrupt the relations among the organisms that are (on the set-theoretic view) members of it? I believe that a case can be made for an affirmative answer. If an endangered species becomes scattered so that human intervention is required if its remaining members are to reproduce, then there remains a chance of preserving the species: that, of course, is what motivates efforts that people sometimes make. Provided that there is a set of organisms belonging to the species, the species persists. Here we have a clear disanalogy with the relationship between organisms and cells, and Hull's argument is blocked.

However, if it is suggested that species are as dependent on the interactions among organisms as organisms are on the relations among cells, it is possible to make a different reply to Hull. Waiving qualms about obligatorily asexual species, let us suppose that it is crucial to the persistence of a species that some of its member organisms be combining their genes in the production of progeny.[8] Now we can say that a species is a set-theoretic entity, to wit a set of organisms subject to a particular relation (or, more precisely, the ordered pair of a set and a relation) where the relation obtains just in case there is that kind of reproductive behavior that is supposed to be crucial to the persistence of species. Could we conceive of organisms after the same fashion, treating them as sets of cells and pieces of extracellular matrix subject to relational conditions? Perhaps. However, at the present state of our knowledge, we can only guess at the complexity of the relations that would have to be adduced. We have not the slightest idea how to define organisms as sets of cells and pieces of matrix (whereas the specification of the relational properties that are required in the case of species seems relatively straightforward). Two points follow. First, the organization of organisms appears much more intricate than that of species—another disanalogy between organisms and species. Second, there is no firm basis for saying that organisms could not be identified with sets subject to a complex of relations (a complex which encapsulated *all* the intricacy of organization), since we have no idea what the explicit specification of the organization of organisms would look like.

I conclude that Hull's argument does not tell against the claim that species are sets. For, depending on your views about what is essential for the persistence of species, it is possible either to find a relevant disanalogy or to find a defensible version of the conclusion that organisms (better: organism-stages) are sets.

On to issues of greater biological significance. The traditional species problem was to delimit the species category by saying which superorganismal entities count as species taxa. If we decide the first question by saying that species are sets, then we can formulate this second problem as that of explaining which sets whose members are organisms are species taxa. Alternatively, if the first question is answered by claiming that organisms are individuals then the second task is to specify which individuals with organisms as parts are species taxa. Notice that it is not a consequence of the set-theoretic view of the ontology of species that *any* set with organisms as members counts as a species. Nor is it a consequence of the mereological approach to species that *any* individual with organisms as parts counts as a species.[9] There are numerous sets with organisms as members and numerous individuals with organisms as parts, and the vast majority of these sets and individuals are of no biological interest whatsoever. To solve the traditional species problem, further specification is needed.

As I interpret them, both Hull and Ghiselin disguise an interesting answer to the second question as a thesis about the ontology of species. The significant point is that species are "historical individuals," chunks of the genealogical nexus. What makes an individual historical? In general I think that this is a hard question to answer, but, in the case of interest, it seems fairly clear that historical connectedness is critical. So, talking in the mereological idiom, we conceive of an individual with organisms as parts to be historically connected just in case for any organismal parts x and y such that x precedes y and for any organism z, if z belongs to a population that descended from a population containing x and that is ancestral to a population containing y then z is also part of the same individual as x and y. Note that the criterion for historical connectedness can easily be reformulated as a condition on sets. A set of organisms is historically connected just in case it satisfies the following condition: for any organisms x, y and z, if x and y are in the set and if z belongs to a population that is descendant from a population which has x as a member and that is ancestral to a population that has y as a member then z is in the set. Hull and Ghiselin *might* have expressed their proposal by saying that species are historically connected entities and shown a studied neutrality on the question whether they are individuals or sets.[10] In my view, of course, this reformulation would have avoided considerable confusion and would have forestalled attempts to give a priori arguments for significant biological theses.[11]

In its neutral version the Hull-Ghiselin proposal is still at odds with Ernst Mayr's biological species concept. For Mayr's account allows for the possibility of species that are not historically connected. Imagine that a species A splits into two parts at t_0, one part consisting of almost all organisms in A and the other of a small isolated population. A (or the bulk of A) persists unmodified, but the peripheral isolate evolves so that, at t_1, it has descendants that are reproductively isolated from A and constitute a new species B. However, the evolutionary change consists in a small genetic modification that is reversed in an isolated population that descends from B, so that, at time t_2, there are descendants of B that make up a population C that is reproductively compatible with A (see figure 6.1). On Mayr's account, the organisms in C are conspecific with the organisms in A. But now it is clear that A is not historically connected. For there are organisms—those in B—that belong to a population ancestral to a population of A and descendant from a population of A but that are not themselves included in A. Hence the biological species concept does not require species to be historically connected.

However, even though the Hull-Ghiselin proposal diverges from the most celebrated answer to the traditional version of the species problem, that proposal does not constitute a complete rival answer to the traditional question. Saying that species are historical entities narrows the range of candidate species taxa but still allows us different ways of splitting up the genealogical nexus. The whole of life—past, present, and future—is one very big historical entity, and, at the opposite extreme, timeslices of particular populations also count as historical entities. Somewhere between these extremes are the species, and, in "A Matter of Individuality," Hull (1978) canvasses some possibilities for delineating them. The diagrams that he presents (see figure 6.2) are persuasive devices for leading us to think that the problem of breaking up the nexus has been solved—or can be solved relatively easily. But I want to urge that the diagrams conceal deep and important problems, that there are serious questions about what the lines and branch points actually mean.[12] The rest of this essay will be devoted to explaining what needs to be done to complete the Hull-Ghiselin account and why the task strikes me as formidable. I hope that

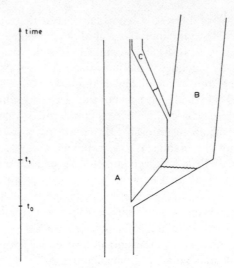

FIGURE 6.1 Hull and Ghiselin *versus* Mayr.

the neutrality of the formulation of the ontological issue (sets versus individuals) will be apparent throughout.[13]

2. The Trouble with Populations

On the account of historical connectedness that I offered above, the historical connectedness of a species depends on the holding of certain relations among popu-

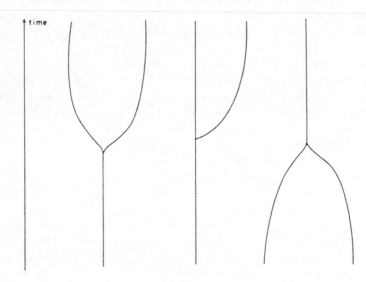

FIGURE 6.2 Three modes of phylogenetic change. After D. Hull, "A Matter of Individuality," *Philosophy of Science* 45 (1978): 335–360.

lations. This reference to populations in accounts of species is as necessary on the Ghiselin-Hull approach as it is to Mayr's well-known biological species concept. But what is a population?

One way to define the standard biologist's notion of *local population* is to take a local population to comprise all the organisms of a chosen species that are present in a particular place at a particular time.[14] There is no objection to using this definition for purposes of exposition, supposing that the notion of species can be taken as already well-understood, but it is useless in a context in which we are trying to use the notion of population to analyse the concept of species. However, Mayr has attempted to do better. He offers the following characterization:

> All members in a local population share in a single gene pool, and such a population may be defined also as 'a group of individuals so situated that any two of them have equal probability of mating with each other and producing offspring', provided, of course, that they are sexually mature, of opposite sex and equivalent with respect to sexual selection. The local population is by definition and ideally a panmictic (randomly interbreeding) unit. An actual local population will, of course, always deviate more or less from the stated ideal.[15]

This passage offers a compelling picture that seems to encapsulate the ways in which many naturalists and theoretical biologists think about populations. Start with a particular sexual organism a. Consider all the organisms in the same region as a (where "region" is defined as a specified function of the distance that a can be expected to travel to mate). Call the totality of these organisms T. Within T we are going to pick out those organisms b such that for any organism c in T, the probability that a mates with b is greater than or equal to the probability that a mates with c. (Here mating requires both copulation and the production of viable offspring). Let S_1 consist of all the b's that meet the condition just stated; intuitively, S_1 comprises the opposite sex of a within the local population. We now assume that, for any b belonging to S_1 there is a unique totality S_2 within T consisting of b's most probable mates (i.e., of those organisms c such that for any d in T the probability that b mates with c is greater than or equal to the probability that b mates with d), that S_2 is the same for each b in S_1, and that a belongs to S_2. Subject to these assumptions, the total local population to which a belongs consists of the organisms in S_1 and S_2.

There are a number of obvious worries that we might have about this picture. In some cases there will be organisms that are not among the probable mates of their most probable mates. If a male bird of paradise has dull plumage, his potential mates will include females who do not include him among their potential mates. Examples like this—and it is easy to see that they are legion—prompt Mayr's suggestion that we treat the notion of population as an ideal, abstracting from the actual differences in sexual selection. Of course, once we demand that mating must involve not only copulation but production of viable offspring, then we encounter troubles with those organisms carrying alleles that are not concordant with the alleles found in members of the opposite sex. If we do not make the demand, then we shall have trouble with populations in which males have the propensity to copulate with females of closely related species as well as with their conspecifics.

For present purposes, however, let us restrict our attention to the difficulty posed by sexual selection, and ask what is meant by claiming that various mating probabilities are equal or unequal. Imagine that a male organism a_1 actually mates with a female a_2 and does not actually mate with another female a_3. Assume that all these organisms are, from the naturalist's standpoint, members of the same population. If Mayr's account of population is to be accepted, then we need to defend the claim that the probability that a_1 mates with a_2 is the same as the probability that a_1 mates with a_3, In making this claim we are obviously expanding our horizons from contemplation of the actual situation alone. We envisage a range of possible situations and suggest that the proportion of situations in which a_1 mates with a_2 is the same as the proportion of situations in which a_1 mates with a_3. The actual world was, as it were, "selected" from this range of situations, and the "selection" produced a situation in which a_1 mates with a_2 and not with a_3. Now what determines the appropriate range of situations, the situations that we tacitly envisage in making our judgment of equiprobability? Or, to put the point another way, what features of the organisms do we allow to vary across this range of possibilities, and which do we hold constant?

Plainly if *all* the features of the actual situation are held fixed, then our consideration is limited to a unique situation, so that the proportion of cases in which a_1 mates with a_2 is 1, and the proportion of cases in which a_1 mates with a_3 is 0. If *no* features of the actual situation are held constant, then we shall be confronted with a range or possibilities so vast that it seems that the proportion of cases in which a_1 mates with any particular organism will be effectively 0. Our probability judgment rests on our striking just the right balance between these two extremes, in abstracting from some features of the actual situation and holding others fixed, so that the probability judgments made in applying Mayr's picture will identify "local populations of conspecific organisms."

A full account of what a population is must tell us how to strike this balance. It must explain how the conception of probability is to be applied here, specifying the class of possible situations that are to fall under our consideration. What properties of the organisms should be held fixed? Which features can be idealized? To see how difficult these questions are, let us consider some cases, which I describe from the perspective of a naturalist who uses the concept of species without analysis.

1. A local population of a social species with a dominance hierarchy in which smaller, weaker males rank lower, contains some males—the smallest and weakest—who do not mate at all. In judging that they have a nonzero probability of mating with high-ranking and low-ranking females, we abstract from the size of these males (i.e., from the characteristic on which their position in the dominance hierarchy depends).
2. Populations of two species, one of which is the dwarf form of the other, inhabit the same region. In judging that dwarf-dwarf matings are more probable than dwarf-normal matings, we do not abstract from considerations of size.
3. Two small populations of related species occur in a marginal habitat at the peripheries of the ranges of both species. In this region, hybridization occurs

as frequently as mating between conspecifics. We avoid lumping the two populations by judging that each organism has a greater probability of mating with an organism from its own species. This judgment rests on abstracting from the composition of the fauna of the region. We distinguish the populations by considering what would happen if the region were not so sparsely populated, which would raise the relative frequency of mating among conspecifics.

4. In a highly polytypic species, showing a continuously distributed range of morphological types, individuals of each type may have a greater propensity to mate with one another than with individuals of different types. Consider a region in which a small number of organisms of the species, exhibiting different types, meet and mate freely. We judge this group of organisms to be a single population, taking the probabilities of cross-type mating to be equal to those of intra-type mating, because, in this case, we do not abstract from the composition of the local fauna.

5. Two species may be reproductively isolated from one another by differences in the times at which they are active. (The differences can consist in differences between the daily cycles of activity and rest or in differences between breeding seasons.) If two such species occur in a given region, we judge the probabilities of various types of mating by holding fixed the times of activity of the organisms concerned. Were we to abstract from the differences in these times, the probability of interspecific matings would be as great as that of intraspecific matings.

6. In some cases, a species may include organisms with a broad ranges of times of activity. Extreme individuals may be debarred from mating because their times of activity do not overlap. Yet we may count these organisms as belonging to the same local population, by abstracting from the differences in times of activity, so that the probabilities of mating become equal across the species.

I claim that if "species" is used as naturalists and theoretical biologists alike use it, then there are numerous examples answering to the descriptions 1–6. What these examples show is that properties of the organisms in question which are held constant in arriving at probability judgments in some cases are allowed to vary in other cases. In other words, the collection of possible situations, with respect to which the probabilities of mating are judged, cannot *obviously* be characterized in any uniform way.[16] If Hull and Ghiselin hope to deploy the concept of population to articulate the idea that species are historical individuals, then they need to articulate the principles we use in setting up the space of "real possibilities" that underlies our probability judgments.

Since Hull has differed with Mayr's use of modal notions (*viz.* the *possibility* of gene exchange) and has insisted that our delineation of species should be based on the pattern of *actual* matings, it is worth exploring briefly whether there is any plausibility to the idea that we can avoid talk of possibilities and probabilities, either explicating the notion of a population in a nonmodal way or bypassing it and building up the concept of historical connectedness from the actual matings among organisms. One obvious trouble results from the fact that, in many species, vast numbers of organisms belonging to the same population do not mate at all. This difficulty could be overcome by supposing that organisms whose parents belong to the same population and that inhabit the same region belong to the same population. Unfortunately, that supposition would debar *by fiat* the possibility of instant speciation, and would yield counterintuitive results in the known cases in which polyploidy results from a single generation event.

Another worry stems from the fact that hybridization does occur in nature, and it is quite probable that there are some organisms that only mate with members of different species. Not only will such instances draw the boundaries of populations in the wrong ways, but, if they are accompanied by instances of relatives that engage in some matings with conspecifics, there is the obvious possibility that the transitivity of the relation *belonging to the same population* will lead to identifications of "populations" that are assemblages of members of different species—perhaps even species that are quite distantly related but connected by a chain of close relatives.

Although both problems are serious, the most fundamental trouble for those who hope to avoid the modal intricacies of Mayr's concept of population seems to me to be a consequence of the fact that populations may have significant internal structure and may fall into groups that have been reproductively disconnected for a number of generations. In some instances in which this occurs there may be incipient speciation; in others not. I deny that we can distinguish the two types of case by appealing to the pattern of actual matings.

Let's consider two examples in which the conspecifics in a region are fragmented into reproductively disconnected groups. The first is an idealization of what actually occurs among the Serengeti lions. Imagine that the females of a species divide into small groups, that these females mate with one or two males who become associated with a group for short periods, and that each male only has one chance to become associated with a group. Under these conditions there is no chain of animals in the population such that *a* mates with *b* who also mates with *c* who also mates with *d* . . . , so that ultimately every member of the species in the region is connected to every other member. Moreover, if there is a large number of groups, and if there is a strong tendency for males to take over groups including offspring of the females in their mothers' groups, then there are likely to be males and females "in the same population" who have no common ancestor in recent generations.

The second example is focused on our own species. It is all too familiar that there have been groups with very strong taboos or laws against various kinds of miscegenation. There are probably some instances in which these taboos have been and are still effective, so that, within a given region, people with different phenotypes have been reproductively disconnected for many generations. I doubt that we want to classify these eases as examples of incipient speciation or to declare that the people concerned belong to different populations. Instead, we want to talk of an extreme of assortative mating *within* a single population.

The moral of this section should by now be apparent. If the Hull-Ghiselin account is to be developed as a reply to the traditional problem of delineating the species taxa, then there is a serious task of analysing the notion of population or of devising some surrogate. If we are even to *understand* the thesis that species are "historical entities," the difficulties that I have indicated must be faced and overcome.

3. The Idiosyncrasies of Isolating Mechanisms

The breaks in the genealogical nexus that are depicted in branching diagrams and that Hull uses to indicate the views about species he regards as serious contenders are typically connected with the attainment of reproductive isolation between populations. Two populations are said to be reproductively isolated from one another if there are mechanisms that prevent interbreeding between their members where they occur together in nature or that would prevent interbreeding between their members if they did occur together in nature. Of course, organisms from populations that are reproductively isolated from one another may produce hybrid progeny in captivity, in the laboratory, or even in places where disturbances of the habitat have produced a large disruption of the normal way of life.[17] Moreover, it is possible for there to be some gene flow between reproductively isolated populations, for example, across stable hybrid zones. Introgression is not precluded, but it must not proceed on so wide a scale that the evolutionary autonomy of either species is threatened.

There is an apparent tension within those accounts that make reproductive incompatibility central to speciation, whether they do so in the classic way of Mayr (the biological species concept) or whether they pursue the idea that species are "historical entities" whose boundaries are marked by episodes of speciation that involve the attainment of reproductive incompatibility. The tension arises from ideas about evolutionary autonomy, specifically:

(a) A small amount of introgression is compatible with reproductive isolation between populations.

(b) A low rate of migration between spatially separated populations (of the same species) is sufficient to ensure that these populations are not (effectively) isolated from one another.

Simultaneous acceptance of (a) and (b) seems problematic. If limited migration between spatially separated populations serves as the "glue" that binds those populations together, making them parts of a single (scattered) species, why does limited gene exchange between populations that are classified as belonging to different species not serve equally effectively to bind those populations into the same kind of genetic/evolutionary unit?[18]

Notice that it won't do to try to solve the problem by insisting that *whenever* there's limited gene exchange the populations in question belong to different species—for, as we saw in the last section, we want to allow for assortative mating within a single species and for island populations of a continental species. There may well be limited gene flow among some subgroups of *Homo sapiens*, among some subgroups of Serengeti lions, among some groups of anoline lizards in the Caribbean, and among oaks in California and Quebec.[19] What we need is a principle for drawing the species boundary, and, in defining the task, it is helpful to pose the issue of what exactly we are attempting to map by speaking of species in the first place.

Remarks by Dobzhansky, and subsequently by Mayr, make plain the motivation for insisting on reproductive isolation. In a famous passage, Dobzhansky introduces the notion of reproductive isolation as the key to the understanding of local diversity.[20] Without the attainment of reproductive isolation, he suggests, gene flow would be uninterrupted, so that, in any locale, the effects on one group of organisms would be felt by the rest of the living residents. Dobzhansky's case for the importance of reproductive isolation presupposes a thesis about the homogenizing effects of gene flow. Even small amounts of gene exchange are taken to threaten the obliteration of genetic differences. Hence the principled division that we sought one paragraph back should explain just how much gene flow can be tolerated without making one group's "evolutionary tendencies and fate" felt by the other.

But it is possible to question the presupposition on which the connection between speciation and the attainment of reproductive isolation depends. As several empirical studies have shown, gene flow in some groups of organisms is far weaker than orthodox evolutionary theorists had supposed: for example, detailed research on dispersal of pollen by insects and by wind has supported the conclusion that "[p]ollen and seed dispersal are either exclusively local or highly leptokurtic."[21] Given this result, it is not easy to see how reproductive community serves as an explanation for the genetic (morphological, ecological) uniformity found in some widely distributed species. As two of the most influential critics conclude: "Our suspicion is that, eventually, we will find that, in some species, gene flow is an important factor in keeping populations of the same species relatively undifferentiated, but that in most it is not. As this becomes widely recognized we will see the disappearance of the idea that species, as groups of actually or potentially interbreeding populations are evolutionary units 'required' by theory."[22]

Theoretical considerations also reveal that reproductive isolation is not a sine qua non for the development and maintenance of diversity. It is at least theoretically possible for considerable differences to evolve within an interbreeding population: even in the absence of barriers to gene flow, sharp differences in the frequencies of alternative alleles can be maintained.[23] Combining the theoretical study of clines with empirical results about gene flow, it becomes hard to sustain the thesis that attainment of reproductive isolation is necessary and sufficient for two groups of organisms to be subject to distinct evolutionary "fates."

Hull's account of the historical individuals that count as species is, I have suggested, incomplete, but his remarks about species fission (and possible fusion) seem wedded to the notion that the genealogical nexus is broken into species at those points at which reproductive isolation is attained.[24] Not only is this approach vulnerable to the familiar objections about the status of species in nonsexual organisms, but, given the considerations that I have been raising here, we need a serious defense of the view that reproductive isolation is necessary and sufficient for the integrity of historical individuals (or for those historical entities that constitute species). It is not just that Hull and Ghiselin have failed to say which among several proposals for splitting the genealogical nexus they are inclined to favor, but we are owed an account of why *any* proposal involving the interruption of gene flow among populations should be seen as theoretically crucial to species diversity.

Let me extend the point by taking note of a response that proponents of the biological species concept have offered to suggestions that gene flow may be insufficient to promote the cohesion of "conspecific" populations. Mayr writes:

> Physiologists and embryologists, likewise, have published evidence for a remarkable uniformity of physiological constants through the range of most species. The essential genetic unity of species cannot be doubted. Yet the mechanisms by which this unity is maintained are still largely unexplored. Gene flow is not nearly strong enough to make these species anywhere nearly panmictic. It is far more likely that all the populations share a limited number of highly successful epigenetic systems and homeostatic devices which place a severe restraint on genetic and phenotypic change.[25]

This response threatens the priority of the concept of reproductive isolation by hinting at a quite different approach to the delimitation of species taxa. Each species taxon is to be associated with an epigenetic system (or a small family of such systems). The persistence of uniform phenotypes across the broad range of a species is to be explained by the difficulty of introducing new alleles that perturb the phenotype, and, by the same token, the distinctness of species is grounded in their having distinct epigenetic systems. No mention need be made of reproductive isolation. It might turn out that the distinctness of epigenetic systems coincided with the possession of isolating mechanisms, or that it did so in most cases, but the division of organisms into species (on this approach) would not rest on the fact of reproductive isolation. What would make organisms belong to different species would be their possession of different epigenetic systems.

If Mayr's account of the persistence of uniformities in phenotype through the prevalence of imperturbable epigenetic systems were correct, then not only would the biological species concept fail to identify the crucial features on which species identity and species difference rest but, more to our present point, species would not need to be characterized as historical entities. Species taxa would be individuated by (families of) epigenetic systems. Of course, we could impose the *additional* requirement that organisms sharing epigenetic systems (of the same family) belong to the same species only if they belong to populations that are historically connected. However, if one believed that it is the presence of the epigenetic systems themselves that explains uniformities and differences, then it would be hard to see this additional requirement as anything other than an *ad hoc* salvaging of the Hull-Ghiselin thesis. Why should we care about reproductive connections if evolutionary fates are fixed by the (family of) epigenetic systems?

I shall conclude my worries about reliance on the notion of reproductive isolation, by considering a disturbing possibility. One of the intuitive, pre-theoretical, ideas that we might have about species is that organisms are either conspecific or not, and that, in either case, there is no relativization to any third factor. Given the organisms, their intrinsic properties and the relations between them, the answer to the question "Are they conspecifics?" is fixed. I do not wish to claim that this pre-theoretical idea is entirely precise, or that it is sacrosanct. However, if we appeal to reproductive isolation as a criterion for species distinctness (or as a criterion for the occurrence of a speciation event that has split the genealogical nexus) then it seems quite possible that there will be a necessary relativization to the environment. This could occur in numerous cases where there are actual or potential disruptions of the habitat with consequences for the cycles of activity of organisms that do not normally overlap. However, I want to consider a pure example in which a mechanical barrier to gene flow might be breached by the environment.

Fertilization in sea urchins involves three fusions between sperm and egg. The first of these involves the acrosome (at the head of the sperm) and a jelly that surrounds the egg: a receptor molecule on the surface of the sperm responds to glycoproteins in the jelly and the result is a change in the pH of the acrosome, a change that allows for release of actin and (ultimately) for the penetration of the egg by the sperm. Two species of sea urchins are distinguished by different glycoproteins at the egg surfaces and by different molecules that bind the glycoproteins to the sperm. The result is that sperm of *Strongylocentrotus purpuratus* cannot fertilize eggs of S. *franciscanus* because the reaction is blocked at the first stage. However, in the presence of trypsin, the glycoproteins *will* bind to the sperm, and, in consequence, hybrid progeny are produced.

To the best of my knowledge, S. *franciscanus* and S. *purpuratus* are isolated only by the mechanism just described. But it is plain that the isolation is environment-relative. In a trypsin-rich environment, there would be no barrier to gene flow between the two species. Now it is doubtful that there are any such environments inhabited by sea urchins—at least outside the laboratory. However, the example[26] points to a general possibility: populations may be reproductively isolated simply because a particular reaction in the formation of a zygote is blocked; however, the presence of certain molecules in the environment—perhaps as a result of abiological features, perhaps because of the presence of further organisms—might allow the reaction to go forward; thus it is quite possible that there are organisms that are reproductively isolated in one environment and not isolated in another (slightly different) environment. If the *very same* organisms had been situated slightly differently, the question whether they are conspecifics would have received a different answer. But perhaps the appropriate moral to draw here is that our initial view about the nonrelativity of species relationships is faulty, and that, in our normal speech, we tacitly relativize to the kinds of environments that actually occur.

4. Segmentation and Serendipity

Imagine that the problems of previous sections have been overcome and that we have successfully made sense of the concept of a population and of a principled notion of reproductive isolation. I'll suppose that we have understood a *lineage* to consist of organisms in some original population (the *founding* population) plus all their descendants, and that our residual task is to segment lineages by using the notion of reproductive isolation to characterize *separation events*. When a separation event occurs, some stages of the lineage just after the event belong to a different species than stages of the lineage from which they descended. The problem is to articulate the idea, specifying exactly how reproductive isolation relates to segmentation.

One proposal is to allow for speciation by anagenesis. Two lineage stages belong to different species if, had they coexisted, they would have been reproductively isolated. This proposal, essentially Simpson's, faces certain obvious difficulties of application—especially within the context of a gradualistic approach to evolutionary change. Notoriously, it has inspired some systematists to express their gratitude for the incompleteness of the fossil record, on the grounds that the gaps allow the delimitation of species taxa![27]

For many contemporary systematists, there is no hope of finding a principled division of lineages while allowing for anagenesis. Instead, we should recognize that the genealogical nexus is broken at those points where speciation produces two contemporary populations that are reproductively isolated from one another. Cladogenetic speciation is completed when the postspeciation descendants of the stages of the lineage preceding the speciation event divide into two groups that are reproductively isolated from one another. For Hennig, a species comprises the organisms on a branch of a lineage bounded by consecutive speciation events: "The limits of the species in a longitudinal section through time would consequently be determined by two processes of speciation: the one through which it arose as an independent reproductive community, and the other through which the descendants of this initial population ceased to exist as a homogeneous reproductive community".[28] Hennig is committed to two claims that distinguish his account from Simpson's: (1) speciation by anagenesis cannot occur; (2) ancestral species cannot survive the events in which they give rise to daughter species.

Wiley[29] has amended Hennig's approach to avoid one source of controversy, and his formulation of an evolutionary conception of species is explicitly designed to wed Simpsonian and Hennigian insights. On Wiley's account each species comprises the organisms on a branch of a lineage bounded by speciation events (not necessarily consecutive). Thus Wiley takes over (1), but does not commit himself to (2). He writes: "Ancestral species may become extinct during speciation events if they are subdivided in such a way that neither daughter species has the same fate and tendencies as the ancestral species."[30] It is fairly clear what Wiley has in mind. If speciation occurs by geographical isolation of a very small population of the ancestral species, so that the full range of antecedent genetic (behavioral, ecological, morphological) variation is retained in that portion of the ancestral species that is *not* isolated, then, in a very obvious sense, the evolutionary history of the branch of the lineage containing the unisolated moiety is unaffected by what occurs on the branch that contains the isolate. Had a cataclysm simply eliminated the organisms that were actually geographically isolated, the subsequent evolution of the unisolated organisms would (at least initially) have been no different. But in this case, there would have been no speciation event, and hence no principled splitting of the lineage into two 'sibling' species that succeed one another temporally. Wiley proposes that ancestral species can survive speciation events if their range of variation is not substantially depleted, and this eminently reasonable idea enables him to cope with cases that Hennig finds troublesome.

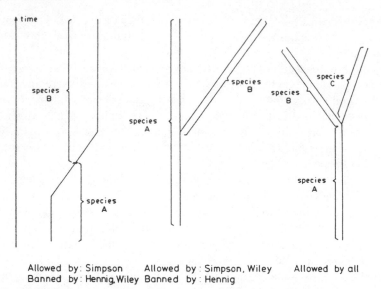

FIGURE 6.3 Three proposals for splitting lineages.

The differences among Simpson, Hennig, and Wiley are easily displayed diagrammatically (see figure 6.3). Hull[31] reproduces similar diagrams, and points out, quite correctly, that it is a significant and difficult issue to choose among the corresponding positions. In the remainder of this section, I want to undersore the difficulties. The Hull-Ghiselin thesis that species are historical entities is committed to the view that there is some principled way of segmenting the genealogical nexus. I hope to show how each of the available principles of segmentation is problematic.

Here is the strategy. In motivating Wiley's departure from Hennig, I developed an argument that contrasts the actual course of evolution with a slightly different possible situation. In the transition from the actual history of the world to this possible situation, the intrinsic properties of and direct relations among stages of one branch of the lineage were left unmodified. Yet Hennig's criterion for species delineation was found to yield different conclusions for the organisms on this branch in the two cases. What discredits the criterion is our acceptance of the following principle:

(*) A proposal to count lineage-stages as stages of the same species should depend only on the intrinsic properties of and direct relations among those stages. It should give the same results in cases which differ only in the existence or properties of organisms occupying a different branch of the lineage.

I shall now try to show how appeals to (*) cast doubt on some of the most basic features of the idea that species are segments of the genealogical nexus.[32]

Let us begin with the thesis that Wiley shares with Hennig, (2), the ban on anagenesis. There is an old worry about this thesis. It is apparently possible that a lineage should evolve quite dramatically without splitting: imagine a world in which the lineage is founded by a population of protists and then evolves into *Homo sapiens* by the sequence of genetic changes that actually link us to our protist ancestors. On the Hennig-Wiley criterion, all the organisms in this lineage would belong to a single species.[33] This strikes many people as counterintuitive (even insane). I shall defend the example and develop it so as to make clear the source of the trouble.

Notice first that the Hennig-Wiley criterion cannot be protected by dismissing the imagined possibility as unreal. It will not do to protest that, in any world in which there was an undivided lineage linking the protists to humans, the laws of nature would have to be very different so that the Hennig-Wiley criterion would be inapplicable. We can describe a world, like our own in certain critical respects, in which the lineage is realized. At each point corresponding to a speciation event in the actual world the same kind of thing happens. Part of the ancestral population takes the first step toward speciation, and, as it does so, the relict of the ancestral population is wiped out. Objection: the story cannot be quite parallel, because the organisms that were eliminated would have exerted selection pressures on the evolving lineage, and, in their absence, the course of evolution cannot be the same. Reply: the selection pressures have to be made up in other ways; one possibility is to suppose that another (distinct) group of protists gives rise to a branching lineage in which organisms evolve to exert the right kinds of pressures on the unbroken lineage.

The heart of the problem can be understood by beginning with the hypothetical situation of the last paragraph and tracing a continuous path back toward actuality. Choose any of the actual branching points along the protist-human lineage — say the event in which the first mammalian species originated from part of the ancestral population. In the hypothetical world, we assume that the first mammalian species survived a cataclysm in which the rest of the ancestral population was wiped out. Now let us suppose that the time of the cataclysm is slightly postponed — the avalanche comes or the river floods a day later than before. As we delay the time of the catastrophe, we finally obtain a situation in which the relict branch achieves reproductive isolation from the main lineage. At this point, the Wiley-Hennig criterion demands that the original lineage is to be split into two distinct species.

The thought-experiment is easier to grasp by reference to figure 6.4. Here W_1 is a world in which A and B (and C, for that matter) are lineage segments belonging to the same species. In W_2, by contrast, A and B (at least) count, by the Hennig-Wiley criterion, as distinct species. To defend the 'Simpsonian intuition' that lineage splitting is forced even in unbranching lineages, one should focus on cases like those contrasted here, and invoke (*). In W_1 and W_2, the intrinsic properties of the organisms in the $A + B$ lineage are the same: the same ranges of genetic, morphological, behavioral, and ecological variation occur at each stage. The same reproductive connections hold along the lineage. All that differs is the timing of a catastrophe that affects only organisms on a *different* branch. Appealing to (*), I claim that the difference is extraneous to the organisms in $A + B$, and that a proper division of the organisms of $A + B$ into species ought to yield the same result in each case.

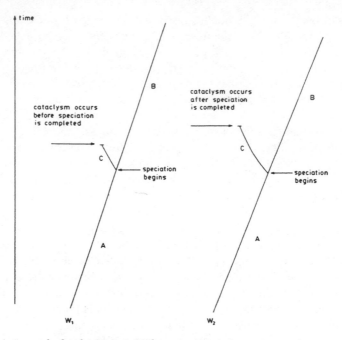

FIGURE 6.4 A puzzle for the Hennig-Wiley criterion.

Allowing for anagenesis would, of course, leave us with the puzzle of *how* to allow for anagenesis. That topic deserves a paper of its own, and I shall not pursue it here. Instead, I want to pose a problem, of the same general form, that strikes at all versions of the thesis that species are historical entities — including those that articulate the thesis along the lines indicated by Simpson. Unlike the argument just offered, we do not have to countenance any exotic possibilities to appreciate the force of the puzzle. It arises form the simple possibility of "dumbbell allopatry" as a mode of speciation.

Imagine an evolving lineage which, at time *t*, is divided into two roughly equal halves by the interposition of a geographical barrier. Assume that, at *t'*, the descendant populations on each branch of the lineage have diverged to a sufficient extent that each behaves as a good species with respect to the other. The criterion of species distinctness can be reproductive isolation — or something different, provided only that termination of speciation should conform to a familiar biological fact, to wit that speciation need not be instantaneous and that it is possible to talk of lineages as undergoing events of speciation (not necessarily at a uniform rate). Suppose, further, that from *t'* a condition of stasis prevails, so that the two lineage branches persist unmodified for a million years, until they become extinct. Finally, let us add the condition that the divergence of both branches is minimal for complete speciation. If the ancestral lineage had persisted unchanged beyond the point of geographical bifurcation, its subsequent stages would not be sufficiently distinct from the stages on either branch to count as a separate species. In other words, each incipient branch retained the full range of variation present in the ancestral lineage,

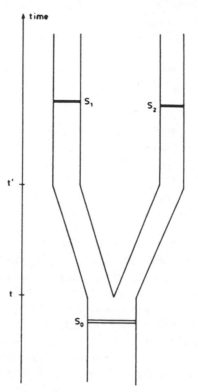

FIGURE 6.5 Species survival and identity.

there are evolutionary changes along both branches, and these are, together but not separately, sufficient for speciation.

The envisaged situation is represented in figure 6.5, where the horizontal axis represents whatever kind of change is taken to be relevant to speciation. By the criterion for speciation S_1 and S_2 are stages of different species. However, by the same criterion, S_1 and S_0 count as conspecific, and so do S_0 and S_2. Hence we face an apparent paradox: there is a species that embraces S_1 and S_0 and a species that embraces S_0 and S_2, but no species that embraces S_1 and S_2.

Formally, this is not a paradox. No contradiction arises unless one holds that for any organism there is at most one species to which it belongs. But if one retains that principle then one must decide which of the judgments about conspecificity to abandon. One approach (Hennig's) is to declare that the ancestral species becomes extinct at t, at which point two daughter species are born. This response falls foul of the argument given in motivating Wiley's departure from Hennig. Had either branch become extinct shortly after t, we would be happy to count the residual branch as a continuation of the ancestral species. Moreover, the situation is symmetrical. Worlds in which either branch survives and the other terminates are happily seen as worlds in which an evolving lineage gets about halfway through what looks like anagenesis—and then stops.

Whatever condition is proposed for guaranteeing the persistence of the ancestral species—retention of full range of genetic variation, for example—can be built in to the scenario. Once again our judgments of conspecificity are grounded in (*). Because the time of extinction of one branch does not make a difference to the intrinsic properties of stages on the other branch, or to the direct relations among them, whether or not those stages belong to the same species cannot depend on whether or not the first branch persists.

If cases like this were to occur (and perhaps they do) a purely formal solution to the problem could be obtained by allowing the same lineage-stage to belong to two different species.[34] Just as two different roads can overlap on the same piece of tarmac, so, we might say, the same lineage-stage can be included in two different species. Biologists, I suspect, will not find this formal solution attractive. A more plausible way of treating such instances is to let one's judgments about division into species conform to the current needs of biological research. For someone investigating the acquisition of reproductive isolation, it might be appropriate to count both branches as distinct daughter species. On the other hand, a biologist concerned with the developmental constraints imposed upon organisms by the facts of their ancestry might prefer to view the branches and the ancestral lineage as constituting an (unusual) single species. Judgments about such cases seem to rest on whether one is more interested in the distinctness of the descendant branches or in their kinship with their common ancestor. I believe that there is no single, objectively right, way to segment the entire lineage into species. Various ways of proceeding offer partial solutions, emphasizing some biological features of the situation and downplaying others. I propose (once again) that we take a pluralistic view of species, allowing that there are equally legitimate alternative ways of segmenting lineages—and indeed legitimate ways of dividing organisms into species that do not treat species as historical entities at all.

5. Conclusions

To say more about pluralism and its virtues would change the focus of this paper. What I have been attempting to show is the extent of the work that needs to be done if Hull's conception of species as historical entities is to cope with the diversity of organisms. Pluralism enters the discussion only because it offers a way out of an apparent difficulty in segmenting lineages. Hull (1987) has complained that pluralism is the counsel of despair, and that monistic proposals for understanding species deserve a run for their money. There is surely a sound point here. Numerous instances from the history of science reinforce the judgment that theories need time to overcome apparently devastating objections. However, what concerns me about the proposal that species are historical entities is that the *difficult* problems about delimiting species taxa seem to have become invisible. As I read the recent literature,[35] an unarticulated version of the proposal seems to be serving as the basis for suspiciously a-priori–looking arguments about evolutionary processes, while issues about the articulation of the proposal are ignored. My aim here has been to bring *some* of the problems back to center stage.

Of course, not all apparent puzzles deserve serious and sustained attention. Most philosophers are familiar with the dismaying degeneration that characterizes fields in which energy is lavished on counterexamples of no theoretical importance. One of the great merits of David Hull's approach to philosophy of biology has been his plea for the use of realistic examples and his dismissal of worries based on unconstrained philosophical fictions. I hope that the examples I have constructed are simply pure types of actual biological situations, so that they will strike him as the kinds of puzzles that his account of species will have to overcome. In this spirit I offer him, not a knockdown argument for pluralism, but just some puzzles about species.

Notes

Thanks to Michael Ruse for conceiving the idea of the volume in which this essay originally appeared and for inviting me to contribute to it.

1. The debate about 'the ontology of species' begins with M. Ghiselin ("A Radical Solution to the Species Problem," *Systematic Zoology* 23 [1974]: 536–544) and D. Hull ("Are Species Really Individuals?" *Systematic Zoology* 25 [1976]: 174–191; Hull, "A Matter of Individuality," *Philosophy of Science* 45 [1978]: 335–360). Criticisms are launched in D. B. Kitts and D. J. Kitts ("Biological Species as Natural Kind," *Philosophy of Science* 46 (1979): 613–622) and A. Caplan ("Back to Class: A Note on the Ontology of Species," *Philosophy of Science* 48 [1981]: 130–140). Hull replies to these in "Kitts and Kitts and Caplan on Species," *Philosophy of Science* 48 (1981): 141–152. My own objections to the Hull-Ghiselin view are presented in Philip Kitcher, "Species," *Philosophy of Science* 51 (1984): 308–333 (republished as chapter 5 in this book); Kitcher, "Against the Monism of the Moment: A Reply to Elliott Sober," *Philosophy of Science* 51 (1984): 616–630; and Kitcher, "Ghostly Whispers: Mayr, Ghiselin and 'the Philosophers' on the Ontology of Species," *Biology and Philosophy* 2 (1987): 184–192. Some of my concerns are addressed in D. Hull, "Genealogical Actors in Historical Roles," *Biology and Philosophy* 2 (1987): 168–184. Sober replies to my 1984 "Species" (E. Sober, "Sets, Species and Evolution: Comments on Philip Kitcher's "Species," *Philosophy of Science* 51 [1984]: 334–341), and my "Against the Monism of the Moment" attempts to rebut Sober's objections.

2. Ghiselin, "Species, Concepts"; Hull, "Are Species Really Individuals?"; and Hull, "A Matter of Individuality."

3. Kitcher, "Species"; Kitcher, "Against the Monism of the Moment"; and Kitcher, "Ghostly Whispers."

4. Hull, "Are Species Really Individuals?"; Hull, "A Matter of Individuality."

5. Kitcher, "Species"; Kitcher, "Against the Monism of the Moment."

6. The following argument is briefly presented in Hull, "Genealogical Actors," but Hull has offered a more extensive version of it in conversation.

7. Talk of the persistence of sets all of whose members are physical objects is tricky. In one obvious sense, any set is an abstract object and therefore exists atemporally. But there is another notion of persistence that underwrites the intuition about the organization of organisms that I am attempting to articulate here. According to this notion, a set of physical objects persists just so long as all of its members exist. When I speak of the persistence of sets of physical objects, I shall be employing this latter notion.

8. This is the type of organization that Hull appears to emphasize in "A Matter of Individuality" (see, for example, p. 342). He has made the point even more explicitly in conversation.

9. This elementary logical point seems to have been very difficult to grasp; see, for example, Mayr's confession of bewilderment in "The Ontological Status of Species: Scientific Progress and Philosophical Terminology," *Biology and Philosophy* 2 (1987): 145–166, and Hull's acknowledgment that he shares Mayr's bewilderment in "Genealogical Actors." My "Ghostly Whispers" tries to forestall the confusion, but perhaps I can make the point even more obvious by noting that the claim "All A's are B's" ("All species are sets," "All species are individuals") doesn't entail "All B's are A's" ("All sets are species," "All individuals are species").

10. It is worth noting that the concept of historical connection that I have just intro-duced isn't strong enough to generate Hull's conceptual point ("A Matter of Individuality," 349) that species cannot reevolve. To require that species can't become extinct and then reap-pear, one needs a condition of *complete historical connection*: an entity with organisms as parts (or members) is completely historically connected just in case for any two organisms belonging to the entity there is a sequence of populations, all of the organisms in which belong to the entity, such that each population in the sequence is either an immediate ances-tor or an immediate descendant of its predecessor and such that the organisms in question belong to the first and last members of the sequence, respectively. I don't hold Hull and Ghiselin to this requirement, because it seems to me to be incorrect. In my "Species," I offered a hypothetical case based on what we know about species of the genus *Cnemidophorus* to suggest that the same species might have a discontinuous career. Examples of a similar kind are probably legion among microorganisms, and the fact that we are eukaryotes shouldn't prevent us from seeing the need for a species concept that can be applied to bacteria and viruses as well.

11. A paradigm example seems to me to be that in N. Eldredge, *Unfinished Synthesis* (New York: Oxford University Press, 1985), where I think that some very important ideas are obscured by developing and defending them in quite the wrong way.

12. A cautionary note: it is easy to draw branching diagrams and to canvass possibilities by appealing to them. But it is always worth asking how we link the organisms that the natu-ralist observes to the branching diagram. By this I do not simply mean to pose the cladists' central question of "retrieving phylogeny" (in Elliott Sober's apposite phrase) but also to point out that we need to be told *precisely* what the phylogenetic branching diagram is supposed to represent. As Hull himself noted in a classic essay ("The Operational Imperative: Sense and Nonsense in Operationism," *Systematic Zoology* 17 [1968]: 438–457), there is all the dif-ference in the world between asking how we obtain evidence for a classificatory judgment and what the classificatory judgment means. So we should wonder not only how we are to find branch points and how we are to assign organisms to lineages, but what "branch point" and "lineage" *mean*.

13. Exercise for the reader: show that any statement made in subsequent pages that uses set-theoretic notions (e.g., *is a member of*) can be replaced without modification of empiri-cal content by a mereological expression (e.g., *is a part of*), provided that the translation is done systematically and that the converse is also true. Ambiguous expressions (*belongs to*) can be read either way.

14. See, for example, D. Futuyma, *Evolutionary Biology* (Sunderland, Mass.: Sinauer, 1979).

15. E. Mayr, *Animal Species and Evolution* (Cambridge, Mass.: Harvard University Press, 1963), 136; Mayr, *Population Species and Evolution* (Cambridge, Mass.: Harvard University Press, 1970), 182.

16. The point that I have been making is an attempt to articulate what I regard as a deep insight in the critique of the biological species concept advanced by R. Sokal and T. J. Crovello in "The Biological Species Concept: A Critical Evaluation," *American Naturalist* 104 (1970): 127–153. The worry is not just that we cannot obtain *evidence* for reproductive community without introducing considerations of phenotypic similarity, but that the *concept* of reproductive community presupposes a projection from the events of actual mating, a projection itself defined by continuities and discontinuities in biological properties. Sokal and Crovello present their case in terms of the decision procedure of a field naturalist, but many of the observations they make can be freed from the emphasis on "operational criteria for application" and reformulated as points about the concept of reproductive community.

17. A classical example is that of the towhees (*Pipilo erythrophtalmus* and *Pipilo ocai*), which occur together in some places without interbreeding, but which hybridize freely where agricultural disturbance has disrupted their habitat. For concise discussion, see Mayr, *Population Species and Evolution*, 73–74, and Mayr, *Animal Species and Evolution*, 121.

18. J. Roughgarden, *Theory of Population Genetics and Evolutionary Ecology* (New York: Macmillan, 1979), chapter 12 provides a review of some basic results about conditions under which migration of alleles among spatially separated populations is likely to be an evolutionarily significant force. He considers the relative strength of effects of migration and drift, and the interplay between migration and a spatially variable selection pressure. However, I have not seen any detailed discussion of the question when introgression in peripheral populations of a species becomes sufficient to break down the autonomy of the species. The conditions for introgression to be important can be expected to depend on (a) the spatial distribution of the populations of both species, (b) the migration rate among conspecific populations, (c) the selection pressures operating on different loci in different populations, and (d) the effective sizes of the populations in question. It seems possible that the phenomenon of introgression in hybrid zones should be asymmetrical: in other words, that genetic changes in one population should spread into the other population, but that there should be no significant flow in the opposite direction. Were this to occur, there would be a fundamental difficulty talking about reproductive isolation as a symmetrical relation between populations.

19. The last example is due to Leigh van Valen. See his extremely interesting—and underread—"Ecological Species, Multispecies, and Oaks," *Taxon* 25 (1976): 233–239.

20. T. Dobzhansky, *Genetics and the Origin of Species* (New York: Columbia University Press, 1937), 311–312.

21. D. Levin and H. Kerster, "Gene Flow in Seed Plants," *Evolutionary Biology* 7 (1974): 139–220; cited at p. 202.

22. P. R. Ehrlich and P. H. Raven, "Differentiation of Populations," *Science* 165 (1960): 1228–1232; cited at p. 1231.

23. See J. Endler, *Geographic Variation, Speciation, and Clines* (Princeton, N.J.: Princeton University Press, 1977), and Roughgarden, *Theory of Population Genetics*, 240–254. My formulation here is deliberately conservative. Many writers would insist that morphological differences and isolating mechanisms can evolve without the interposition of a geographical barrier. See, for example, G. L. Bush, "Modes of Animal Speciation," *Annual Review of Ecology and Systematics* 6 (1975): 339–364, and M. J. D. White, *Modes of Speciation* (San Francisco: Freeman, 1978). For a response, see D. Futuyma and G. Mayer, "Non-allopatric Speciation in Animals," *Systematic Zoology* 29 (1980): 254–271. If they are correct, then the case for making reproductive isolation crucial to the understanding of species diversity is even weaker than I have portrayed it as being.

24. See Hull, "A Matter of Individuality," 344–349.

25. Mayr, *Animal Species and Evolution*, 523.

26. For information and discussions relevant to this example, I am extremely grateful to William Loomis. For details on sea urchin fertilization, see V. Vacquier, "The Interactions of Sea Urchin Gametes during Fertilization," *American Zoologist* 19 (1979): 839–849, and S. Podell and V. Vacquier, "Wheat Germ Agglutin Blocks the Acrosome Reaction in *Strongylocentrus purpuratus* Sperm by Binding a 210,000 mol. wt. Membrane Protein," *Journal of Cell Biology* 99 (1984): 1598–1604.

27. See, for example, a first-rate textbook in palentology by D. Raup and S. Stanley, *Principles of Paleontology* (San Francisco: Freeman, 1978), 111. I should note that Raup and Stanley also lament the "valuable information" that has been lost, so they do perceive the gappiness of the record as a *mixed* blessing.

28. W. Hennig, *Phylogenetic Systematics* (Urbana: University of Illinois Press, 1966), 58.

29. E. O. Wiley, *Phylogenetics* (New York: Wiley, 1981).

30. Ibid., 25.

31. Hull, "A Matter of Individuality."

32. Note that (*) is connected with the intuition that I canvassed in the last section in discussing the possible environment-relativity of species divisions. I don't rule out the possibility that we might ultimately want to reject (*), but I think that any such rejection would have to be based on a thorough scrutiny of the central ideas, aims and presuppositions of systematics, the type of investigation for which I have argued in "Against the Monism," 628–630. In a line, philosophical discussion of species concepts ought to return to the issue of what we are after in devising a scheme for mapping organismic diversity, the issue that was originally posed by Mayr and Dobzhansky in the 1930s and 1940s, but that has become lost in subsequent discussions. The puzzles of this paper are intended as motivational preludes for the return.

33. One can't duck the issue by claiming that all the organisms belong to a single *taxon*, *Eucaryota*. For, if we believe that every organism belongs to a species, then trivially for each organism in the lineage there is a species to which it belongs. By the Wiley-Hennig criterion, any two organisms in the lineage are conspecific. Hence there is a unique *species* to which every organism in the lineage belongs. It makes little difference whether one calls this species *Eucaryota integra* or *Homo sapiens*.

34. Aficionados of the literature on personal identity will recognize the problem I have posed and the line of solution I indicate here. A similar puzzle and similar canvassing of possible solutions is given by Laurance Splitter in a forthcoming paper.

35. For example, Eldredge, *Unfinished Synthesis*; Mayr, "The Ontological Status of Species"; M. Ghiselin, "Species Concepts, Individuality, and Objectivity," *Biology and Philosophy* 2 (1987): 127–143; Hull, "Genealogical Actors in Historical Roles."

7

Function and Design (1993)

I

The organic world is full of functions, and biologists' descriptions of that world abound in functional talk. Organs, traits, and behavioral strategies all have functions.[1] Thus the function of the *bicoid* protein is to establish anterior-posterior polarity in the *Drosophila* embryo; the function of the length of jackrabbits' ears is to assist in thermoregulation in desert environments; and the function of a male baboon's picking up a juvenile in the presence of a strange male may be to appease the stranger, or to protect the juvenile, or to impress surrounding females. Ascriptions of function have worried many philosophers. Do they presuppose some kind of supernatural purposiveness that ought to be rejected? Do they fulfil any explanatory role? Despite a long, and increasingly sophisticated, literature addressing these questions, I believe that we still lack a clear and complete account of function-ascriptions. My aim in what follows is to take some further steps toward dissolving the mysteries that surround functional discourse.

I shall start with the idea that there is some unity of conception that spans attributions of functions across the history of biology and across contemporary ascriptions in biological and nonbiological contexts. This unity is founded on the notion that the function of an entity S is *what S is designed to do*. The fundamental connection between function and design is readily seen in our everyday references to the functions of parts of artifacts: the function of the little lever in the mousetrap is to release the metal bar when the end of the lever is depressed (when the mouse takes the cheese) for that is what the lever is designed to do (it was put there to do just that). I believe that we can also recognize it in pre-Darwinian perspectives on the organic world, specifically in the ways in which the organization of living things is taken to reflect the intentions of the Creator: Harvey's claim that the function of the heart is to pump the blood can be understood as proposing that the wise and beneficent designer foresaw the need for a circulation of blood and assigned to the heart the job of pumping.

Now examples like these are precisely those that either provoke suspicion of functional talk or else prompt us to think that the concept of function has been altered in the course of the history of science. Even though we may retain the idea of the "job" that an entity is supposed to perform in contexts where we can sensibly speak of systems fashioned and/or used with definite intentions—paradigmatically machines and other artifacts—it appears that the link between function and design must be broken in ascribing functions to parts, traits, and behaviors of organisms. But this conclusion is, I think, mistaken. On the view I shall propose, the central common feature of usages of function—across the history of inquiry, and across contexts involving both organic and inorganic entities—is that the function of S is what S is designed to do; design is not always to be understood in terms of background intentions, however; one of Darwin's important discoveries is that we can think of design without a designer.[2]

Contemporary attributions of function recognize two sources of design, one in the intentions of agents and one in the action of natural selection. The latter is the source of functions throughout *most* of the organic realm—there are occasional exceptions as in cases in which the function of a recombinant DNA plasmid is to produce the substance that the designing molecular biologist intended. But, as I shall now suggest, the links to intentions and to selection can be more or less direct.

II

Imagine that you are making a machine. You intend that the machine should do something, and that is the machine's function. Recognizing that the machine will only be able to perform as intended if some small part does a particular job, you design a part that is able to do the job. Doing the job is the function of the part. Here, as with the function of the whole machine there is a direct link between function and intention: the function of X is what X is designed to do, and the design stems from an explicit intention that X do just that.

It is possible that you do not know everything about the conditions of operation of your machine. Unbeknownst to you, there is a connection that has to be made between two parts if the whole machine is to do its intended job. Luckily, as you were working, you dropped a small screw into the incomplete machine and it lodged between the two pieces, setting up the required connection. I claim that the screw has a function, the function of making the connection. But its having that function cannot be grounded in your explicit intention that it do that, for you have no intentions with respect to the screw. Rather, the link between function and intention is much less direct. The machine has a function grounded in your explicit intention, and its fulfilling that function poses various demands on the parts of which it is composed. You recognize some of these demands and explicitly design parts that can satisfy them. But in other cases, as with the luckily placed screw, you do not see that a demand of a particular type has to be met. Nevertheless, whatever satisfies that demand has the function of so doing. The function here is grounded in the contribution that is made toward the performance of the whole machine and in the link between the performance and the explicit intentions of the designer.

Pre-Darwinians may have tacitly relied on a similar distinction in ascribing functions to traits and organs. Perhaps the Creator foresaw all the details of the grand design and explicitly intended that all the minutest parts should do particular things. Or perhaps the design was achieved through secondary causes: organisms were equipped with abilities to respond to their needs, and the particular lines along which their responses would develop were not explicitly identified in advance.

So the Creator intended that jackrabbits should have the ability to thrive in desert environments and explicitly intended that they should have certain kinds of structures. However, it may be that there was no explicit intention about the length of jackrabbits' ears. Yet, because the length of the ears contributes to the maintenance of roughly constant body temperature, and because this is a necessary condition of the organism's flourishing (which is an explicitly intended effect), the length of the ears has the function of helping in thermoregulation.

Understanding this distinction enables us to see how earlier physiologists could identify functions without engaging in theological speculation.[3] Operating on the presupposition that organisms were designed to thrive in the environments in which they are found, physiologists could ask after the necessary conditions for organisms of the pertinent types to survive and multiply. When they found such necessary conditions, they could recognize the structures, traits, and behaviors of the organisms that contributed to satisfaction of such conditions as having precisely such functions—without assuming that the Creator explicitly intended that those structures, traits, and behaviors perform just those tasks.

I have introduced this distinction in the context of machine design and of pre-Darwinian biology because it is more easily grasped in such contexts. I shall now try to show how a similar distinction can be drawn when natural selection is conceived as the source of design, and how this distinction enables us to resolve important questions about functional ascriptions.

III

We can consider natural selection from either of two perspectives. The first, the organism-centered perspective, is familiar. Holding the principal traits of members of a group of organisms fixed, we investigate the ways in which, in a particular environment or class of environments, variation with respect to a focal trait, or cluster of focal traits, would affect reproductive success. Equally, we can adopt an environment-centered perspective on selection. Holding the principal features of the environment fixed, we can ask what selective pressures are imposed on members of a group of organisms. In posing such questions we suppose that some of the general properties of the organisms do not vary and consider the obstacles that must be overcome if organisms with those general properties are to survive and reproduce in environments of the type that interests us.

So, for example, we might consider the selection pressures on mammals whose digestive systems are capable of processing vegetation but not meat (or carrion) in an environment in which the accessible plants have tough cellulose outer layers. Holding fixed the very general properties of the animals that determine their need to take in food and the more particular features of their digestive systems, we recognize that they will not be able to survive to maturity (and hence not able to reproduce) unless they have some means of breaking down the cellulose layers of the plants in their environments. Thus the environments impose selection pressure to develop some means of breaking down cellulose. Organisms might respond to that pressure in various ways: by harboring bacteria that can break down cellulose or by having molars that are capable of grinding tough plant material. If our mammals do not have an appropriate colony of intestinal bacteria, but do have broad molars that break down cellulose, we may recognize the molars as their particular response to the selection pressure and ascribe them the function of processing the available plants in a way that suits the operation of their digestive systems. At a more fine-grained level, we may hold fixed features of the dentition, and identify properties of particular teeth as having functions in terms of their contributions to the break-down of cellulose.

This illustration can serve as the prototype of a style of functional analysis that is prominent in physiology and in general zoological and botanical studies. One starts from the most general evolutionary pressures, stemming from the competition to reproduce and concomitant needs to survive to sexual maturity, to produce gametes, to identify and attract mates, and so forth. In the context of general features of the organisms in question and of the environments they inhabit, we can specify selection pressures more narrowly, recognizing needs to process certain types of food, to evade certain kinds of predators, to produce particular types of signals, and so forth. We now appreciate that certain types of complex structures, traits, and behaviors enable the organisms to satisfy these more specific needs. *Their* functions are specified by noting the selection pressures to which they respond. The functions of their constituents are understood in terms of the contributions made to the functioning of the whole. Here, I suggest, we have a mixture of evolutionary and mechanistic analysis. There is a link to selection through the environment-centered perspective from which we generate the selection pressures that determine the functions of complex entities, and there is a mechanistic analysis of these complex entities that displays the ways in which the constituent parts contribute to total performance.

I claim that understanding the environment-centered perspective on selection enables us to draw an analogous distinction to that introduced in section 2, and thus to map the diversity of ways in which biologists understand functions. However, before offering an extended defense of this claim, two important points deserve to be made.

First, the environment-centered perspective has obvious affinities with the idea that organisms face selective "problems," posed by the environment, an idea that Richard Lewontin has recently criticized.[4] According to Lewontin, there is a "dialectical relationship" between organism and environment that renders senseless the notion of an environment prior to and independent of the organism to which "problems" are posed. Lewontin's critique rests on the correct idea that there is no specifying which parts of the universe are constituents of an organism's environment, without taking into account properties of the organism. In identifying the environment-centered perspective, I have explicitly responded to this point, by proposing that the selection pressures on organisms arise only when we have held fixed important features of those organisms, features that specify limits on those parts of nature with which they causally interact. Quite evidently, if we were to hold fixed properties that could easily be modified through mutation (or in development), we would obtain an inadequate picture of the organism's environment and, consequently, of the selection pressures to which it is subject. If, however, we start from those characteristics of an organism that would require large genetic changes to modify—as when we hold fixed the inability of rabbits to fight foxes—then our picture of the environment takes into account the evolutionary possibilities for the organism and offers a realistic view of the selection pressures imposed.

Second, as we shall see in more detail below, recognizing a trait, structure, or behavior of an organism as responding to a selection pressure imposed by the environment (in the context of other features of the organism that are viewed as inaccessible to modification without severe loss of fitness) we do not necessarily commit ourselves to claiming that the entity in question originated by selection or that it is maintained by selection. For it may be that genetic variation in the population allows for alternatives that would be selectively advantageous but are fortuitously absent. Thus the entity is a response to a genuine demand imposed on the organism by the environment even though selection cannot be invoked to explain why it, rather than the alternative, is present. In effect, it is the analogue of the luckily placed screw, answering to a real need, but not itself the product of design. I shall be exploring the consequences of this point below.

IV

The simplest way of developing a post-Darwinian account of functions is to insist on a direct link between the design of biological entities and the operation of natural selection. The function of X is what X is designed to do, and what X is designed to do is that for which X was selected. Since the publication of a seminal article by Larry Wright, etiological accounts of function have become extremely popular.[5] Wright claimed that the function of an entity is what explains why that entity is there. This simple account proved vulnerable to counterexamples: if a scientist conducting an experiment becomes unconscious because gas escapes from a leaky valve, then the presence of the gas in the room is explained by the fact that the scientist is unconscious (for otherwise she would have turned off the supply), but the function of the gas is not to asphyxiate scientists.[6] Such objections can be avoided by restricting the form of explanations to explanations in terms of selection, so that identifying the function of X as that for which X was selected enables us to preserve Wright's idea that functions play a role in explaining the presence of their bearers without admitting those forms of nonselective explanation that generate counterexamples.[7] However, this move forfeits one of the virtues of Wright's analysis, to wit, its recognition of a common feature in attributions of functions to artifacts and to organic entities.

There are other issues that etiological analyses of functional ascriptions must confront, issues that arise from the character of evolutionary explanations. First is the question of the *time* at which the envisaged selection regime is supposed to act. Second we must consider the *alternatives* to the entity whose presence is to be explained and the extent of the role that selection played in the singling out of that entity.[8] If these issues are neglected—as they frequently are—the consequence will be either to engage in highly ambiguous attributions of function or else to fail to recognize the demands placed on functional ascription.

Selection for a particular property may be responsible for the original presence of an entity in an organism or for the maintenance of that entity.[9] In many instances, selection for P explains the initial presence of a trait *and* the subsequent maintenance of that trait: the initial benefit that led to the trait's increase with respect to its rivals also accounts for its superiority over alternatives that arose after the original process of fixation. But as a host of well-known examples reveals, this is by no means always the case. To cite one of the most celebrated instances, feathers were apparently originally selected in early birds (or their dinosaur ancestors) for their role in thermoregulation; after the development of appropriate musculature (and other adaptations for flight) the primary selective significance of feathers became one of making a causal contribution to efficient flying.

Faced with examples in which the properties for which selection initially occurs are different from those for which there is selection in maintaining a trait, behavior, or structure, the etiological analysis must decide which of the following conditions is to govern functional attributions:

1. The function of X is Y only if the initial presence of X is to be explained through selection for Y.
2. The function of X is Y only if the maintenance of X is to be explained through selection for Y.
3. The function of X is Y only if both the initial presence of X and the maintenance of X are to be explained through selection for Y.

But deciding among these three conditions is only the beginning of the enterprise of disambiguating the etiological analysis of function. Just as the properties important in initiating selection may not be those that figure in maintaining selection, it is possible that an entity may be *maintained* by selection for different properties at different times. Hence, both (2) and (3) require us to specify the appropriate period at which the maintenance of X is to be considered. I believe that there are two plausible candidates with respect to (2), namely the present and the recent past, and that the most well-motivated version of (3) requires that the character of the selective regime is constant across all times. Thus we obtain:

2a. The function of X is Y only if selection of Y has been responsible for maintaining X in the recent past.
2b. The function of X is Y only if selection for Y is currently responsible for maintaining X.
3. The function of X is Y only if selection for Y was responsible for the initial presence of X and for maintaining X at all subsequent times up to and including the present.

A consequence of adopting (1)—which effectively takes functions to be *original* functions—is that two of Tinbergen's famous four why-questions are conflated: there is now no distinction between the "why" of evolutionary origins and the "why" of functional attribution.[10] In those biological discussions in which an etiological conception of function is most apparent (ecology, and especially behavioral ecology), Tinbergen's distinction seems to play an important role. Thus I doubt that an etiological analysis based on (1) reflects much that is significant in biological practice.

Etiological analyses clearly based on (3) can sometimes be found in the writings of those who are critical of unrigorous employment of the notion of function. So, for example, Stephen Jay Gould's and Elisabeth Vrba's contrast between functions and "exaptations" seems to me to thrive on the idea that specification of functions must rest on the presupposition that selection has been operating in the same way in originating and maintaining traits (and, indeed, that traits maintained by selection were originally fashioned by selection).[11] Because there is frequently no available evidence for this presupposition, adoption of etiological conception based on (3) can easily fuel skepticism about ascriptions of function.

I suspect that some biologists do tacitly adopt an etiological conception of function founded on (3), and that their practice of ascribing functions is subject to Gould's strictures. Others plainly do not. Thus, Ernst Mayr explicitly recognizes the possibility of change of function over evolutionary time, suggesting that he acknowledges *two* notions of function, one ("original function") founded upon (1) and another ("present function") based on some version of (2),[12] For biologists who draw such distinctions, Gould's criticisms will seem to claim novelty for a point that is already widely appreciated. (Of course, one of the most prominent features of the debates about adaptationism is the opposition between those who believe that the criticisms tiresomely remind the evolutionary community of what is already well known and those who contend that what is professed under attack is ignored in biological practice.)[13]

The most prevalent concept of function among contemporary ecologists is, I believe, an etiological concept founded on some version of (2). Claims about functions are founded on measurements or calculations of fitness, and the measurements and calculations are made on *present* populations. Faced with the question, "Do you believe that the properties for which selection is now occurring are those that originally figured in the fixation of the trait (structure, behavior)?" sophisticated ecologists would often plead agnosticism. Their concern is with what is currently occurring, and they are happy to confess that things may have been different in a remote past that is beyond their ability to observe and analyze in the requisite detail. Hence the concept of function they employ is founded on the link between functions and contemporary processes of selection that maintain the entities in question, a link recorded in (2).

But which version of (2) should they endorse? Here, I believe, philosophical analyses reveal unresolved ambiguities in biological practice. An account of functions that effectively endorses (2b) has been proposed by John Bigelow and Robert Pargetter (who, idiosyncratically it seems to me, attempt to distance themselves from Wright and other etiological theorists).[14] My own prior discussions of functional ascriptions presuppose a concept based on (2a), and this notion of function has been thoroughly articulated by Peter Godfrey-Smith.[15] On what basis can we decide among these accounts?

As Godfrey-Smith rightly notes, a "recent history" notion of function, committed to (2a), gives functional ascriptions an explanatory role. Identifying the function of an entity outlines an explanation of why the entity is now present by indicating the selection pressures that have maintained it in the recent past. Arguing that philosophers ought to identify a concept that does some explanatory work, he concludes that (2a) represents the right choice. But this seems to me to be too quick. The conception of function defended by Bigelow and Pargetter, founded on (2b), is perhaps most evident in those biological discussions in which the recognition that a trait is functional supports a prediction about its future presence in the population. Yet the "forward-looking" conception also allows ascriptions of function to serve as explanations of why the trait will continue to be present. There is still an explanatory project, but the *explanandum* has been shifted from current presence to future presence.

Biological practice seems to me to be too various for definitive resolution of these differences. Sometimes attributions of function outline explanations of current presence, sometimes offer predictions about the course of selection in the immediate future, sometimes sketch explanations of the presence of traits in succeeding generations. Moreover, since it is often reasonable to think that the environmental and genetic conditions are sufficiently constant to ensure that the operation of selection in the recent past was the same as the selection seen in the present, it will be justifiable to combine the main features of the "recent past" and "forward-looking" accounts to found a notion of function on a combination of (2a) and (2b)

2c.   The function of X is Y only if selection of Y is responsible for maintaining X both in the recent past and in the present.

In situations in which there is reason to think that the action of selection has been constant across the relatively short time periods under consideration, use of a notion of function founded on (2c) will allow functional attributions to play a role in all the explanatory and predictive projects I have considered.

If biological practice overlooks potential ambiguities with respect to the timing of the selection processes that underlie attributions of function, it is even more silent on issues about the competition involved in such processes. What are the alternatives to the biological entity whose presence is due to selection? And to what extent is selection the *complete* explanation of the presence of that entity?

Ecologists working on pheromones in insects or on territory size in birds can sometimes specify rather exactly the set of alternatives they consider. Holding fixed certain features of the organisms they study, features that would, they suppose, only be modifiable by enormous genetic changes that render rivals effectively inaccessible, they can impose necessary conditions that define a set of rival possibilities: pheromones must have such-and-such diffusion properties, territories must be able to supply such-and-such an amount of food, and so forth. In light of these constraints, they may be able to construct a mathematical model showing that the entity actually found in the population is optimal (or, more realistically, "sufficiently close" to the optimum).[16] A different strategy is to consider alternatives that arise by mutation in populations that can be observed and to measure the pertinent fitness values. Either of these approaches will support claims about selection processes that have occurred/are occurring in the recent past or the present. In both instances there may be legitimate concern that unconsidered alternatives might have figured in historically more remote selection processes, either because the organisms were not always subject to the constraints built into the mathematical model or because the genetic context in which mutations are now considered is quite different from the genetic contexts experienced by organisms earlier in their evolutionary histories. So far this simply underscores our previous conclusions about the greater plausibility of analyses based on some version of (2).

But now let us ask how exactly selection is supposed to winnow the alternatives. Suppose we ascribe a function to an entity X, basing that function on a selection process with alternatives $X_1, \ldots, X_n$. Must it be the case that organisms with X have higher fitness than organisms with any of the X_i? On a strict etiological analysis of functional discourse, this question should be answered affirmatively: where selection is the *complete* foundation of the design that underlies X's function, X is favored by selection over *all* its rivals. Thus on the strongest version of an etiological conception, functional ascriptions should be based either on recognition that X has greater fitness than all the alternatives arising by mutation in current populations, or on an analysis that shows X to be strictly optimal. I believe that some biologists—particularly in ecology and behavioral ecology—make functional claims in this strong sense and attempt to back them up with careful and ingenious observations and calculations.[17] Nonetheless, there is surely room for a less demanding account of biological function.

Consider two possibilities. First, our optimality analysis shows that, while X is reasonably close to the optimum, it is theoretically suboptimal. We do not know enough about the genetics and developmental biology of the organisms under study to know whether mutations providing a genetic basis for superior rivals could arise in the population. Under these circumstances, one cannot claim that the presence of X is entirely due to the operation of selection. It may be that X is present because theoretically possible mutants have not (recently) arisen, and selection, acting on a limited set of alternatives, has fixed X. Second, we may be able to identify actual rivals to X that are indeed superior in fitness but that have fortuitously been eliminated from the population. During the period that concerns us (present or recent past) organisms bearing some entity X_i have arisen, and these have had greater fitness than organisms bearing X. By chance, however, such organisms have perished. Here, we can go further than simply recognizing an inability to support the strong claim about optimality—we recognize that X is definitely suboptimal, and that its presence is not the result of selection alone.

Nevertheless, many biologists would surely be uninterested in these possibilities or actualities, regarding X as having the function associated with the selective process, even if it were possibly, even definitely, suboptimal. There are various ways of weakening the requirement that X's fitness be greater than those of alternatives. We might demand that X be fitter than *most* alternatives, that it be fitter than the *most frequently occurring* alternatives, and so forth. It requires only a little imagination to devise scenarios in which an entity is inferior in fitness to most of its rivals and/or to its most frequently occurring rivals, even though it may still be ascribed the function associated with the selection process.

Imagine that there is a species of moth that is protected from predatory birds through a camouflaging wing pattern that renders it hard to perceive when it rests on a common environmental background. We observe the population and discover a number of rival wing colorations, none of which ever occurs in substantial numbers. Less than half of these alternatives are absolutely disastrous, and organisms with them are vulnerable to predation, and quickly eliminated. Investigating the others, we find, to our surprise, that they prove slightly superior to the prevalent form, in affording improved camouflage, without any deleterious side effects. However, as the result of various events that we can identify—disruptions of habitat, increased concentrations of predators in areas in which there is a high frequency of the mutants—these alternatives are eliminated as the result of chance. Nonetheless, although it is somewhat inferior to most of its rivals, the common wing pattern still has the function of protecting the moth from predation.

I think that it is obvious what we should say about this and kindred scenarios. The impulse to recognize X as having a function can stem from recognition that X is a response to an identifiable selection pressure, *whether or not the presence of X is completely explicable in terms of selection.* Thus, instead of trying to weaken the conditions on etiological conceptions of function, I suggest that we can accommodate cases that prove troublesome by drawing on the distinctions of sections 2 and 3. I shall now try to show how this leads to a rich account of functional ascriptions that will cover practice in physiology as well as in those areas in which the etiological conception finds its most natural home.

V

Entities have functions when they are designed to do something, and their function is what they are designed to do. Design can stem from the intentions of a cognitive agent or from the operation of selection (and, perhaps, recognizing how unintuitive the notion of design without a designer would have seemed before 1859, from other sources that we cannot yet specify). The link between function and the source of design may be direct, as in instances of agents explicitly intending that an entity perform a particular task, or when the entity is present because of selection for a particular property (that is, its presence is completely explained in terms of selection for that property). Or the link may be indirect, as when an agent intends that a complex system perform some task and a component entity makes a necessary causal contribution to the performance, or when organisms experience selection pressure that demands some complex response of them and one of their parts, traits, or behaviors makes a needed causal contribution to that response. As noted in the previous section, there are also ambiguities about the time period throughout which the selection process is operative. It would be easy to tell a parallel story about agents and their intentions.

I have noted that the strong etiological conception—that based on a direct link between function and the underlying source of design (in this case, selection)—is very demanding. While some ecologists undoubtedly aim to find functions in the strong sense, much functional discourse within ecology, as well as in other parts of biology is more relaxed. Imagine practicing biologists accompanied by a philosophical Jiminy Cricket, constantly chirping doubts about whether selection is *entirely* responsible for the presence of entities to which functions are ascribed. Many biologists would ignore the irritating cavils, contending that the attribution of function is unaffected by the possibilities suggested by philosophical conscience. It is enough, they would insist, that genuine demands on the organism have been identified and that the entities to which they attribute functions make causal contributions to the satisfaction of those demands. What is wrong with the relaxed attitude?

Functional attributions in the strong sense have clear explanatory work to do. They indicate the lines along which we should account for the presence of the entities to which functions are ascribed. To say that the function of X is F is to propose that a complete explanation of the presence of X (at the appropriate time) should be sought in terms of selection for F. Once we relax the demands on functional ascriptions, the role of selection is no longer clear; indeed, a biologist may explicitly allow that selection has not been responsible for maintaining X (or, at least, not completely responsible). But there is a different type of explanatory project to which the more lenient attributions contribute. They help us to understand the causal role that entities play in contributing to complex effects.

Here we encounter a central theme of the main philosophical rival to the etiological conception, lucidly articulated in an influential article by Robert Cummins.[18] For Cummins, functional analysis is about the identification of constituent causal contributions in complex processes. This style of activity is prominent in physiological studies, where the apparent aim is to decompose a complex "organic function" and to recognize how it is discharged. I claim that Cummins has captured an important part of the notion of biological function, but that his ideas need to be integrated with those of the etiological approach, not set up in opposition to it.

When we attribute functions to entities that make a causal contribution to complex processes, there is, I suggest, always a source of design in the background. The constituents of a machine have functions because the machine, as a whole, is explicitly intended to do something. Similarly with organisms. Here selection lurks in the background as the ultimate source of design, generating a hierarchy of ever more specific selection pressures, and the structures, traits, and behaviors of organisms have functions in virtue of their making a causal contribution to responses to those pressures.

Without recognizing the background role of the sources of design, an account of the Cummins variety becomes too liberal. Any complex system can be subjected to functional analysis. Thus we can identify the "function" that a particular arrangement of rocks makes in contributing to the widening of a river delta some miles downstream, or the "functions" of mutant DNA sequences in the formation of tumors—but there are no genuine functions here, and no functional analysis. The causal analysis of delta formation does not link up in any way with a source of design; the account of the causes of tumors reveals *dysfunctions*, not functions.

Recognizing the liberality of Cummins-style analyses, proponents of the etiological conception drag evolutionary considerations into the foreground. In doing so they make *all* projects of attributing functions focus on the explanation of the presence of the bearers of those functions. However, important though the theory of evolution by natural selection undoubtedly is to biology, there are other biological enterprises, some even continuous with those that occupied pre-Darwinians, that can be carried out in ignorance of the details of selective regimes. Thus the conscience-ridden biologists who offer more relaxed attributions of function can quite legitimately protest that the niceties of selection processes are not their primary concerns: without knowing what alternatives there were to the particular valves that help the heart to pump blood, they can recognize both that there is a general selection pressure on vertebrates to pump blood and that particular valves make identifiable contributions to the pumping. Selection, they might say, is the background source of design here, but it need not be dragged into the foreground to raise questions that are irrelevant to the project they set for themselves (understanding the mechanism through which successful pumping is achieved).

I believe that the account I have offered thus restores some unity to the concept of function through the recognition that each functional attribution rests on some presupposition about design and a pertinent source of design. But it allows for a number of distinct conceptions of function to be developed, based on sources of design (intention versus selection), time relation between source of design and the present, and directness of connection between source of design and the entity to which functions are ascribed. This pluralism enables us to capture the insights of the two main rival philosophical conceptions of function, and to do justice to the diversity of biological projects.

Does it go too far? In their original form, etiological accounts were vulnerable to counterexample, and the resolution invoked selection ad hoc. Am I committed to supposing that the leaky valve that asphyxiates the scientist has the function of so doing? No. For there is no explaining the presence of the valve in terms of selection for ability to asphyxiate scientists, nor is there any selection pressure on a larger system to whose response the action of the valve makes a causal contribution. Even though the account I have offered is more inclusive than traditional etiological conceptions, it does not seem to fall victim to the traditional counterexamples.

VI

I have tried to motivate my account of function and design by alluding to some quickly sketched examples. This strategy helps to elaborate the approach, but invites concerns to the effect that a more thorough investigation of biological practice would disclose less ambiguity than I have claimed. To alleviate such concerns I now want to look at some cases of functional attribution in a little more detail.

I shall start with two examples that are explicitly concerned with evolutionary issues. The first concerns a "functional analysis of the egg sac" in golden silk spiders.[19] The orb-weaving spider *Nephila clavipes* lays its eggs under the leaf canopy, covers them with silk, and weaves a loop of silk around twig and branch which holds the sac in place. The authors of the study, T. Christenson and P. Wenzl, investigate the functions of components of the egg-laying behavior. I shall concentrate on the spinning of the loop.

Christenson and Wenzl write:

The functions of the silk loop around the attachment branch were assessed by examining clutches that fell to the ground. We found 19 of the 59 egg sacs that fell due to naturally occurring twig breakage; 84.2% (16) failed to produce spiderlings, 13 because of ground moisture and subsequent rotting, and 3 because of predation. . . . The remaining three sacs had fallen a few weeks prior to the normal time of spring emergence; the spiderlings appeared to disperse and inhabit individual orbs.[20] In contrast to those that fell, sacs that remained in the tree were dry and appeared relatively safe from predation. Only 4.5% (15 of 353) showed unambiguous signs of predation, that is, some damage to the silk such as a tear or a bore hole.[21]

I interpret this passage as demonstrating a marked fitness difference between spiders who perform the looping operation that attaches the egg sac to twig and branch and those who fail to do so. Christenson and Wenzl are tacitly comparing the normal behavior of N. *clavipes* with mutants whose ability to weave an attachment loop was somehow impaired. Their emphasis on evolutionary considerations is evident not only in their detailed measurements of survivorships, but also in the framing of their analysis and in their final discussion. The authors begin by noting that "[f]unctional analyses of behaviours are often speculative due to the difficulty of demonstrating that the behavior contributes to the individual's reproductive success, and what the relevant selective agents might be."[22] They conclude by contending that "Female *Nephila* maximize their reproductive efforts, in part, through the construction of an elaborate egg sac."[23] This study is thus naturally interpreted as deploying the strong etiological conception of function, linking function directly with selection and proposing that the entities bearing functions are optimal.

Similarly, a study of the function of roaring in red deer by T. Clutton-Brock and S. Albon explicitly connects the attribution of function to claims about selection.[24] The authors begin by examining a traditional proposal:

> A common functional explanation is that displays serve to intimidate the opponent. . . . This argument has the weakness that selection should favour individuals which are not intimidated unnecessarily and which adjust their behaviour only to the probability of winning and the costs and benefits of fighting.[25]

Here it seems that a necessary condition on the truth of an ascription of function is that there should not be possible mutants that would be favored by selection. The same strong conception of function is apparent later in the discussion, when Clutton-Brock and Albon consider the hypothesis that roaring serves as an advertisement enabling stags to assess others' fighting ability. Although their careful observations indicate that stags rarely defeat those by whom they have been out-roared, they recognize that their data leave open other possibilities for the relation between roaring and fighting ability. They suggest that fighting and roaring may both draw on the same groups of muscles, so that roaring serves as an "honest advertisement" to other stags. But they note that this depends on assuming that "selection could not produce a mutant which was able to roar more frequently without increasing its strength or stamina in fights."[26] I interpret the caution expressed in their discussion to be grounded in recognition of the stringent conditions that must be met in showing that a form of behavior maximizes reproductive success, and thus their reliance on the strong version of the etiological conception.

I now turn to two physiological studies in which the connection to evolution is far less evident. Here, there are neither detailed measurements of the fitnesses (or proxies such as survivorships) of rival types of organism (as in the study of golden silk spiders) or connections with mathematical models of a selection process (as in the investigation of the roaring of stags). Instead, the authors undertake a mechanistic analysis of the workings of a biological system. Consider the following discussion of digestion in insects,

> Food in the midgut is enclosed in the peritrophic membrane, which is secreted by cells at the anterior end of the midgut in some insects or formed by the midgut epithelium in most. It is secreted continuously or in response to a distended midgut, as in biting flies. It is likely that the peritrophic membrane has several functions, although the evidence is not conclusive. It may protect the midgut epithelium from abrasion by food or from attack by microorganism or it may be involved in ionic interactions within the lumen. It has a curious function in some coleopterous larvae, where, in various ways, it is used to make the cocoon.[27]

The interesting point about this passage is that it could easily be accepted by a biologist ignorant of or hostile to evolutionary theory. So long as one has a sense of the overall life of an insect and of the conditions that must be satisfied for the insect to thrive, one can view the peritrophic membrane as making a causal contribution to the organism's flourishing. Of course, Darwinians will view these conditions as grounded in selection pressures to which insects must respond, but physiology can keep this Darwinian perspective very much in the background. It is enough to recognize that insects must have a digestive system capable of processing food items, that the passage of food through the system must not abrade the cells lining the gut, and so forth. I suggest that this, like so many other physiological discussions, presupposes a background picture of the selection pressures on the organisms under study and analyzes the causal mechanisms that work to meet those pressures, without attending to the fitness of alternatives that would have to be considered to underwrite a claim about the operation of selection.

Finally, I turn to a developmental study of sexual differentiation in *Drosophila*.[28] The problem is to understand simultaneously how an embryo with two X chromosomes becomes a female, how an embryo with one X chromosome becomes a male, and how the organism compensates for the extra chromosomal material found in females. The author (M. Kaulenas) summarizes a complex causal story, as follows:

> The primary controlling agent in sex determination and dosage compensation is the ratio between the X chromosomes to sets of autosomes (the X:A ratio). This ratio is "read" by the products of a number of genes; some of which function as numerator elements, while others as denominator elements. Two of the numerator genes have been identified [*sisterless a (sis a)* and *sisterless b (sis b)*] and others probably exist. The denominator elements are less clearly defined. The end result of this "reading" is probably the production of DNA-binding proteins, which, with the cooperation of the *daughterless (da)* gene product (and possibly other components) activate the *Sex lethal (sxl)* gene. This gene is the key element in regulating female differentiation. One early function is autoregulation, which sets the gene in the functional mode. Once functional, it controls the proper expression of

the *doublesex (dsx)* gene. The function of *dsx* in female somatic cell differentiation is to suppress male differentiation genes. *Dsx* needs the action of the *intersex (ix)* gene for this function. Female differentiation genes are not repressed, and female development ensues.[29]

Here is a causal story about how female flies come to express the appropriate proteins in their somatic cells. The elements of the story concern the ways in which particular bits of DNA code for proteins that either activate the right genes or block transcription of the wrong ones. In the background is a general picture of how selection acts on sexually reproducing organisms, a picture that recognizes the selectively disadvantageous effects of failing to suppress one set of genes (those associated with the distinctive reactions that occur in male somatic cells) and of failing to activate the genes in another set (those whose action is responsible for the distinctive reactions of female somatic cells). The functions of the specific genes identified by Kaulenas are understood in terms of the causal contributions they make in a complex process. There is no attempt to canvass the genetic variation in *Drosophila* populations or to argue that the specific alleles mentioned are somehow fitter than their rivals. The discussion takes for granted a particular type of selection pressure — thus adopting the environment-centered perspective on evolution — and considers only the causal interactions that result in a response to that selection pressure. The causal analysis is vividly presented in a diagram (reproduced in figure 7.1), which shows the kinship between the type of mechanistic approach adopted in this study and the analysis of complex systems designed by human beings. Selection furnishes a context in which the overall design is considered, and, within that context, the physiologist tries to understand how the system works,

I offer these four examples as paradigmatic of two very different types of biological practice offering ascriptions of function. I hope that it is evident how introducing the strong etiological conception within the last two would distort the character of the achievement, rendering it vulnerable to skeptical worries about the operation of selection that are in fact quite irrelevant. By the same token, it is impossible to appreciate the line of argument offered in the explicitly evolutionary studies without recognizing the stringent requirements that the strong etiological conception imposes. There are undoubtedly many instances in which the notion of function intended is far less clear. I believe that keeping our attention focused on paradigms will be valuable in the work of disambiguation.

VII

Philosophical discussions of function have tended to pit different analyses and different intuitions against one another without noting the pluralism inherent in biological practice.[30] On the account I have offered here, there is indeed a unity in the concept of function, expressed in the connection between function and design, but the sources of design are at least twofold and their relation to the bearers of function may be more or less direct. This means, I believe, that the insights of the main competitors, Wright's etiological approach and Cummins's account of func-

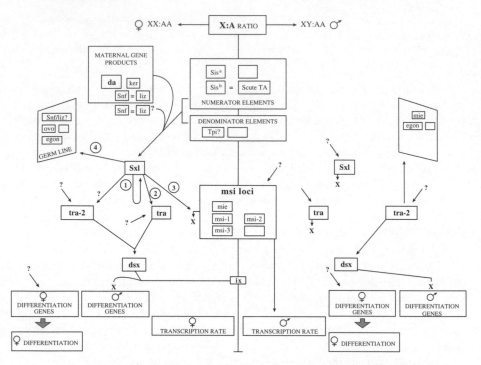

FIGURE 7.1 Diagram illustrating the interrelationships of the genes involved in the control of sexual differentiation and dosage compensation in *Drosophila*. From M. Kaulenas, *Insect Accessory Reproductive Structures: Function, Structure and Development* (New York: Springer, 1992), 18.

tional analysis, can be accommodated (and, as the discussion in section 4 indicates, variants of the etiological approach can also be given their due).

The result is a general account of functions that covers both artifacts and organisms. I believe that it can also be elaborated to cover the apparently mixed case of functional ascriptions to social and cultural entities, in which both explicit intentions and processes of cultural selection may act together as sources of design. But working out the details of such impure cases must await another occasion.

Notes

I am extremely grateful to the Office of Graduate Studies and Research at the University of California–San Diego for research support, and to Bruce Glymour for research assistance. My thinking about functional attributions in biology has been greatly aided by numerous conversations with Peter Godfrey-Smith. Despite important residual differences, I have been much influenced by Godfrey-Smith's careful elaboration and resourceful defense of an etiological view of functions.

1. I shall sometimes identify the bearers of functions simply as "entities," sometimes, for stylistic variety, talk of traits, structures, organs, behaviors as having functions. I hope it will be obvious throughout that my usage is inclusive.

2. This aspect of Darwin's accomplishment is forcefully elaborated by Richard Dawkins in *The Blind Watchmaker* (London: Longmans, 1987). Although I have reservations about Dawkins's penchant for seeing adaptation almost everywhere in nature, I believe that he is quite correct to stress Darwin's idea of design without a designer.

3. The fact that the intentions of the Creator are in the remote background in much pre-Darwinian physiological work is one of the two factors that allow for continuity between pre-Darwinian physiology and the physiology of today. As I shall argue later, appeals to selection as a source of design are kept in the remote background in contemporary physiological discussions.

4. Richard Lewontin, "Organism and Environment" (manuscript), and Lewontin and Richard Levins, *The Dialectical Biologist* (Cambridge, Mass.: Harvard University Press, 1987).

5. Wright, "Functions," *Philosophical Review* 82 (1973): 139–168. For further elaboration, see Ruth Millikan, *Language, Thought, and Other Biological Categories* (Cambridge, Mass.: 1984); Karen Neander "Functions as Selected Effects: The Conceptual Analyst's Defense," *Philosophy of Science* 58, 168–184; and Peter Godfrey-Smith, "A Modern History Theory of Functions," *Noûs* 28 (1994): 344–362.

6. This example stems from Christopher Boorse, "Wright on Functions" *Philosophical Review* 85 (1976): 70–86.

7. This way of evading the trouble is due to Millikan, *Language*.

8. These issues are broached by Godfrey-Smith in "A Modern History Theory of Functions." He and I are in broad agreement about questions of timing and diverge in our approaches to the second cluster of questions.

9. Here, and in the ensuing discussion, I permit myself an obvious shorthand. In speaking of the origination of an entity in an organism I do not, of course, mean to refer to the mutational and developmental history that lies behind the emergence of the entity in an individual organism but in the process that culminates in the initial fixation of that entity in members of the population. I hope that this abbreviatory style will not cause confusions.

10. See N. Tinbergen, "On War and Peace in Animals and Man," *Science* 160 (1968): 1411–1418.

11. S. J. Gould and E. Vrba, "Exaptation—A Missing Concept in the Science of Form," *Paleobiology* 8 (1982): 4–15.

12. E. Mayr, "The Emergence of Evolutionary Novelties," in Mayr, *Evolution and the Diversity of Life* (Cambridge, Mass.: Harvard University Press, 1976).

13. The point that biologists often ignore in practice the strictures on adaptationist claims that they recognize in theory is very clearly expressed in S. J. Gould and R. Lewontin, "The Spandrels of San Marco and the Panglossian Paradigm: A Critique of the Adaptationist Programme," *Proceedings of the Royal Society* B 205 (1979): 581–598.

14. J. Bigelow and R. Pargetter, "Functions," *Journal of Philosophy* 84 (1987): 181–196.

15. See his "Functions." For my own commitments to a similar view, see Philip Kitcher, "Why Not the Best?" in John Dupr, *The Latest on the Best: Essays on Evolution and Optimality* (Cambridge, Mass.: MIT Press, 1988), and Kitcher, "Developmental Decomposition and the Future of Human Behavioral Ecology," *Philosophy of Science* 57 (1990): 96–117 (reprinted as chapter 14 of this volume).

16. See, for example, the discussion of Geoffrey Parker's ingenious and sophisticated work on copulation time in male dungflies in Kitcher, *Vaulting Ambition: Sociobiology and the Quest for Human Nature* (Cambridge, Mass.: MIT Press, 1985), chapter 5.

17. See the examples given in section 6.

18. Robert Cummins, "Functional Analysis," *Journal of Philosophy* 72 (1975): 741–765.

19. T. Christenson and P. Wenzl, "Egg-Laying of the Golden Silk Spider, *Nephila clavipes* L. (Araneae, Araneidae): Functional Analysis of the Egg Sac," *Animal Behaviour* 28 (1980): 1110–1118.

20. I should note here that spiderlings typically overwinter in the egg sac, so that the period of a few weeks represents a fall only a *short* time before the usual time of emergence. Thus the successful instances are those in which the normal course of development is only slightly perturbed.

21. Christenson and Wenzl, "Egg Laying" 1114.

22. Ibid., 1110.

23. Ibid., 1115.

24. T. Clutton-Brock and S. Albon, "The Roaring of Red Deer and the Evolution of Honest Advertisement," *Behaviour* 69 (1979): 145–168.

25. Ibid., 145.

26. Ibid., 165.

27. J. McFarlane, "Nutrition and Digestive Organs," in M. Blum, *Fundamentals of Insect Physiology* (Chichester: Wiley, 1985), 59–90. The quoted passage is from p. 64.

28. M. Kaulenas, *Insect Accessory Reproductive Structures: Function, Structure, and Development* (New York: Academic Press, 1992), section 2.3, "Genetic Control of Sexual Differentiation."

29. Ibid., 17.

30. As I have argued elsewhere, biological practice is pluralistic in its employment of concepts of gene and species and in its identification of units of selection. See Kitcher, "Genes," *British Journal for the Philosophy of Science* 33 (1982): 337–359; Kitcher, "Species," *Philosophy of Science* 51 (1984): 308–333; and Kim Sterelny and Kitcher, "The Return of the Gene," *Journal of Philosophy* 85 (1988): 335–358, reprinted as chapters 4 and 6 of this book.

8

The Evolution of Human
Altruism (1993)

The problem of altruism has loomed large in evolutionary biology ever since Charles Darwin. How do tendencies to kindly, even self-sacrificial, behavior evolve in an unkind, Darwinian world? More exactly, conceiving altruistic behavior, as biologists do, as behavior that promotes the fitness of another organism at costs in fitness to the agent, how can propensities to engage in such behavior originate and be maintained under natural selection? There are, of course, two problems here: we must not only explain how altruistic tendencies are sustained once they are prevalent but also how they spread when they are rare.

During the past thirty years, standard answers to these problems have emerged. Some actions that are altruistic in the biologists' sense involve benefits conferred upon kin. In these instances, natural selection favors propensities that will promote the spread of underlying genes irrespective of the bodies in which those genes are housed. If the relationship is sufficiently close, then the probability that the beneficiary will share the relevant allele and will receive a large enough gain in fitness may offset the fitness costs to the agent.[1] When the recipient is not a relative, the agent's short-term fitness losses may be made up in long-term gains, provided that the altruistic action is reciprocated. In recent years, intense exploration of the evolutionary fortunes of strategies for playing iterated prisoner's dilemma has shown how a particular altruistic strategy, *Tit-for-Tat*, might be maintained under natural selection.[2] Of course, both proposed selection regimes may be combined in cases where an organism's behavior to a relative exceeds the demands of kinship and where the extra cost is made up through reciprocity.

So far, so good, at least for biology. But is this all there is to understanding the evolution of what is commonly called altruism, those forms of human behavior which most inspire admiration (and which, though they may initially be identified in our own species, might also be found in other species)? Some biologists have thought so, claiming that the evolutionary explanations I have mentioned account for *all* altruistic behavior, revealing the human actions we prize to be "ultimately selfish."[3] It is not hard to rebut the debunking arguments, showing that there are important differences between the biologists' conception of altruism and the everyday notion.[4] But simply identifying argumentative flaws leaves a complex of problems: What is *human* altruism—that sort of altruism we take to have moral significance—and how might *it* evolve?[5]

The goal of this paper is to begin a line of solution to this complex of problems. I shall start by modifying the currently standard treatment of reciprocal altruism, developing a biological model that might provide a more convincing account of the origin and maintenance of altruistic behavior toward nonrelatives in cognitively sophisticated organisms.[6] Section 2 offers an account of the special features of human altruism in terms of the psychological mechanisms that underlie behavior, and section 3 shows how, given the model of section 1, a specific type of mechanism might evolve. The final section is devoted to surveying various ways of making the problem more realistic and of refining my analysis.

I

Orthodox treatments of the evolution of reciprocal altruism suppose that organisms repeatedly face situations that can be modeled as prisoner's dilemmas played against the same opponent a large, but indefinite, number of times. Standard prisoner's dilemma is a *compulsory* game: players have no choice about the partner/opponent and they have no way of opting out. I suggest that for many interactions among cognitively sophisticated organisms, both features might be modified. An *optional* game is a situation in which a player has the opportunity to engage in interaction with others, with the possibility of signaling willingness to play with some and unwillingness to play with others, but also has a possibility of opting out and foreswearing social interaction altogether. I shall explore the evolution of altruism in the context of repeated optional games (in this chapter, in the context of repeated optional prisoner's dilemmas).

Imagine a group of primates, a baboon troop, for example. Each animal needs to remove parasites that become concealed in its fur. One way for the job to be done is by mutual grooming, but any such interaction takes the form of a standard prisoner's dilemma. If both partners cooperate, doing a thorough job, both will receive a high payoff in that both animals will be virtually parasite-free. If one receives thorough grooming but does a quick, sloppy job in return, then that animal will do even better, having well-groomed fur and extra time to invest in other activities. If both groom quickly and sloppily both will fare poorly, but they will be better off than animals who are exploited, who lavish time on others without receiving benefits in return. But, of course, any animal in the troop could elect to groom itself, receiving a benefit greater than that provided by a sloppy grooming from another but less than that yielded by a thorough grooming from a partner (others can clean areas that the organism itself cannot reach).

I have obviously simplified a small part of primate life, but, granted the simplifications, the situation has a structure we can represent as follows:

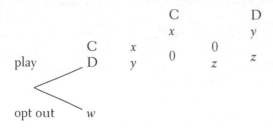

where $y > x > w > z > 0$ (and $y \leq x + z$), and C and D represent the strategies *cooperate* and *defect*.

I suggest that situations with this structure might occur repeatedly in the lives of savannah-dwelling primates (including our hominid ancestors). In seeking food, an animal might have the option of foraging by itself (thus restricting itself to prey of lesser quality) or might enter into a hunting partnership with another (allowing for the possibility of obtaining higher quality prey but laying it open to the possibility of exploitation). Similarly, males might try to win copulations with females by themselves or might enter into partnerships with another male to displace a third male from the vicinity of an estrous female. By the same token, females might try to choose mates by acting by themselves or by forming coalitions to drive away unwanted males. In each of these situations, the interactive option can be regarded as a prisoner's dilemma, and there is a noninteractive behavior ("opting out") that yields a payoff intermediate between DD and CC.

I shall suppose that opportunities for optional games might have arisen repeatedly for our hominid ancestors. Imagine, then, that we have a population of size N (which I shall assume to be relatively small, fifty at most) and that organisms in the population have, on average, M opportunities to play the same game, whose structure is given above. In accordance with the usual assumptions of evolutionary game theory, I shall suppose that the payoffs are in units of evolutionary fitness.[7] I shall also make strong assumptions about the epistemic abilities of the organisms: each is able to represent what others have done in past interactions with it. Divide strategies for playing the repeated optional game into those which are discriminating and those which are undiscriminating. Discriminating strategies adjust behavior to the past performances of other organisms. Undiscriminating strategies always direct the same behavior.

Organisms playing discriminating strategies recognize some organisms in the population as potential partners and classify others as nonadmissible. There is no coercion. Interaction only takes place if each player counts the other as a potential partner. Here is a fantasy about how coordination is accomplished. Imagine that the opportunities for playing arise for all members simultaneously. Those who are willing to interact with others signal their willingness, pairing off in much the way that children do when they choose partners for games. A strategy for a player in the repeated optional game is specified by identifying the conditions on potential partners and the action to be performed in interaction with each partner.

Consider the following simple strategies. *Discriminating altruists* (DAs) are prepared to play with any organism that has never defected on them, and, when they play, they always cooperate. *Willing defectors* (WDs) are always prepared to play, and always defect. *Solos* (SOs) always opt out. *Selective defectors* (SDs) are prepared to play with any organism that has never defected on them, and always defect.

1. DA invades a population fixed for SO. For, in a population of SOs (individuals whose lives are "*solitary*, poor, nasty, brutish, and short"), a single mutant DA is indistinguishable from other members of the population. (Since there are no other organisms prepared to interact, DA is forced to opt out.) Thus, DA can drift into the population. But, as soon as there are

two DA mutants, they recognize one another, and interact successfully for a lifetime payoff of Mx, which is, ex hypothesi, greater than Mw.

2. SO invades a population of SDs. For the payoff to SD will be $rz + (M - r)w$, where r is the expected number of occasions on which an SD encounters a hitherto unknown SD, so that $r > 0$. The payoff to SO is Mw, and, since $w > z$, this payoff is greater than the payoff to SD.

3. SD cannot invade a population of DAs. For, in a population consisting predominantly of DAs, with a small number, n, of SDs, the worst that can befall a DA is to be exploited by all the SDs and interact with DAs on other occasions. So the payoff to DA is bounded below by $(M - n)x$. The best that can happen to an SD is to exploit all the DAs, after which, debarred from further interaction with the DAs, its maximal payoff comes from opting out. So the payoff to SD is bounded above by $(N - n)y + (M - N + n)w$. Provided that M is large in relation to N and that x is not too close to w, SD cannot invade. The exact condition is:

$$M(x - w) > N(y - w) + n(x + w - y)$$

and, since n is small, this is effectively

$$M/N > (y - w)/(x - w)$$

4. If there are two mutant DAs, then DA can invade a population of SDs. For the analysis is exactly that given above in (3) except that now we assume that n is no longer small but $N - 2$. So DA spreads if

$$M(x - w) > (N - 2)x + 2(y - w)$$

which is approximately

$$M/N > x/(x - w)$$

Taken together, (1)–(4) offer considerable encouragement for the possibility of the evolution of altruism (at least in the limited biological sense). Provided that there are genuine benefits from cooperation (i.e., $x - w$ is sufficiently large) and the number of opportunities for playing the optional game is large in relation to the population size (M is significantly bigger than N),[8] discriminating altruism can be sustained once it is prevalent.[9] Moreover, provided that we have a *pair* of DA mutants, discriminating altruism can invade a population of defectors. (4) shows this in the case of selective defectors, and willing defectors are even worse off. Now, it may be thought that I have cheated by allowing DA mutants to arise in pairs.[10] The point of (1) and (2) is to defuse this worry. If antisocial populations (SDs or WDs) are doomed to collapse into asocial populations (SOs), and if SOs allow mutant DAs to enter and, with nontrivial probability, to leave a pair of DA offspring, then discriminating altruism can become directly established from asociality and indirectly from antisociality.

As has become abundantly clear in recent discussions, evolutionary origination of reciprocal altruism requires that altruists interact more with one another than they do with other members of the population.[11] There are various ways of achieving this nonrandom interaction: the population may be divided into family groups or spatially organized. The heart of my approach is to focus on organisms that have the cognitive capacities for recognizing others and reacting to their past behavior, and to base the nonrandom interactions on these capacities.[12] An obvious objection is that reciprocal altruism is apparently more widespread in nature than my analysis would allow, occurring among organisms that seem to lack the cognitive capacities upon which I have relied. One way to respond would be to allow for different types of propensities for altruistic behavior, so that, in cognitively sophisticated organisms, the disposition to discriminating altruism (in my sense) might overlay a more primitive propensity to play *Tit-for-Tat* in repeated compulsory games. But, as recent reviews of alleged cases of reciprocal altruism have made abundantly clear, there are few, if any, instances of reciprocal altruism among organisms lacking the abilities central to the account I have given (roughly a capacity to identify other individuals and to respond to actions that we would classify as defections).[13]

The model I have outlined cries out for refinement, modification, and extension in many ways, some of which I shall mention in the final section. For the present, I shall briefly respond to one apparent difficulty. I have effectively assumed that each DA recognizes each other DA as a potential partner, and have concluded that, on any occasion on which an optional game arises, all the DAs will engage in pairwise interactions. This is easily subverted if there are preferences for certain partners or errors in identifying the performances of others (both issues I shall take up briefly in the last section), but it also fails as a simple matter of arithmetic if the number of DAs is odd. If there are $2k + 1$ DAs in the population, then the lower bound on the payoff to a DA should be adjusted to $2k/2k + 1$ the amount given in my analysis. The worst case is obviously when $k = 1$, when the correct value of the payoff is $\frac{2}{3}$ that which I have given, but, even in such cases, a sufficient frequency of decision opportunities (M sufficiently large) enables discriminating altruism to become established.

II

The concept of an optional game and the approach to repeated optional games begun in the last section can be applied to understand the evolution of altruism in the limited biological sense as well as to study the behavior of calculating economic agents. But, as I emphasized at the beginning, my aim is to understand ways in which *human* altruism might evolve. The next task is to develop a representation of the type of altruism that is of most interest to philosophy (and to ordinary moral reflection).

Thomas Nagel[14] is surely right to understand altruism as involving "not abject self-sacrifice, but merely a willingness to act in consideration of the interests of other persons, without the need of ulterior motives" (ibid., 79). One way to develop this characterization is to think of an altruist as an individual with a psychological disposition that involves modification of desires that might otherwise lead to action so that the desires that ultimately cause action take into account the interests of others. Conceive of human agents (and perhaps other organisms) as subjects with incompatible desires, so that *different* payoff matrices may represent a subject's valuations.

Genuine altruists must feel the tug of selfish actions, so that, in some sense, a purely selfish payoff matrix represents their desires. But altruists are disposed to transform this payoff matrix to a different matrix, one which takes into account the well-being of another (or of others), and these transformed matrices move them to action.[15]

The idea of a divided self in which valuation and motivation come apart is not new. Discussions of human free will and of weakness of the will recognize the possibility of conflicting desires: the recovering alcoholic (to cite a hackneyed example) both wants and does not want the proffered drink.[16] I suggest that altruism involves analogous problems to those which arise in connection with *akrasia* and compulsion: for unless the altruist is genuinely moved to value what is given up in performing the selfish action *and* genuinely wants to help the other(s), there is no altruism of the inspiring—human—kind. The failures of attempts to extrapolate biological treatments of altruism are, I believe, directly attributable to the neglect of this point.[17]

I propose that we think of decision situations as involving *two* payoff matrices. The *primary* matrix represents the agent's selfish desires, those desires which would have moved it to action if there had been no effects on the well-being of another.[18] The *secondary* matrix is obtained by transforming the primary matrix by taking into account the well-being of another (or, possibly, of others). Completely selfish agents may perform a degenerate transform so that primary and secondary matrices are identical. But, in any case, the agent's actions are determined by the structure of the secondary matrix.

Suppose that the primary matrix for a game between A and B is as follows

$$
\begin{array}{cc}
b_{11} & b_{12} \\
a_{11} & a_{12} \\
b_{21} & b_{22} \\
a_{21} & a_{22}
\end{array}
$$

I assume that A and B are completely aware of the values that the other attaches to particular outcomes. A's secondary matrix

$$
\begin{array}{cc}
a^*_{11} & a^*_{12} \\
a^*_{21} & a^*_{22}
\end{array}
$$

is formed by setting $a^*_{ij} = \theta a_{ij} + (1 - \theta)b_{ij}$. Here θ represents the weight A gives to A's own interests, $1 - \theta$ the weight given to B's. Based on A's assessment of B, A may also assign subjective values to B.[19]

Complete selfishness corresponds to $\theta = 1$, complete self-denial to $\theta = 0$, and "golden-rule" altruism (treating the other as oneself) to $\theta = 0.5$. A golden-rule altruist would transform the prisoner's dilemma matrix

to

9   5
5   1

and, if the game is compulsory, the choice is completely determinate (cooperating dominates).[20]

I have said that action is determined by the desires represented in the secondary matrix. I have not explained, however, how payoffs to agents are actually determined. From the evolutionary perspective, what counts is reproductive success, and, the *objective payoff matrix* will register the consequences of combinations of actions in terms of units of fitness. The evolution of human altruism is only problematic when the entries in the primary matrix are well-correlated with those in the objective matrix, for, when this occurs, basing action on a different matrix will normally reduce fitness.[21] I want to illustrate my way of representing human altruism by showing how some types of altruistic disposition might evolve quite easily.

Suppose that a parent has acquired some item of food and that there are two options, to devour it whole or to share it with one of its young. There is a selfish desire to devour the food which we can represent by the matrix

Devour   10   (0)
Share   7   (9)

where the values to the child are given in parentheses. (Since the child has no way of affecting the outcome, the organisms are not playing a game.) The secondary matrix takes the form

Devour 10θ
Share $9 - 2\theta$

so that when $\theta < \frac{3}{4}$ (the parent is modestly altruistic), the parent will share. Suppose now that the objective matrix, reflecting the inclusive fitness payoffs, takes the form

Devour   0.7
Share   1.0

Then it is easy to recognize that altruistic tendencies ($\theta < \frac{3}{4}$) are favored by selection.[22]

But how can this be altruism, real altruism, if the organism ends up gaining by its action? The question is as natural as it is misconceived. The alleged "gains," in terms of spread of genes, are outcomes that all but a minute fragment of cognitively sophisticated organisms are unable to represent and, even for those organisms who can represent the outcomes, only the volitionally disordered would be moved to action by the representation. What is important is that the organism fights desires that in the absence of effects on others, would have led unproblematically to action, and that the desires that cause behavior are formed by recognizing the consequences for another's welfare. We are inclined to retract our admiration for an apparently altruistic act when we suspect that the agent might have seen forthcoming benefits. But there is no reason to take a similar stance when we are confident that the causal explanation of the action involves recognition and response to the needs of another.[23]

On my account, there is no essential connection between human altruism and reduction of evolutionary fitness, although altruists typically lose something that, at some level, they value (and that is represented in the primary matrix). I want explicitly, however, to disavow any claim to the effect that this analysis of altruism is complete.[24] To see its shortcomings, we must recognize that altruists often recognize how potentially altruistic actions may affect those who are altruistically disposed toward them: in contemplating sacrifices toward others, we are sometimes moved by the thought that their concerns for us will be ignored if we act in the envisaged way. What I have so far described is *paternalistic* altruism: the agent modifies the primary matrix to take into account the other's entries in the *primary* matrix. *Mutualistic* altruists would recognize that those with whom they interact also transform the primary matrix, and would be disposed to produce a *tertiary* matrix based on taking into account the other's interests in the *secondary* matrix. From the matrix whose entries are a_{ij}, b_{ij}, A would construct the secondary matrix, a_{ij}^*, b_{ij}^* in the fashion described above, and the tertiary matrix a_{ij}^{**}, b_{ij}^{**} such that

$$a_{ij}^{**} = \theta a_{ij}^* + (1-\theta)b_{ij}^* = \theta[\theta a_{ij} + (1-\theta)b_{ij}] + (1-\theta)[\theta' b_{ij} + (1-\theta')a_{ij}]$$

where θ' is the value assigned by A in connection with B's transformation of the primary matrix. How does the process terminate and so lead to action? Perhaps the agent simply stops transforming matrices at some arbitrary point. Or perhaps there is convergence to an equilibrium value. It is not hard to show that the latter will occur if and only if $\theta + \theta' = 1$, and transformation beyond the secondary matrix makes no difference. So, in the special case in which both players are golden-rule altruists ($\theta = \theta' = 0.5$) and recognize one another as such, mutualistic and paternalistic altruism coincide.

There are further complexities that must eventually be explored. Sometimes the recognition that an outcome has been brought about by altruistic behavior on the part of two individuals can confer, for each, special value on the outcome.[25] Perhaps the primary matrix is sometimes transformed by taking into account the mutual sacrifices that both parties have made to attempt to satisfy the desires of the other, so that an outcome that might initially seem to have the lowest value eventually is assigned the highest value. For the rest of this essay, however, I shall operate with the representation of the simplest form of altruism, paternalistic altruism, seeking to understand how altruistic propensities of this elementary kind might evolve.

III

Let us now return to interactions among unrelated individuals, and to the treatment of reciprocation in terms of optional games, developed in section I. Altruists of different degrees (measured by different values of θ) can be differentiated by the PD games in which they find cooperation dominant. Consider the spectrum of optional PDs, for which the objective matrix and the primary matrix is

$$\begin{array}{c c c} & x & y \\ x & 0 & \\ & 0 & z \\ y & z & \end{array}$$

and the payoff for opting out w. $y = x + z = 10$. For G_i $(i = 1 - 5)$, $z = i$, $w = i + 1$. Consider four types, for whom $\theta = 1, 0.8, 0.5,$ and 0.3, respectively. The associated secondary matrices are:

θ	G₁		G₂		G₃		G₄		G₅	
1	9	0	8	0	7	0	6	0	5	0
	10	1	10	2	10	3	10	4	10	5
0.8	9	2	8	2	7	2	6	2	5	2
	8	1	8	2	8	3	8	4	8	5
0.5	9	5	8	5	7	5	6	5	5	5
	5	1	5	2	5	3	5	4	5	5
0.3	9	7	8	7	7	7	6	7	5	7
	3	1	3	2	3	3	3	4	3	5

The values for opting out remain the same: 2 for G_1, 3 for G_2, 4 for G_3, 5 for G_4, and 6 for G_5. Note that the decision making of the golden-rule altruist is completely determinate for all the games: playing G_1–G_4 and cooperating is preferable either to defecting or to opting out *whatever the other does*; opting out is preferable in G_5 *whatever the other does*. For the less altruistic, $\theta = 0.8$, defecting is preferred to cooperating for G_3–G_5. The hyperaltruist, $\theta = 0.3$, would prefer to cooperate—and be defected on—in G_5.

Now, let us enrich our earlier discussion of optional games by considering the psychological mechanisms and allowing for situations in which the payoff matrices differ. Earlier, we considered two generic strategies for responding to repeated optional game situations: the undiscriminating and the discriminating. The first operates by having a fixed value of θ and using this to guide action in the optional game. The other adjusts values of θ, depending on the past behavior of a candidate partner.

We can now consider various psychological types, which refine our original notion of a discriminating altruist. $DA_{0.8}$ initially sets $\theta = 0.8$ for anyone who has never defected on it, but sets $\theta = 1$ for those who have defected once. $DA_{0.5}$ initially sets $\theta = 0.5$ for those who have never defected, while also setting $\theta = 1$ for defectors. Both strategies will behave like our original DAs with respect to G_1, but they will respond differently when confronted by G_3 or G_4.

Distinguish between games of our normal form those which are *tempting* and those which are *refusable*. A game is tempting only if $x > w$—that is, if the cooperative payoff exceeds what one could expect to achieve by opting out; otherwise, it is refusable. For $DA_{0.5}$, the subjective value matrix always takes the form

$$\begin{array}{c c} x & 5 \\ 5 & z \end{array}$$

and, since $x + z = 10$, $x > z$, cooperating will always dominate defecting. Opting out will also be dominated if $w < 5$ ($z < 4$). We can differentiate two further types: $DA_{0.5C}$ and $DA_{0.5R}$. The former opts out if $x > w > 5$, the latter plays and cooperates. Thus, $DA_{0.5R}$ plays all and only tempting games, provided that there is a potential partner who has not defected.

It should be clear that $DA_{0.5R}$ can invade a population of $DA_{0.5C}$ provided that two $DA_{0.5R}$ mutants arise together. For, with respect to some tempting games, the $DA_{0.5R}$ mutants achieve the payoff x while the $DA_{0.5C}$ only receive w.

Imagine that we have, as before, a population of size N, and that a spectrum of optional games $\{G_i\}$ of our normal form repeatedly arises for members of this population. Suppose that there are n games in the spectrum, that $x_n = z_n = 5$, and that they are ordered so that, if $i < j$, then $x_i > x_j$ (and, consequently, $z_i < z_j$). G_1–G_m are tempting, the other games are refusable. The expected number of times that an individual confronts G_i is f_i.

Consider two undiscriminating strategies, both of which fix $\theta = 1$. SOs always have a secondary payoff matrix of form

$$\begin{matrix} x & 0 \\ 10 & z \end{matrix}$$

but they prefer the sure thing w to the risk involved in aiming for 10 and possibly getting only z. WDs have the same secondary payoff matrix but prefer to take the risk (they always play and always defect). Thus, in obvious analogy to the notation already introduced, SO is UA_{1C} and WD is UA_{1R}. The SDs of our earlier analysis are DA_{1R}. We thus have a three-dimensional representation of strategies in optional PD games, where the dimensions are measured by (i) whether or not one assesses track records, (ii) the values assigned to θ, and (iii) how one responds to situations in which the preferred action in the game does not dominate opting out.

We can now give a more fine-grained analysis of the original problem. First, a population of SOs, WDs, or SDs is invadeable by any DA_{sC} if $0.5 \le s < 1$, subject to the proviso that $x_1 > 10s$ (this means that there is one game for which the cooperative payoff exceeds the secondary value of the maximum payoff) and that the f_i and $x_i - w_i$ are large enough. For imagine that two DA mutants DA_{sC} and $DA_{s'C}$ arise together, where $0.5 \le s' < s < 1$. Let G_k be the first game that DA_{sC} is not willing to play, and G_l be the first game that $DA_{s'C}$ is not willing to play ($k \le l$). The payoff to SOs is

$$\sum_1^n f_i w_i.$$

The payoff to WD is less than

$$10 + (f_1 - 2)z_1 + \sum_1^n f_i z_i$$

The payoff to SD is less than

$$10 + (f_1 - 2)w_1 + \sum_2^n f_i w_i$$

The payoff to DA_{rC} and to $DA_{r'C}$ is greater than

$$(f_1 - N + 2)x_1 + \sum_2^{k-1} f_i x_i + \sum_k^n f_i w_i$$

(assuming that $f_1 > N$; if this assumption does not hold, the algebra is more complicated, but a qualitatively similar conclusion can be reached). Both mutants have greater fitness than all the noncooperatives if

$$(f_1 - N + 2)x_1 + \sum_2^{k-1} f_i(x_i - w_i) > 10 + (f_1 - 2)w_1$$

A similar analysis readily shows that DA_{sC} is invadeable by $DA_{s'C}$ if $0.5 \leq s' < s$, provided that there are enough opportunities for playing the optional games. Moreover, $DA_{s'C}$ will sweep to fixation against DA_{sC}. Thus we can expect selection to move the population to $DA_{0.5C}$. As already noted, $DA_{0.5C}$ is invadeable by—and replaceable by—$DA_{0.5R}$.

$DA_{0.5R}$ is stable relative to the strategies so far considered: SO, WD, SD, and DA_s ($s > 0.5$). For, intuitively, the higher DA_ss do not play some tempting games, and thus lose the higher payoff x_i, settling for the return for opting out, w_i. Hence, we seem to have shown that selection will favor "golden rule" altruism of a discriminating kind (treating the other as oneself so long as one has no basis for thinking that the other will not do the same).

But might selection go further, introducing the hyperaltruists who initially set $\theta < 0.5$? Hyperaltruists behave like $DA_{0.5}$ with respect to tempting games. But, in refusable games, they will value most the outcome in which they cooperate and the other defects, regarding this as the only outcome preferable to opting out. There are various ways to understand hyperaltruistic behavior, but the highest payoffs come to those who are only prepared to play with another hyperaltruist once: they hope for the outcome $\langle C, D \rangle$, and, when this is not forthcoming, they henceforth opt out. Such extremely conservative hyperaltruists are at a slight selective disadvantage with respect to $DA_{0.5}$. Hyperaltruists who are prepared to play refusable games more frequently do worse. Those who pass up tempting games except with partners who defect on them do even worse. Hence hyperaltruism does not displace $DA_{0.5}$.

IV

The foregoing sections provide a representation of human altruism and show how a recognizable, if rather minimal, type of human altruism might evolve under natural selection. I conclude by explaining how the ideas so far introduced might be extended and refined.

Other Types of Altruism

The approach in terms of primary, secondary (and possibly higher-order) matrices needs to be developed to capture the more subtle forms of altruism portrayed by

historians and writers of fiction. It should also be linked more directly to a substantive psychological account, so that the "divided mind" approach to motivation
could be more clearly elaborated. But even within the relatively simple system of
representation I have used, it is possible to consider a broader set of altruistic strategies. The forms of DA are all unforgiving. In a group in which organisms can obtain
information about the performances of others in interactions with third parties,
however, it would be possible both to be even less merciful (never interact with
anyone who has ever defected on anybody)[26] or more flexible, allowing θ to decline
from 1 as a past defector is observed to cooperate with others.[27] The restricted space
of strategies considered in the present paper is analytically tractable. Investigating
all the possibilities for varying θ in response to the behavior of others almost certainly requires computer simulation.

Error

Two kinds of error need to be considered. A discriminating altruist might mistakenly choose to play with someone who had already defected on it. Alternatively, a
discriminating altruist might fail to recognize that there are perfectly good partners
available for interaction. If one assigns fixed probabilities to the two types of error,
then, provided that the number of decision opportunities is big enough relative to
the size of the population, it is not hard to show that fairly substantial error rates
can be tolerated. Yet, while this probabilistic treatment may be adequate as a way
of representing the first type of mistake, it seems to me highly unrealistic as an
approach to the second. For it assumes that each situation is treated de novo, and
that the organism has some small probability of failing to recognize all the potential partners in the population. A more adequate model would suppose that the
organism acts as if it had an internal chart on which the members of the population were divided into the good (those with whom it would be prepared to interact) and the not-so-good (those with whom it would refuse to interact). Lapses of
memory might cause a one-way flow away from the good side of the chart: once an
organism is crossed off the list of potential partners, it stays beyond the pale. If the
"mutation rate" at which lapses occur is sufficiently high, the altruist may quickly
move into a state in which it refuses to play with any other organism in the population. This effect can be mitigated, however, by the kinds of reputation effect discussed in the previous paragraph.

Variation in Partner Quality

Some organisms in the population may be more valuable as potential partners than
others. (So, for example, probability of success in cooperative hunting may depend
on the speed or strength of the prospective partner.) A simple way of modeling this
aspect of cooperative situations is to suppose that the population divides into two
groups, the strong and the weak. Weak-weak interactions receive the original
payoffs, weak-strong interactions receive the payoffs multiplied by a factor k ($k > 1$),
and strong-strong interactions receive the payoffs multiplied by k^2. It is easy to show,
given this representation, that the strong will interact preferentially with one

another. Interestingly, if the population is spatially (or otherwise) structured so that weak-strong interactions often represent the only available choices for prospective players, then it is possible for a population of discriminating altruists to be invaded by a pair of symbiotic mutants—strong organisms that occasionally defect on the weak, and weak organisms that tolerate a small amount of defection by strong partners.

Dynamics and Population Genetics

The analyses offered above are somewhat casual about possible ways in which either the order of entry of mutants or the underlying genetics might affect the outcome of a selection process. It is not hard to envisage scenarios in which, despite my results about the fitness relations, discriminating altruism fails to become established: if, in a population of Solos, selective defector mutants arise whenever discriminating altruist mutants arise, discriminating altruism will be unable to invade. Computer simulations of the dynamics, under various assumptions about mutation and the underlying genetic structure, would reveal how seriously we should take such possible subversions.

Optional Games and Compulsory Games

It is possible to mix together the Axelrod framework and that which I have developed here. Suppose that organisms face M_1 optional games (in which they can play only with those with whom they willingly interact) and M_2 compulsory games (in which they are forced to play with a specific partner/opponent). Preliminary analyses suggest that discriminating altruism can resist invasion by antisocial strategies provided that the number of optional games, M_1 is sufficiently high. Interestingly, if this condition is met, the size of the load of compulsory games, M_2 does not seem to matter. A more interesting, and more realistic, problem is to consider a spectrum of game situations ranging from the completely optional, through situations with restricted choice, to situations of compulsory interaction, and to take into account both variation in partner quality and the possibility of error. I do not yet know what the fate of discriminating altruism within this richer context would be.

Multiple Games

The most important, and difficult, extension of the approach begun here involves recognizing that the decision opportunities for organisms might involve many different types of game. Not only the objective matrices but also the number of players may vary.[28] Here is a general statement of the problem that an account like mine must ultimately address: given a population whose members have repeated opportunities for interactions of many different kinds, what types of psychological mechanisms, conceived along the lines of section 2, can we expect to be favored by natural selection? Ideally, the problem must be tackled in the light of the various complications I have mentioned in this section. At this point, there seems no alternative to wide-ranging and intricate computer simulations, and I must rest content with placing my idealized treatment in proper perspective and offering a formulation of the residual issues.

Notes

An earlier version of these ideas was presented at the 1992 meeting of the Society for Philosophy and Psychology. I am extremely grateful to Ronald de Sousa, who prepared insightful comments in record time. Subsequent versions were given as a Rothman Lecture in Cognitive Science at UCLA and at Philosophy Colloquia at the University of California/Berkeley and at UCSD. I am grateful to many friends and colleagues whose questions and comments have helped to shape the present paper: Richard Arneson, Michael Bratman, Patricia Churchland, Paul Churchland, Vincent Crawford, Michael Devitt, Rochel Gelman, Bruce Glymour, Peter Godfrey-Smith, Alison Gopnik, Patricia Kitcher, William Loomis, Sandra Mitchell, Michael Rothschild, Gila Sher, Brian Skyrms, Joel Sobel, Elliott Sober, Kyle Stanford, Barry Stroud, Charles E. Taylor, Stephen White, and David Sloan Wilson.

1. Here I simplify W. D. Hamilton's intricate treatment of inclusive fitness and the theory of kin selection to which it has given rise. For Hamilton's original work, see "The Genetical Evolution of Social Behavior" (most easily accessible in G. Williams, ed., *Group Selection* [Chicago: Aldine, 1971], 23–43). Introductory accounts are given in Richard Dawkins, *The Selfish Gene*, 2nd ed. (New York: Oxford University Press, 1991), and in Philip Kitcher, *Vaulting Ambition: Sociobiology and the Quest for Human Nature* (Cambridge, Mass.: MIT, 1985).

2. The idea of reciprocal altruism was first bruited by Robert Trivers in "The Evolution of Reciprocal Altruism" (repr. in T. Clutton-Brock and P. Harvey, eds., *Readings in Sociobiology* [San Francisco: Freeman, 1979], 189–226). Trivers's ideas were developed in the context of iterated prisoner's dilemma by Robert Axelrod and Hamilton, "The Evolution of Cooperation," *Science* 211: 1390–1396; and by Axelrod, *The Evolution of Cooperation* (New York: Basic, 1984). The work of Axelrod and others is most clearly successful in solving the problem of how altruistic behavior is *maintained*—although Axelrod and Hamilton outlined an account of how reciprocal altruism might initially spread from interactions among relatives. For an interesting recent suggestion about the origination of altruism in spatially structured populations, see M. A. Nowak and R. M. May, "Evolutionary Games and Spatial Chaos," *Nature* 359 (1992): 826–829.

3. See, e.g., E. O. Wilson, *On Human Nature* (Cambridge, Mass.: Harvard University Press, 1978), 155. Also David Barash, *The Whisperings Within* (London: Penguin, 1979), 135, 167.

4. See Kitcher, *Vaulting Ambition*, 396–406.

5. As I have already hinted, I take human altruism to be that admirable behavior which is most readily identified in our own species but which may also be present in the actions of other organisms. As will become apparent, I want to resist the suggestion that human altruism is restricted to *Homo sapiens*.

6. It will also become clear that this account bears on attempts to understand the emergence of cooperative behavior among rational egoists, and thus connects with projects in economics and in moral philosophy. I should emphasize, from the beginning, that the modal language of my formulations is deliberate: the task is to explain how certain tendencies to behavior *might* evolve. As I argue in *Vaulting Ambition* (pp. 74–75), it is crucial not to leap, without evidence, from possibility to actuality.

7. See John Maynard Smith, *Evolution and the Theory of Games* (New York: Cambridge, 1982). I should note that I am making the usual simplification of evolutionary game theory to the effect that "like begets like." Since the organisms with which I am concerned are not haploid, it will ultimately be necessary to show that there are models of the underlying genetics that will sustain my conclusions. For purposes of this essay, such genetic complications will be ignored.

8. Wilson has suggested to me that requiring that N be small and M large eliminates much of "the real-world parameter space." But the evolutionary conditions that concern me are precisely those which might have affected savannah-dwelling primates (and other mammals) who live in small groups (or whose *effective* populations are small), who are relatively long-lived, and who have relatively frequent opportunities for cooperation (food-sharing, grooming, foraging, may be useful almost every day). Thus I believe that the assumption that M is large in relation to N is innocuous.

9. Strictly, DA is no more stable than is TFT in the Axelrod-Hamilton model. For a population of DAs (or TFTs) can be invaded by cooperators, through drift, and this then provides an opportunity for antisocial strategies to increase in frequency. It should be noted, however, that my account makes this possibility less worrisome, since, as shown above, DA can reinvade antisocial populations. We can tentatively conclude that altruism can be expected to be present—and prevalent—*most* of the time. Firmer conclusions must await detailed simulations of possible scenarios.

10. The worry here would be exactly analogous to concerns raised about evolution by macromutation. As critics of Richard Goldschmidt's ideas have repeatedly emphasized, where does the "hopeful monster" find a mate? Similarly, a single mutant DA would be bereft of cooperative partners.

11. See Elliott Sober, "The Evolution of Altruism: Correlation, Cost, and Benefit," *Biology and Philosophy* 7 (1992): 177–187.

12. This is the core of attempts to solve formally similar problems, such as the rationality of cooperation among economic entrepreneurs or the rationality of cooperative behavior among Hobbesian egoists. Both Robert Frank ("If *Homo Economicus* Could Choose His Own Utility Function, Would He Want One with a Conscience?" *American Economic Review* 77 [1987]: 593–604); and David Gauthier (*Morals by Agreement* [New York: Oxford, 1986], chapter 6) resolve the difficulty of showing that cooperative behavior can benefit the agent by allowing the agent to identify likely partners. The general notion of an optional game, which I have sketched in this section, allows for systematic treatment and refinement of such proposals. In the version I have developed in the text, the recognition of others is based on identification of past performance, but there is no reason not to combine this with other signals of the types discussed especially by Frank.

13. For a thorough review of the main instances of reciprocal altruism, see the special edition of *Ethology and Sociobiology* 9, 2–4 (1988), and, in particular, the articles by Gerald Wilkinson, Craig Packer, and Robert Seyfarth and Dorothy Cheney. As the editors, Charles E. Taylor and Michael T. McGuire, note, "it is surprising that so few instances of reciprocity (humans excluded) have been satisfactorily documented" (p. 68). The most likely cases seem to be food sharing in vampire bats and coalition formation in baboons, both of which involve organisms that could engage in the kinds of recognition and response tasks demanded by my analysis. (There is now considerable evidence that primates can perform far more complex cognitive tasks than any I have presupposed: see Cheney and Seyfarth, *How Monkeys See the World* [Chicago: University of Chicago Press, 1990]). The one potentially challenging example, involving egg trading in hermaphroditic fish (of the family Serranidae), seems to me to involve a misrepresentation of the payoff structure in the game that the organisms are playing, in that the alleged "defect" strategy brings no benefits to the "defector."

(For a contrary view of the example, see the article by Eric Fischer in *Ethology and Sociobiology*.)

14. Thomas Nagel, *The Possibility of Altruism* (New York: Oxford University Press, 1970). Not all examples of extreme self-sacrifice are abject, however, and the account I offer will ultimately have to come to terms with those types of human action that strike us as especially noble. See the brief discussion in note 19.

15. This formulation adopts a rather Kantian approach to moral psychology: the moral agent overcomes baser desires in being moved to action by the valuation that takes into account the interests of others. I had originally hoped to remain neutral between this approach and a more Humean conception of moral agency, but to do so would require a somewhat different treatment of the conflicting desires from that adopted in the text. Instead of seeing the primary desires (matrix) as being present—and being conquered through the exercise of moral reason—we might see those desires as representing the values that the agent *would have attached to the situation if no other's interests had been involved*. When others are present and are affected, we can envisage that the agent forms *directly* the transformed matrix through some process of empathy. I am indebted to Barry Stroud for a valuable discussion of this point.

16. For lucid treatment, to which I am much indebted, see Gary Watson, "Free Agency," *Journal of Philosophy* 77, 8 (April 24, 1975): 205–220; and Watson, "Skepticism about Weakness of the Will," *Philosophical Review* 87 (1977): 316–339. Similar ideas are broached in a different context by Thomas Schelling, *Choice and Consequence* (Cambridge, Mass.: Harvard University Press, 1984); and the idea of different payoff matrices within the same subject is suggested by Howard Margolis, *Selfishness, Altruism and Rationality* (Chicago: University of Chicago Press, 1984), in a discussion of social duties. Watson traces the fundamental idea to Plato's separation of rational desire from the brute appetites, but that specific way of distinguishing multiple schemes of valuation will not serve my purposes here.

17. See the references in note 3.

18. A rough-and-ready procedure for the construction of the primary matrix is to ask what the organism would have valued if it had had the same set of possible outcomes in isolation from conspecifics.

19. After developing this approach to the representation of human altruism, I discovered that similar ideas about the transformation of matrices had been put to work in social psychology for the exploration of interpersonal relationships. A groundbreaking text in this area is Harold H. Kelley and John W. Thibaut's *Interpersonal Relations: A Theory of Interdependence* (New York: Wiley, 1978). The approach I suggest is not only akin to the account offered by Kelley and Thibaut but also concordant with other psychological discussions of altruism. See, in particular, Nancy Eisenberg, *Altruistic Emotion, Cognition, and Behavior* (Hillsdale, N.J.: Erlbaum, 1986); Eisenberg, ed., *The Development of Prosocial Behavior* (New York: Academic Press, 1982); and Valerian Derlega and Janusz Grzelak, eds., *Cooperation and Helping Behavior* (New York: Academic Press, 1982). It does not seem impossible that the abstract approach in terms of matrices may ultimately be given more psychological substance.

20. As Sobel pointed out to me, cooperation is not unambiguously recommended if the payoff for defection is increased sufficiently (to more than twice the cooperative payoff). But if this is done, we violate the conditions for prisoner's dilemma, and the appropriate *cooperative* solution should be that of coordinating a pattern of alternating defections. I shall not explore how my discriminating altruists might resolve this different problem.

21. Let me make explicit the point that organisms do not represent the genetic consequences of their actions and assign values to those consequences by measuring the prolifer-

ation of their genes. But it may well be that selection has favored the presence of desires—for certain types of food, certain types of cover, certain types of mates—whose satisfaction will typically work to spread our genes.

22. Sober has reached similar conclusions by a different route. See his "Did Evolution Make Us Psychological Egoists?" in Sober, *From a Biological Point of View* (Cambridge: Cambridge University Press, 1994), 8–27.

23. Nor is there any reason to underrate human altruism by recognizing that human propensities for responding to the well-being of others might be causally determined. The apparent force of this worry stems from neglect of the possibility of compatibilist analyses of human freedom (such as that developed by Watson in "Free Agency" or by Harry Frankfurt in "Freedom of the Will and the Concept of a Person," *Journal of Philosophy* 68 [January 14, 1971]: 6–20). For further discussion of the idea that human altruism is impossible because people lack the requisite type of freedom, see Kitcher, *Vaulting Ambition*, 404–417.

24. One obvious concern is that my simple weighting of interests does not take into account the *distribution* of payoffs. But altruists might not simply be concerned to maximize their joint well-being but to have a fair distribution. This could be achieved by setting $a^*_{ij} = \theta a_{ij} + (1 - \theta)b_{ij} - \tau(a_{ij} - b_{ij})^2$. Transformations of more complex forms that correspond to desire for fairness in interpersonal relations are considered by Kelley and Thibaut, *Interpersonal Relations*.

25. A wonderful example of this occurs in O. Henry's famous story "The Gift of the Magi." I was reminded of the story by Roger Bingham and am grateful to Sher for helpful discussion of it.

26. I originally set the problem up in precisely these terms. By doing so, the selective advantage for discriminating altruism is increased.

27. My discriminating altruists are akin to the *retaliators* who played in Axelrod's computer tournaments (see *The Evolution of Cooperation*). Although this strategy did fairly well within Axelrod's version of the problem (repeated *compulsory* games) it proved too unforgiving to win. My suggestions in the text that θ might adjust to the reputations others earn point toward the possibility of analogs for *Tit-for-Tat* (and its kin) within the framework of optional games.

28. Ultimately, the account of psychological mechanisms developed in section 2 must be extended to cope with multiplayer interactions. White has emphasized the importance of the many-person prisoner's dilemma (in a currently unpublished manuscript, "Constraints on an Evolutionary Explanation of Morality"). I hope that the account I have offered can be developed to accord with White's demands.

9

Evolution of Altruism in Optional and Compulsory Games (1995)

COAUTHORED WITH JOHN BATALI

1. Introduction

Contemporary biological discussions of the evolution of altruism define altruistic behavior as that which increases the fitness of another animal at a cost in fitness to the animal engaging in the behavior. In the past decades, study of the evolution of genetic dispositions to altruistic behavior in this sense has been advanced by considering two special instances: (i) cases in which the animal's own reproductive losses are made up through increases in the reproductive successes of kin, and (ii) cases in which the short-term losses of altruistic actions are made up through a system of reciprocation. Separating these instances is heuristically helpful, but artificial, since it is possible that interactions with relatives might involve short-term fitness losses that are made up through contributions from both sources.

Our concern is with cases of type (ii) and with a modification of what has become the standard way of dealing with such cases. Reciprocal altruism was originally introduced in a seminal paper by Robert Trivers.[1] Following the work of Axelrod and Hamilton,[2] reciprocal altruism has been explored by considering strategies for playing iterated Prisoner's Dilemma (henceforth PD). Initially it appeared that the strategy *Tit-for-Tat* would be evolutionarily stable,[3] but further investigation has shown that it is not so.[4] Indeed, the many discussions of the Axelrod-Hamilton approach have revealed unsuspected complexities in the selection of strategies for playing the iterated Prisoner's Dilemma.[5] Our aim in this essay is to suggest a modification of the Axelrod-Hamilton scenario and to explore the dynamics of the selection of altruistic strategies.

We begin by presenting a new version of the Prisoner's Dilemma in which players have the option of not participating in interactions. Analytical evaluations of strategies for this game are presented to show that populations playing the optional games will achieve states of high cooperation more reliably than will populations playing the standard, compulsory, Prisoner's Dilemma. We then present the results of computer simulations of these games that confirm the analytic results, and also illustrate the dynamics of strategy changes in different kinds of game.

1.1. Compulsory and Optional Prisoner's Dilemmas

In each round of the standard Prisoner's Dilemma, each of two players must choose one of two actions: C, (Cooperate) or D (Defect). After each player has chosen, a payoff for each is computed from a payoff matrix as follows:

	C	D
	R	T
C R.		S
	S	P
D T		P

with R = reward for mutual cooperation; T = "traitor" payoff to defector if other cooperates; S = "sucker" payoff to cooperator if other defects; P = payoff for mutual defection, where $T > R > P > S$ and $2R > T + S$.

This standard game is compulsory in the following sense: first, the players have no choice of partner but are forced to interact with the assigned individual; second, they have no choice but to play: there is no individual asocial behavior available to them. In focusing on the iterated Prisoner's Dilemma, Axelrod and Hamilton considered the evolution of altruism in a population of animals that is forced into social activity, so that the problem is posed in terms of the victory of cooperation over anti-social behavior. As we shall explain below, we believe that it may be more realistic to consider the evolution of altruism in situations in which the possibility of asociality is present, and that this will make a difference to the evolutionary scenarios.

Various authors have considered the possibility that animals facing repeated gamelike situations might either have an asocial option[6] or be able to discriminate partners,[7] but we believe that the concept of an optional game has not previously been clearly articulated.[8] In an optional game, individuals have the possibility of signaling willingness to play to other individuals in the population. When signals of mutual willingness are given, the individuals play the game together. An animal that is unwilling to play with any of the animals willing to play with it is forced to opt out, performing some asocial behavior. We allow that such asocial behavior might have some intermediate benefit to the player, which we represent by the payoff W. In the cases of interest, $T > R > W > P > S$—that is, the opting-out payoff is less advantageous to both players than is mutual cooperation, but is more advantageous to both players than is mutual defection. Optional games of this sort will be referred to as "fully optional" versions of the iterated Prisoner's Dilemma.

Intermediate between fully optional and compulsory games are "semi-optional" games. These are of two types. In one version, individuals have no choice of partners, but they do have the ability to opt out. In the other, they cannot opt out, but can signal willingness to play with particular partners. The latter type of game will be of no concern to us, and we shall henceforth use "semi-optional" game to refer only to those games in which there is the possibility of opting out, but no choice of partners. We hope that the distinction between compulsory, fully optional, and semi-optional games is now sufficiently clear, and proceed with an illustrative example.

1.2. An Optional-Game Model of Grooming

Mutual grooming among primates serves the function of removing parasites from the fur. An animal that grooms another incurs a cost, depending on the length of time taken up in grooming, an animal that is groomed receives a benefit, which we can also assume to be proportional to the amount of time spent in grooming. Assume that there are two possibilities for each animal: to groom for a long period (C) or to groom for a short period (D). Let the cost of grooming for a short period be c_0, the benefit from a short period of grooming be b_0, the cost of a long period of grooming be c_2, and the benefit from a long period of grooming be b_2. Then, in a single interaction between two animals, the payoff matrix will be

	C	D
C	$b_2 - c_2$ $\quad$ $b_2 - c_2$	$b_2 - c_0$ $\quad$ $b_0 - c_2$
D	$b_0 - c_2$ $\quad$ $b_2 - c_0$	$b_0 - c_0$ $\quad$ $b_0 - c_0$

Since $b_2 > b_0$ and $c_2 > c_0$, it is clear that $b_2 - c_0$ is the highest payoff and $b_0 - c_2$ is the lowest. We may reasonably assume that a more thorough grooming is worth the extra time spent, and thus that $b_2 - c_2 > b_0 - c_0$. Given this inequality, it follows that $2(b_2 - c_2) > b_2 - c_2 + b_0 - c_0 = (b_2 - c_0) + (b_0 - c_2)$. Thus the payoff matrix may be rewritten as:

	C	D
C	R $\quad$ R.	T $\quad$ S
D	S $\quad$ T	P $\quad$ P

with: $R = b_2 - c_2$; $T = b_2 - c_0$; $S = b_0 - c_2$; $P = b_0 - c_0$; and so $T > R > P > S$ and $2R > (T + S)$. These are the conditions for a compulsory Prisoner's Dilemma.

But our primates surely have another option—they can groom themselves. Presumably the payoff from self-grooming is less than one would receive from an earnest grooming job from another, but greater than the payoff from a more desultory performance. This can be modeled by assigning an intermediate cost c_1, and an intermediate benefit b_1 for self-grooming. The payoff from self-grooming is thus $b_1 - c_1$, and we will assume that $b_2 - c_2 > b_1 - c_1 > b_0 - c_0$. Given that primates can also signal willingness to engage in grooming interactions with some animals and to forswear grooming interactions with others, we can expect that grooming interactions among primates take the form of an optional Prisoner's Dilemma, where the opt-out payoff W is intermediate between R (the reward for mutual cooperation) and P (the penalty for mutual defection). To understand the evolution of mutual grooming under natural selection, we therefore need to consider the evolution of cooperative behavior in iterated optional Prisoner's Dilemmas.

Our illustration is loosely based on recent discussions of the social lives of primates.[9] Plainly, it would be possible to test our assumptions about the ordering of costs and benefits by assessing the contributions to fitness of various commitments of time and of various states of parasitic infestation. Until such testing is done, we can only suggest that it is plausible that this part of primate social life may be understood in terms of optional Prisoner's Dilemmas. We also think it likely that optional games will prove useful in understanding cooperative hunting, cooperative foraging, mate-seeking coalitions, systems of defense against predation, and cooperation among females in exercising mate choice. But in all these cases the promise of the approach we recommend must be assessed in the light of field studies.

2. Analytic Results for Simple Strategies

Consider the following very simple strategies for playing an iterated optional Prisoner's Dilemma with the standard payoffs T, R, P, S and the opt out payoff W:

Solo: always opt out;

Undiscriminating Altruist: always interact and always play C;

Discriminating Altruist: interact with any animal that is willing to interact with you provided that that animal has never previously defected on you, and cooperate in any such interaction, (or, if there are no such animals, opt out);

Undiscriminating Defector: always interact and always play D;

Discriminating Defector: interact with any animal that is willing to interact with you, provided that that animal has never previously defected on you, and defect in any such interaction (or, if there are no such animals, opt out).

We now consider, from a standard evolutionary game-theoretic perspective, the fitness relationships among some of these strategies under various conditions. Throughout we shall suppose that the population of animals with we are dealing has size N, and that the average number of occasions that an animal has to play the optional Prisoner's Dilemma during its lifetime is M. We assume that M is significantly larger than N, an assumption that we base on the idea that animals have frequent opportunities for playing the game and that they interact with a relatively small number of conspecifics. (Think of the number of times primates groom one another during their lifetimes and the relatively small sizes of the groups of conspecifics with which they interact.) Since our strategies involve the ability of animals to recognize one another, to recognize when another has defected on them, and to remember past defections, we are obviously supposing that our animals have substantial cognitive capacities. We are encouraged both by recent studies of the cognitive lives of primates[10] and by investigations of cooperative behavior in guppies.[11] However, these assumptions about the cognitive capacities of the animals concerned are plainly in tension with the "like begets like" assumption of evolutionary game theory.[12] For the moment, we shall simply suppose that "like begets like" is a good approximate rule of thumb. Later sections will explore the dynamics of the evolutionary game in a less simplistic way.

Result 1. Discriminating Defector Can Be Invaded by Solo

Suppose that the population contains $N - n$ Discriminating Defectors and n Solos ($n > 0$). The payoff to each Solo is MW (on each of the M occasions on which the opportunity arises, a Solo opts out for payoff W). The payoff to each Discriminating Defector is:

$$(N - n - 1)P + (M - N + n + 1)W,$$

since each Discriminating Defector plays with each other Discriminating Defector just once, each attempting to exploit the other and receiving payoff P; when all have been tested, all opt out. Since $W > P$, the payoff to Solos is greater than that to Discriminating Defectors. A population fixed for Discriminating Defector should be invaded by Solo, and Solo should sweep to fixation.

Result 1.1. Solo Invades Undiscriminating Defector

The payoff to Undiscriminating Defector is less than that for Discriminating Defector. The payoff to Solo is the same as in Result 1.

Result 2. Solo Can Be Invaded by Discriminating Altruist

In a population consisting entirely of Solos, a single mutant Discriminating Altruist is indistinguishable from the other members. If there are n (> 1) Discriminating Altruists in a population with $N - n$ Solos, then the payoff to each Discriminating Altruist is MR (provided that n is even) and $(n - 1)Mr/n + MW/n$ (when n is odd). (These provisions are needed since, with an odd number of Discriminating Altruists, if the decision opportunities arise for all simultaneously, arithmetical considerations dictate that one will have to opt out; the probability that any particular Discriminating Altruist is the unlucky one is $1/n$.) Given $n > 1$, it is trivial that the payoff to a Discriminating Altruist exceeds that to a Solo, for all values of n. Hence Discriminating Altruist can be expected to invade a population of Solos, and to sweep to fixation.

Result 2.1 When Both Discriminating Altruists and Discriminating Defectors Enter a Population of Solos, the Discriminating Altruists May Be Driven Out; but It Is Also Possible for Them to Enter, Given the Right Order of Mutations

If a population of Solos containing a single Discriminating Altruist mutant comes to have a single Discriminating Defector mutant, then the Discriminating Altruist mutant will be selected against. The condition for Discriminating Altruists to enter the population when Discriminating Defector mutations are also likely to arise is that the order of mutations be [Discriminating Altruist, Discriminating Altruist] rather than [*Discriminating Altruist, Discriminating Defector*]. If both mutations arise at the same frequency, we can expect that the probability that *Discriminating Altruists* will drift into a population of *Solos* will be $p/2$, where p is the probability that *Discriminating Altruists* would increase in frequency in a population of *Solos* without *Discriminating Defector* mutations (i.e. a population like that described in Result 2).

> Result 3. If the Reward for Cooperation Is Large in Relation to the Payoff for Opting Out, and if the Number of Decision Opportunities Is Large in Relation to Population Size, Discriminating Altruists Can Increase in Frequency Against Discriminating Defectors

Suppose that a population contains n *Discriminating Altruists* and $N - n$ *Discriminating Defectors* where $n > 1$. The worst case for a *Discriminating Altruist* is to be exploited by each *Discriminating Defector* and then to spend the rest of the decision opportunities cooperating with other *Discriminating Altruists* (or, occasionally, opting out if the number of *Discriminating Altruists* is odd). This means that the payoff to a *Discriminating Altruist* is bounded below by

$$(N - n)S + (M - N + n)[(n - 1)R/n + W/n].$$

The best a *Discriminating Defector* can expect to do is to exploit each of the *Discriminating Altruists* once, and opt out on the remaining occasions. Hence the payoff to a *Discriminating Defector* is bounded above by

$$nT + (M - n)W.$$

The condition for *Discriminating Altruists* to increase in frequency under selection is thus

$$(N - n)S + (M - N + n)[(n - 1)R/n + W/n] > nT + (M - n)W.$$

This reduces to

$$M/N > K + Hn/N,$$

where

$$K = [(n - 1)R + W - nS]/(n - 1)(R - W);$$

$$H = [n(T + S) - (n + 1)W - (n - 1)R]/(n - 1)(R - W).$$

When n is large (approximately N), the crucial condition for the maintenance of *Discriminating Altruists* is

$$M/N > (T - W)/(R - W),$$

which is clearly satisfied if the reward for cooperating is significantly larger than the payoff for opting out, the payoff to exploiters is not too big, and the number of decision opportunities is sufficiently large relative to the population size. Hence it will be possible for *Discriminating Altruists* to resist invasion by *Discriminating Defectors*. At the other extreme, when n is small, the worst case is given by

$n = 3$. Since $N \gg 3$, we can ignore the term in H, and approximate the inequality by

$$M/N > (2R + W - 3S)/2(R - W).$$

As before, if the reward for cooperation is large in relation to the payoff for opting out, and if the number of decision opportunities is large in relation to population size, *Discriminating Altruists* can increase in frequency against *Discriminating Defector*.

These results are encouraging, for they suggest that, contrary to our naïve expectations, it might be very hard for antisocial or asocial behavior to be evolutionarily sustainable. By Result 1, antisocial populations are likely to decay into states of asociality. By Result 2, asocial populations are likely to be invaded by *Discriminating Altruists*, and, given Result 3, *Discriminating Altruists* can resist invasion by *Discriminating Defectors*. The only problem for a population of *Discriminating Altruists* is that *Undiscriminating Altruists* can drift in unnoticed, and, once the population has a sufficient number of them it is ripe for invasion by *Discriminating Defector* (or even by *Undiscriminating Defector*). Nevertheless, the combination of Results 1 and 2, and also Result 3, show that *Discriminating Altruists* can stage a comeback. Our analysis reveals that, while altruism may not be stable, the absence of altruism is also unstable. Moreover, when fitness differences are marked, we can expect that populations will spend most of their time in states of high cooperation, with occasional crashes and brief recovery periods. As we shall see later, this optimistic expectation is confirmed by computer simulations.

We now briefly explore the consequences of supposing that interactions are not fully optional. Animals are paired at random, and can either play with their assigned partner or opt out. As before, *Solos* always opt out, *Discriminating Altruists* play if and only if the assigned partner has not previously defected on them (and opt out otherwise) and they cooperate when they play, *Discriminating Defectors* also play if and only if the assigned partner has not defected on them and they defect when they play.

Result 4. Discriminating Altruists Can Invade a Population Fixed for Solo

Initially *Discriminating Altruists* are indistinguishable from *Solos*. Once there are two (or more) *Discriminating Altruists* there is a nonzero probability that they will be paired, and, on such occasions, each will receive R, a payoff that exceeds the opt out payoff W. So the fitness of *Discriminating Altruists* can be written as $M(rR + (1 - r)W)$ where $r > 0$, which exceeds the payoff for *Solos* of MW.

Result 5. Solo Can Invade a Population Fixed for Discriminating Defector

In a population of *Discriminating Defectors*, the payoff to a *Discriminating Defector* will be $M(rP + (1 - r)W)$, where $r > 0$ (r is now the probability that two *Discriminating Defectors* who have never previously met are paired). The payoff to *Solo* is MW, and, since $W > P$, *Solo* has a selective advantage.

Result 6. Although Discriminating Altruists Can Still Increase in Frequency against Discriminating Defector, the Condition for Doing So in a Semioptional Game Is More Stringent Than in the Fully Optional Game

Suppose the population has n Discriminating Altruists and $(N - n)$ Discriminating Defectors. The expected number of encounters of a Discriminating Altruist with a Discriminating Defector is $M(N - n)/N$. The total payoff to a maximally unlucky Discriminating Altruist from these encounters is $(N - n)S + M(N - n)W/N$. The rest of the time Discriminating Altruists are paired with one another for total payoff MnR/N. The total payoff for Discriminating Altruists is thus bounded below by

$$(N - n)S + M[(N - n)W + nR]/N.$$

The total payoff for Discriminating Defector is bounded above by

$$(N - n)T + (M - N + n)W.$$

Discriminating Altruists will have a selective advantage provided that

$$Mn(R - W)/N > (N - n)(T - S - W).$$

When Discriminating Altruists are prevalent (n is close to N), this condition becomes

$$M/N > (T - S - W)/(R - W),$$

which is very similar to the condition in the fully-optional game. When Discriminating Altruists are rare ($n > 1$, but small in relation to N), increase of Discriminating Altruists requires that

$$2M/N^2 > (T - S - W)/(R - W).$$

This is far more exacting than the condition of Result 3, and it is thus only in special cases (M extremely large) that we could expect a small number of Discriminating Altruists to invade a population fixed for Discriminating Defector in the semioptional game.

From Results 4–6, we can expect that the dynamics of the evolution of cooperative behavior in semioptional games will resemble that in the fully-optional case, but that recovery from crashes is likely to be slower and mediated by the presence of Solos. Intuitively, the direct route from Discriminating Defector to Discriminating Altruist is now partially blocked, and, very frequently, only the trajectory from Discriminating Defector to Solo to Discriminating Altruist will be available.

3. Computational Simulations

The above analytical results are based on assuming that the populations are relatively simple in two ways: first, that the populations consist of individuals who use one of a small set of strategies; second, that the strategies are chosen from the set of simple strategies described in section 2. In order to investigate the properties of more heterogeneous populations, and of populations containing individuals following more complex strategies, we performed a number of computational simulations of populations of players who participate in the iterated Prisoner's Dilemma by following inherited strategies. Our simulation results support and expand upon the analytic results, and also illustrate how the genetic representation of strategies can influence the evolutionary dynamics of populations whose members deploy those strategies.

This section describes the algorithms used in the computational simulations. The next section presents the results of those simulations, and describes the dynamics of a few of the runs we performed.

In the simulations, the actions performed by each player are represented as a history sequence. The lengths of histories recorded in our simulations varied from 2 to 4. For example the sequence (C C) indicates that a player cooperated on both of the previous two rounds; the sequence (D C) indicates that a player defected two rounds ago, but cooperated the last round. The symbol N is used when the players haven't played as many rounds as the history records. Thus for a history of length 2, the sequence (N N) indicates that no rounds at all have been played; the sequence (N C) means that a single round was played, and the player cooperated.

Strategies are represented by pairing each possible history of opponent's actions with the action to make in response the next round. This pairing of a history with a response action will be called a "move." The following move represents the response of defecting if the opponent cooperated twice in a row:

((C C) D)

This move represents the action of cooperating in the first round:

((N N) C)

Given a specific history length, a complete strategy contains a move for each possible history sequence of that length. For example this strategy represents the *Tit-for-Tat* strategy in which two steps of history are recorded:

((N N)   C)
((N C)   C)
((N D)   D)
((C C)   C)
((C D)   D)
((D C)   C)
((D D)   D)

In this strategy, the player begins by cooperating and then responds with whatever its opponent did the last round.

A sequence of rounds between two players is simulated by using the strategies of the two players to determine their moves for each round, depending on what the other player did the last rounds. Each player receives an increment to a "fitness" value according to this payoff schedule in table 9.1. In a generation, each player plays against each other player in the population some number of rounds.

TABLE 9.1

Payoff	Explanation	Value
T	Defect if other cooperates	7
R	Both cooperate	5
W	Opt out	3
P	Both defect	2
S	Cooperate if other defects	0

For example consider a simulation of the compulsory game in which one step of history is recorded. Two players with the following strategies are chosen to play the game:

```
((N)  C)    ((N)  D)
((C)  C)    ((C)  D)
((D)  D)    ((D)  C)
Player one    Player two
```

The players begin their interactions with fitness values of 0. In the first round both of the histories are (N), so player one will cooperate and player two will defect. Player one's fitness will remain 0, while player two's fitness will be set to 7. In the second round, player one consults its strategy with the history (D) and defects. Player two uses the history (C) and also defects. Both players fitness values are incremented by 2. On the third round, player one defects again, but player two cooperates. Hence player one receives 7 and player two receives 0. The next round player one cooperates again and so does player two, so they both receive 5 points. The next round player one cooperates but player two defects. After this point the histories are identical to that after the first move, and so the players perform as they did then, and continue to cycle through the same sequence of moves.

In the first generation of a simulation, each of the players is assigned a random strategy—the response action to each possible history sequence is randomly chosen from the available moves. At the end of each subsequent generation, the set of players is sorted in order of decreasing value of the total fitness payoffs each received while playing against the other players. The top third of the players is preserved into the next generation, and those players are also used to create the strategies of the rest of the players in the next generation. Each new strategy is created by mixing the strategies of two of the most successful players—for each possible history, the response action is taken randomly from one or the other parent's strategy. Mixing strategies in this way has the effect of rapidly distributing advantageous moves through the population. A small fraction (for most of our runs: 1 percent) of the moves are then mutated by replacing the action part of the move with a randomly chosen action.

In each generation, a record is kept of the total number of moves of each type: "cooperated," "defect," and for the optional games, "opt out". At the end of a generation the average fitness of the population is also recorded. A sample run of the compulsory game is shown in figure 9.1. As is typical for the runs reported here,

the population moves through a number of states in which the levels of cooperation and defection are fairly stable for tens of generations or longer. Rapid transitions then occur, yielding other stable states.

In this run the population quickly enters a state of high cooperation and fitness.
Around generation 70, it reverts to a state of virtually 100 percent defection and low
fitness. This is followed at generation 150 with a state of 50 percent cooperation,
50 percent defection and an intermediate fitness value. Around generation 310, the
population again finds a state of very high cooperation.

The results of an entire run are summarized with two numerical values:

The "cooperativity" measure is meant to quantify the degree to which cooperative behavior dominated during the run. Cooperativity is defined as the fraction
of generations during a run when the difference between the percentages of cooperative moves and defection moves is greater than a threshold of 25. The cooperativity value for the run shown in figure 9.1 is 0.494. (The precise value of the
threshold for computing the cooperativity value is not crucial. For example, changing the threshold to 70 for the run in figure 9.1 changes the cooperativity value
from 0.494 to 0.488. This is because the runs tend to remain in states where either
cooperation or defection is relatively high, and the other is correspondingly low.)

The "instability" measure is meant to quantify the degree to which the amount
of cooperation varies from generation to generation. This is defined as the average
of the square of the difference between the number of cooperative moves in successive generations. The instability value for the run shown in figure 9.1 is 19.41.
The instability measure increases if a run enters more states, or if the states that it
enters do not have constant values of cooperation. The value of the instability
measure for simulations depends on the simulation parameters. For example in a
set of 20 runs of 500 generations of the compulsory game with 36 players, a history
of length 2, and 10 rounds between each pair of players, increasing the fraction of
moves mutated each generation from 0.1 percent to 10 percent changed the average
instability value from 3.28 to 28.37. As is shown below, the instability value is also
strongly affected by whether the game is compulsory or optional.

Two versions of the optional game were simulated: the "semi-optional" and the
"fully optional" game. In the semi-optional game players are paired off as in the
compulsory game, and each pair plays some number of rounds against each other.
If either player chooses to opt out in any move, both players receive the opt-out
payoff W.

In the fully optional game, players first attempt to locate other players who will
not opt out against them. Such pairs then play one round against each other as in
the compulsory game, and each player keeps a record of what its opponent does
which it consults the next time they are looking for partners. All players who do not
find a willing partner in a given round are awarded the opt-out payoff. The fully
optional game requires more computational overhead to simulate, and the runs take
much longer, than the semi-optional game, because each player must record all
of the histories of its interactions with all of the players it has played against in a
generation; and because the process of pairing off the players is more complicated
than for the semi-optional game.

Runs of the optional games are illustrated and discussed below.

TABLE 9.2

Game	Mean cooperativity	Standard deviation	Mean instability	Standard deviation
Compulsory	0.105	0.160	5.28	3.80
Semi-optional	0.719	0.283	13.2	5.96

TABLE 9.3

Game	Mean cooperativity	Standard deviation	Mean instability	Standard deviation
Compulsory	0.248	0.267	14.8	8.85
Fully optional	0.668	0.292	29.9	5.10

4. Simulation Results

The results of this study can be summarized in tables 9.2 and 9.3. The statistics in table 9.2 are for a set of 27 runs, each of 500 generations. There were 60 players in each run, two steps of history were recorded, and each pair of players played 30 rounds each generation. As can be seen from table 9.2, the compulsory game yields populations which are more stable but less cooperative than those that play the semi-optional game.

Fewer runs of the fully optional game were done, as they take much longer. However the general result is similar. Table 9.3 shows results for 12 runs of each of the indicated games of 500 generations each, with 30 players, two steps of history were recorded and the players played against each other an average of ten rounds.

The reason that the statistics for the compulsory game in table 9.3 are different from those in table 9.2 is that the algorithm for pairing off players in this set of runs corresponded to that used for the fully optional game, except that no player could opt out. In each round of the game, a player was paired with another player in the population at random. The fact that some pairs played fewer (and some played more) rounds than the average of ten is reflected in the higher instability values for these runs as compared with the first table. Again however, the optional version of the game yields more cooperative generations than does the compulsory version of the game, and the instability of the optional game is higher than that of the compulsory game.

4.1. Runs of the Compulsory Game

Runs of the compulsory game tend to become stuck in a small number of states, either with very high cooperation, very high defection, or half cooperation and half defection. Often a run will be stuck in a state for many generations. This is reflected in the relatively low instability value for the compulsory game, and the high ratio of the standard deviation of its cooperativity to the mean value.

For example, when one step of history is recorded, a run often first enters a state where all of the players defect each round. This is because the initial strate-

((N) D)	((N) D)	((N) C)	((N) C)	((N) C)	((N) D)
((C) D)	((C) C)	((C) C)	((C) C)	((C) D)	((C) D)
((D) D)	((D) D)	((D) D)	((D) C)	((D) C)	((D) C)
a	b	c	d	e	f

FIGURE 9.2 Some strategies for the compulsory game, when one step of history is recorded.

gies are random, and so there are a large number of players who cooperate no matter what their opponents do. Hence the defectors receive the high T payoff, and their offspring take over the population.

Subsequent events can be understood by examining the strategies shown in figure 9.2 beginning with the *Undiscriminating Defector* strategy shown in figure 9.2a Two of the single-move mutations of this strategy will be at a disadvantage playing against it because they will cooperate in a round where the original will defect. If everyone in the population is defecting all the time, however, the move ((C) D) is never exercised. So a mutation from the move ((C) D) to ((C) C) will not affect the behavior of (nor the fitness payoffs accumulated by) a player using the strategy.

After several generations of this kind of "genetic drift", a population initially containing only players with strategy 2a can be expected to contain a fraction of players with strategy 2b. From here, a single mutation in the ((C) D) move can change the strategy to *Tit-for-Tat*, as shown in figure 9.2c. Provided that enough of these mutations occur at about the same time, players using *Tit-for-Tat* can dominate the population, which will enter a state of very high cooperation. This is essentially what happens in the run shown in figure 9.1 around generation 10.

However *Tit-for-Tat* is not immune to variation. For example the ((N) C) move could mutate back to ((N) D) and return the population to strategies like 2b and a state of high defection. This is what happens near generation 70 in the run shown in figure 9.1.

Another mutation from *Tit-for-Tat* can yield the *Undiscriminating Altruist* strategy 2d. If the population is in a state where everyone else is following *Tit-for-Tat*, following the *Undiscriminating Altruist* strategy will not affect the fitness of the player, as the ((D) C) move in *Tit-for-Tat* is never exercised. However if defectors appear, they can exploit the player who always cooperates. For example, strategy 2e is one mutation away from *Undiscriminating Altruist*. If it appears, the number of players using it will increase in the population as they prey on the cooperators. A single mutation from 2e is 2f, which increases defection in the population even more.

Strategies like the one shown in figure 9.2f can lead to very stable populations, with relatively low cooperation and fitness values. A pair of players playing this strategy will cooperate 50% of the time and defect 50% of the time. Furthermore there is no possibility of genetic drift with this strategy as each move of it is exercised and each one-mutation variant of this strategy is at a disadvantage against it. The population in the run shown in figure 9.1 enters a state where each member of the population is playing a variant of this strategy around generation 155. Note that

```
((N   N)   D)   •
((N   C)   C)
((N   D)   C)   •
((C   C)   D)   •
((C   D)   C)   •
((D   C)   C)   •
((D   D)   D)
```

FIGURE 9.3 A stable suboptimal strategy for the compulsory game, when two steps of history are recorded. The moves marked * are exercised when a pair of players with this strategy compete against each other.

the state remains steady for around 60 generations, before the population returns to a state of relatively high cooperation.

With longer history lengths, strategies like 2f, in which pairs of players using the strategies alternate between cooperation and defection, can be very stable. For example the strategy shown in figure 9.3 is a variant of 2e for a history length of 2. In figure 9.3, the '*' character indicates moves that are exercised when players using this strategy play against each other. Five of the seven moves are used in such an encounter, and therefore cannot mutate away from those shown without having an immediate effect. Mutations to the other two moves actually serve to reinforce the patterns of interactions seen when the rest of the players are using this strategy. It is important to note that the stability of this strategy is not a matter of any static superiority of the strategy compared with all possible competitors: instead this strategy is stable because its genetic representation is such that there are no better alternatives to it via single mutations. So a population in which all players are following such a strategy constitutes a robust local optimum in the evolutionary search. We have observed runs of the compulsory game in which the population remains stuck in states where populations are using such strategies for thousands of generations, though transitions to states of high cooperation or high defection eventually occur.

4.2. Runs of the Optional Games

With the option of opting out, populations in the optional games can escape from states in which there is a significant amount of defection. Since the payoff for opting out, W, is larger than both S, the payoff for cooperating when the opponent defects, and P, the payoff for mutual defection, the presence of defection in the population makes opting out an advantageous alternative. Thus populations playing the optional games will tend to revert to states of high opting out whenever a number of defectors appear. This fact alone accounts for some of the reason why the optional game leads to higher cooperativity—it just can't become stuck in states of high defection.

A run of the semi-optional game with one step of history recorded is shown in figure 9.4. As is typical for runs of the optional games, the' population enters more states than otherwise equivalent runs of the compulsory game, and the states that it enters are much less stable.

Since the initial strategies are random, and therefore include many defectors, opting out is initially favored, and most runs of the semi-optional game enter states of virtually 100 percent opting out in the early generations, often as early as generation 10. From then one, the populations tend to go through cycles of various sorts. One kind of cycle involves the appearance of defectors, which is usually soon followed by a reversion to high opting out.

Another kind of cycle seen in the semi-optional games involves the appearance and subsequent disappearance of players who always cooperate. If almost every other member of the population is opting out each round, there is no danger to the few players who mutate to strategies that involve some cooperation. Indeed, when these players play against each other, and receive the reward R for mutual cooperation (which is larger than the W opting out payoff), they will increase in the population. However when there are lots of careless cooperators in the population, there is an advantage to be gained by defecting. If mutations occur to create strategies that involve some defection, defectors will rapidly increase, effectively destroying the cooperators. Since the resultant level of defection is high, opting out is now relatively advantageous, and the population reverts to a state where everyone is opting out. This pattern is similar to the "predator–prey" cycles seen in population biology.

A typical run of the semi-optional game will go through a number of cycles. In some cases it is possible that the mutations that increase defection either include or are followed by mutations that increase the discriminatingness of the strategy, either by playing *Tit-for-Tat*, or by opting out when an opponent defects, that is, the *Discriminating Altruist* strategy. If such mutations occur before defection rises significantly, it is possible for the players possessing these strategies to continue to increase in the population even when defection rises temporarily. Thus the population can enter and remain in a relatively stable state of high cooperation.

This process is illustrated in figure 9.5. This shows a portion of a run of the semi-optional game. At generation 60, virtually all of the players are opting out in each game. Around generation 70, a few players begin cooperating. Since they manage to find other cooperators, their numbers increase. Within two generations, however, a few defectors appear. Since these defectors will prevail over the cooperators, their numbers increase rapidly, and by generation 80 or so, the cooperators are gone. A similar but less dramatic pattern of this sort begins almost immediately, and is over by generation 90.

At generation 91, another set of cooperators appears, followed closely by defectors. However in this case, at least some of the cooperators are playing a discriminating strategy, and in fact by generation 108, the defectors begin disappearing from the population. By generation 110, virtually all of the players are cooperating all of the time.

As with the compulsory game, states of high cooperation are not stable either. With all of the members of the population cooperating, genetic drift can set in, and mutate some of the discriminating strategies to their careless versions, providing fodder for defectors when they appear by mutation. As before, the high rate of defection will ultimately be followed by an increase in opting out.

The general dynamics of the fully optional game are similar to those for the semi-optional game. Slightly higher cooperativity is seen, as illustrated in table 9.4,

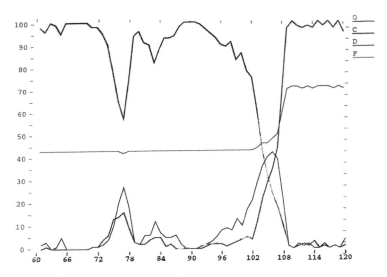

FIGURE 9.5 A portion of a run of the semi-optional game, showing two "predator-prey" cycles, and the beginning of a state of high cooperation.

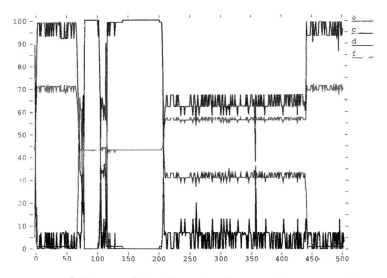

FIGURE 9.6 A run of the fully optional game with two steps of history recorded. The "cooperativity" value for this run is 0.716; its "instability" is 31.7.

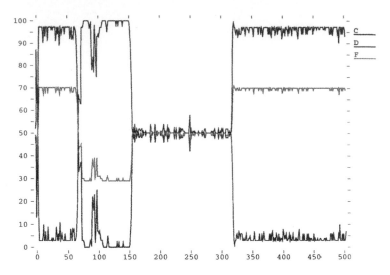

FIGURE 9.1 A run of the compulsory game, with two steps of history recorded. The trace marked "C" records the percentage of "cooperate" moves each generation; the trace marked "D" records the percentage of "defect" moves each generation; the trace marked "F" records the average fitness of the population as a percentage of the maximum possible value. The "cooperativity" value for this run is 0.494; its "instability" is 19.41.

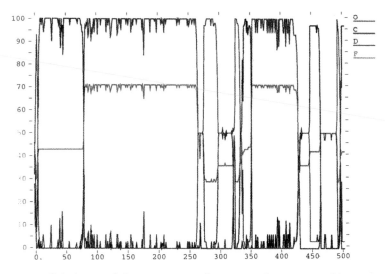

FIGURE 9.4 A run of the semi-optional game, with one step of history recorded. The trace marked "O" records the percentage of "opt out" moves each generation. The "cooperativity" value for this run is 0.581; its "instability" is 45.1.

TABLE 9.4

Game	Mean cooperativity	Standard deviation	Mean instability	Standard deviation
Semi-optional	0.401	0.263	27.8	10.02
Fully optional	0.668	0.292	29.9	5.10

but it is not clear if this is significant. The parameters for these runs are: 30 players, 10 game interactions, history length of 2. The statistics are for 12 runs of 500 generations each.

A run of the fully optional game with 2 steps of history recorded is shown in figure 9.6. The population quickly finds a state of high cooperation without an intervening period of opting out, as predicted by the analysis in section 2. Around generation 70, this state crashes and yields a period of high opting out that lasts (with one short glitch) until generation 205. At this point, the population enters a state where cooperation is still relatively high, but the fraction of defection moves is around 30 percent. This state lasts until generation 440, when a state of high cooperation occurs. Again, the transition to the higher level of cooperation happens without an intervening period of opting out.

5. Conclusion

The superiority of the optional games in reaching states of high cooperation can be demonstrated analytically, and is supported by the dynamic properties that simulations of such games manifest. There is no way for a population playing the compulsory game to escape from a state of high defection, except if several favorable mutations appear simultaneously. In the optional games there are routes out of states of high defection. The option of asocial behavior facilitates the appearance and maintenance of altruistic behavior.

In thinking about the evolution of social behavior it is important to recognize that such behavior occurs against a changing environment consisting of the behaviors of the other members of the populations. Thus such an evolutionary process is a feedback system, and the global properties of such a process should be expected to fluctuate, perhaps chaotically. The relative fitness of a given behavior or strategy cannot be assessed statically, with respect to a specific, or to a fixed, environment. In the long run, the evolutionary dynamical properties of strategies and their genetic representations, may have the most significant effect on the careers of populations using those strategies.

Obviously, in addition to more detailed analysis and simulation of the optional games, it is important to see if the optional games provide a more ethologically valid model of some animal interactions. One would have to be able to distinguish between an animal's refusal to participate in an interaction ("opting out") and its failure to reciprocate an altruistic action of another animal ("defection"). Our model predicts that geographically separate, but genetically equivalent, populations of the same species might differ markedly in their social interactions, with some populations exhibiting high cooperation, some behaving asocially, and others enduring periods of highly antisocial behavior.

There are many ways in which to introduce complications into the study of altruism in optional and semi-optional games. For example, animals may make various types of errors of recognition, potential partners may vary in quality, and different types of game-theoretic situations may arise. Our preliminary analyses reveal that these complications do not markedly affect the results presented herein. A slightly more detailed survey of some possible complications is given by Kitcher,[13] where the evolution of specifically human types of altruism is also addressed.

Notes

1. R. Trivers, "The Evolution of Reciprocal Altruism," *Quarterly Review of Biology* 46 (1971): 33–57.

2. R. Axelrod and W. D. Hamilton, "The Evolution of Cooperation," *Science* 211 (1981): 1390–1396; R. Axelrod, *The Evolution of Cooperation* (New York: Basic Books, 1984).

3. In the sense of J. Maynard Smith and G. R. Price, "The Logic of Animal Conflict," *Nature* 246 (1973): 15–18; see also J. Maynard Smith, *Evolution and the Theory of Games* (Cambridge: Cambridge University Press, 1982).

4. R. Boyd and J. P. Lorberbaum, "No Pure Strategy Is Evolutionarily Stable in the Repeated Prisoner's Dilemma Game," *Nature* 327 (1987): 58–59; J. Farrell and R. Ware, "Evolutionary Stability in the Repeated Prisoner's Dilemma," *Theoretical Population Biology* 36 (1989): 161–168; M. Mesterton-Gibbons, "On the Iterated Prisoner's Dilemma, *Bulletin of Mathematical Biology* 54 (1992): 423–444.

5. L. A. Dugatkin and D. S. Wilson, " 'Rover': A Strategy for Exploiting Cooperators in a Patchy Environment," *American Naturalist* 138, no. 3 (1991): 687–701; J. R. Peck and M. W. Feldman, "The Evolution of Helping Behavior in Large, Randomly Mixed Populations," *American Naturalist* 127 (1986): 209–221; K. Lindgren, "Evolutionary Phenomena in Simple Dynamics," in C. G. Langton, C. Taylor, J. D. Farmer, and S. Rasmussen, eds., *Artificial Life II* (Redwood City, Calif.: Addison-Wesley, 1992), 295–311.

6. R. R. Miller, "No Play: A Means of Conflict Resolution," *Journal of Personality and Social Psychology* 6, no. 2 (1967): 150–156; S. L. Lima, "Iterated Prisoner's Dilemma: An Approach to Evolutionary Stable Cooperation," *American Naturalist* 134 (1989): 828–834.

7. R. Nöe, C. P. van Schaik, and J. van Hoof, "The Market Effect: An Explanation for Pay-off Asymmetries among Collaborating Animals," *Ethology* 87 (1991): 97–118; J. R. Peck, "Evolution of Outsider Exclusion," *Journal of Theoretical Biology* 148 (1990): 563–571; J. J. Bull and W. R. Rice, "Distinguishing Mechanisms for the Evolution of Co-operation," *Journal of Theoretical Biology* 149 (1991): 63–74.

8. The closest approach we know is that of E. A. Stanley, D. Ashlock, and L. Tesfatsion, "Iterated Prisoner's Dilemma with Choice and Refusal of Partners," in C. G. Langton, ed., *Artificial Life III* (Redwood City, Calif.: Addison-Wesley, 1994), 131–175.

9. For example, B. B. Smuts, D. L. Chesney, R. M. Seyfarth, R. W. Wranghan, and T. T. Struhsaker, eds., *Primate Societies* (Chicago: University of Chicago Press, 1987).

10. D. L. Cheney and R. M. Seyfarth, *How Monkeys See the World* (Chicago: University of Chicago Press, 1990).

11. L. A. Dugatkin and M. Alfieri, "Guppies and the Tit for Tat Strategy: Preference Based on Past Interaction," *Behavioral Ecology and Sociobiology* 28 (1992): 243–246.

12. Maynard Smith, *Evolution and the Theory of Games*.

13. P. S. Kitcher, "The Evolution of Human Altruism," *Journal of Philosophy* 90 (1993): 497–516; reprinted as chapter 8 in this volume.

10

Infectious Ideas

Some Preliminary Explorations (2001)

1. Two Prevalent Analogies

Efforts to use concepts from contemporary biology in understanding the dissemination of culture have been inspired by two main analogies. One of these supposes that there are cultural units—memes—that share important similarities with genes, and a number of authors have attempted to exploit this analogy to develop precise theories of cultural transmission.[1] According to a second analogy, the spread of culture is like the infection of a population by a virus (or a bacterium or a parasite). Very often, the two analogies are developed in tandem: Dawkins's introduction of the meme contains both his own assimilation of memes to genes and explicit approval of an elegant formulation of the view of memes as parasites by his then-colleague Nicholas Humphrey;[2] similarly, Daniel Dennett's influential presentation of cultural transmission as the spread of memes[3] appeals to one or other analogy indifferently as if they were equivalent. My aim in what follows is to take a hard look at the second analogy (culture as infection), treating it with a bit more precision than is customary.[4] We'll find that the formal details of transmission processes are by no means equivalent, and that even the kinematics is more complicated than seems to have been appreciated.

2. Infection: Some Basics

Infectious agents typically spread through populations in roughly the following fashion. At any given stage, there's a subpopulation consisting of those who are currently infected. As these individuals come into contact with the rest of the population there's a chance that the virus (or whatever) will be transmitted from the carrier to a hitherto uninfected person. (This chance may be different for different members of the population, for some may have genetically based susceptibilities that others lack; I'll exploit an analog of this important point below, in section 11 and following.) The distribution of the infection in the population, which can be indexed by the carrier subpopulation, grows insofar as the germ is transmitted and decreases insofar as carriers recover (or die). It's not hard to see that there are cultural analogues for the properties and relations that figure in this story. Carriers are those who are in possession of a cultural item; there are occasions of contact among individuals that can result in transmission of cultural items; and forgetting serves as the counterpart of recovery.

3. The General Form of Cultural Contact

Before we go any further, however, it's important to think carefully about how contact among individuals gives rise to cultural transmission. Before the contact, we have two people, exactly one of whom stands in a relation to a cultural item; afterwards the individual who lacked any relation to that cultural item stands in a relation to that cultural item. Let's use the shorthand 'Oxy' to indicate that x has none of the psychological relationships in which we're interested to the cultural item y. Then the fully general form of an occasion of cultural transmission from a to b with respect to the cultural item c involves three psychological relations Ψ_1, Ψ_2, Ψ_3 (so that "Obc" now denies that b stands in Ψ_i to c [for each i]) and can be represented as

$$(3.1) \qquad << \Psi_1 ac, Obc >, < \Psi_2 ac, \Psi_3 bc >>$$

This pedantic formulation is intended to expose the fact that the Ψ-relations need not be the same. Suppose, for example, that an evangelist explains to me the doctrines of his favorite sect, one with which I wasn't previously familiar. Initially, the evangelist firmly believes the doctrines, whereas I know nothing about them; perhaps after the event I entertain them (they belong to my list of daft dogmas), and the evangelist is made somewhat less certain by my resistance (or maybe even more zealous by my recalcitrance). Similarly, Don Basilio's commendation of *la calunnia* envisages spreading through the city of Seville a belief that its ultimate sources (Don Bartolo and himself) know to be false. The analogy between cultural transmission and disease isn't intended to cover heterogeneous cases of cultural contact: after contact, the recipient comes to be in the same state with respect to the pathogen that the carrier was before. So let's narrow our focus to *homogeneous* cultural transmission in which $\Psi_1 = \Psi_2 = \Psi_3$. Although the Ψ-relation can be construed in many different ways—believing a proposition; entertaining a proposition; having a musical item, a story, or a joke in one's repertoire; having the capacity to use a particular tool or style of making an artefact; having a disposition to use a tool or a style; and so forth—we can explore many features of cultural transmission without adopting a particular interpretation. But it's important to be aware of different possibilities because the types of formal models that are relevant depend sometimes on what psychological relations we envisage.

4. Cultural Atoms?

Many critics of the analogy between memes and genes have protested the idea of atomizing culture.[5] Although I believe that many of their points are well-taken, it's not clear that the criticisms extend to the comparison of cultural transmission with the spread of pathogens. The items that figure in the general transmission process described in section 3 don't have to be restricted to some particular small size—chunks of tradition could be passed on as wholes. Yet the fundamental worry is that allowing cultural items to be relatively large will further complicate the form of the transmission process in that the psychological relations of (3.1) will obtain between people and *different* cultural items (b will come to stand in some psychological relation to an item c^*, a modified version of the item c that a is related to). This general issue has been explored most thoroughly by Dan Sperber.[6] To accommodate his concerns, we could either develop the analogy by imagining a virus (or other pathogen) with a significant chance of mutation on each occasion of transmission, or else treat cultural transmission (understood as taking the form of (3.1)) as a special case of a more general phenomenon of cultural contact exemplified by

$$(4.1) \qquad << \Psi_1 ac, Obc >, < \Psi_2 ac, \Psi_3 bc^* >>$$

For the purposes of this essay, I'll explore the former strategy. Readers skeptical of the general idea of the transmission of cultural units may wish to see my account of the complexities of cultural kinematics as complementary to Sperber's concerns.

5. Some Varieties of Cultural Transmission

Before we delve into the formal details, it's important to acknowledge another limitation of the analogy with infection. The account of the transmission of disease sketched in section 2 is importantly different from common examples of cultural transmission. The idea of many centers of infection isn't always appropriate to the cultural analogue. Consider two extreme cases. One replaces many-many transmission with one-many transmission. Imagine that a community living in a harsh climate requires authoritative information about the winter weather, and that there's one reliable source (a particular radio station, say). Each day, every member of the community tunes to that station, and each day, everybody comes to believe exactly the same thing about the imminent weather. Here, in *One-Stage Universal Transmission*, an initial state of a belief held by one is followed by a state in which that belief is shared by all members of the community. The other extreme is *One-One Sequential Transmission*, perhaps best exemplified in the children's game "Telephone" (players sit in a circle; one player whispers a sentence [or a brief story] into the ear of his/her rightmost (or leftmost) neighbor; the message is transmitted around the circle until it reaches the player who began; the point of the game is to enjoy the difference between the initial message and final product). In One-One Sequential Transmission, it takes n-1 transmissions for a community of size n to go from an initial state in which one person has the cultural item to a final state in which all members have (some version of) the item. If the chance of uptake of a modified ("mutated") message is μ at each stage, then the probability that the same version of the message is universally shared at the end is $(1 - \mu)^{n-1}$, so that the chance that different versions are present at the end is $1 - (1 - \mu)^{n-1}$. In taking the many-many approach detailed in section 2 as a way of thinking about cultural transmission, we shouldn't ignore the fact that other processes are far from rare. Rather the situation in which there are multiple centers and random contact can be viewed as a default, against the background of which we can explore the causal factors that produce different modes of transmission. Like the Hardy-Weinberg equilibrium in population genetics, the approach cited in section 2 can help us classify ways in which cultural transmission departs from random contact, either in the extreme forms I've mentioned or in less radical ways.

6. Infection without Competition: The Basic Case

So much for preliminaries. Turn now to the very simplest model of infectious ideas. Imagine a population of size N. Within this population is an infected population of size I. Members of this subpopulation make contact at random with other members of the population. They do so at a rate ρ, and when contact occurs, there's a fixed probability τ that the cultural item will be transmitted. Let $\rho\tau = k$. Assume that, once infected, individuals never lose the cultural item (this is surely more plausible when we take the psychological relation to be something like *having in one's repertoire* rather than *belief*); and that transmission only occurs when contact takes place with a hitherto uninfected individual. In unit time, each infected individual engages in ρ contacts; with probability $(N - I)/N$ the contact is with an uninfected individual; in cases where the contact is with an uninfected individual, transmission occurs with probability τ. So it's not hard to see that the kinematics of transmission is governed by the equation

$$(6.1) \qquad \frac{dI}{dt} = \frac{I(N-I)}{N} \rho\tau$$

that is

$$(6.2) \qquad \frac{dI}{dt} = \frac{kI(N-I)}{N}$$

This is the familiar logistic equation from population ecology and shows asymptotic convergence to N. The explicit solution of (6.2), assuming that we begin from a state in which just one person has the cultural item, is

$$(6.3) \qquad I(t) = \frac{e^{kt}}{(N-1)+e^{kt}} N$$

It is not hard to show that, if the initial population divides into two types, those who are susceptible to the cultural item and those who are impervious to it, the kinematics is just the same except that we must replace the population size, N, with that of the susceptible population, S. Although these results are relatively banal, they do allow us to compare population statistics on the spread of cultural items with the logistic trajectory. There's a relatively good fit, for example, between that trajectory and the sizes of the Christian population of Mediterranean cities for the third and fourth centuries,[7] and this supports the idea that Christian belief may have spread through the Greco-Roman world by random contacts between believers and members of the susceptible population. (This hypothesis allows, of course, for a variety of explanations of why transmission occurred in particular instances.)

7. Infection without Competition: Cultural Loss

The treatment outlined in section 6 supposed a process in which cultural items, once acquired, are permanently retained. This may be appropriate when we're considering such psychological relations as *having within one's repertoire*, but it's quite implausible for belief or a disposition to use some style or artefact. It's not hard to see how to build in the idea of recovery from infection (loss of the cultural item). We now have to keep track of two populations: those who are infected and those who have recovered from infection. Start with the simplest case, in which those who recover are never reinfected. Then, if the recovery rate is r, and the size of the population that has recovered is R, we have:

$$(7.1) \qquad \frac{dR}{dt} = rI$$

$$(7.2) \qquad \frac{dI}{dt} = kI \frac{(N-I-R)}{N} - rI$$

Plainly, if $r > k$, transmission never goes far, but even when $k > r$, the long-term state of the population will be one in which the cultural item is completely forgotten. (This conclusion would have to be modified if we suppose that the population goes through several biological generations, so that new susceptible individuals are introduced. We'll briefly consider the interaction among cultural transmission and biological birth and death below.) A more interesting possibility is to allow for reinfection at a reduced rate. Thus in place of (7.2), We'd have

$$(7.3) \qquad \frac{dI}{dt} = kI \frac{(N-I-R)}{N} - k'R - rI$$

where $k'(<k)$ is the rate of reinfection. Evidently, the population can't reach a steady state until each member is in one of two conditions—either infected with the item or having been infected but now having lost the item. Hence at steady state $N = I + R$, so that the equilibrium is given by

$$(7.4) \qquad I = \frac{k'N}{k'+r}$$

8. Infection without Competition: Resistance

Return now to the simpler case in which cultural items, once acquired, are never lost. Suppose that there's an initial recalcitrant population of size R (in the notation of section 6, $R = N - S$). Assuming that recalcitrance remains constant, whatever the success of the cultural item in invading the population, the kinematics is governed by the logistic equation

$$(8.1) \qquad \frac{dI}{dt} = kI\frac{(N-R-I)}{N}$$

But we don't have to suppose that R is fixed: perhaps resistance to a cultural item drops as it becomes popular. The simplest case is to suppose that

$$(8.2) \qquad \frac{dR}{dt} = -g(I)R$$

where $g(I) = 0$ if $I \le I_c$, $g(I) = q$ if $I > I_c$. Evidently, if $I_c > N - R_0$ is the size of the initial recalcitrant population, this isn't going to make any difference: the final state will show an infected population of size $N - R_0$. But if $I_c < N - R_0$, there'll be a two-stage process: in the first stage, there'll be logistic growth to an infected population of size I_c; at that point; R will begin to decay, and the idea will eventually spread through the entire population. This scenario can easily be extended to a stratified set of sets of individuals with increasing degrees of recalcitrance (measured by increasing critical values), allowing a multi-stage process in which the cultural item sweeps through some initial segment of the ordered sequence of strata, or through the entire population.

9. Competitive Infection

So far I've ignored the most obvious feature of the spread of cultural items, to wit: that they can compete for access to our minds. Once again, this is less pertinent to instances in which the psychological relations concern incorporation within a repertoire—although even then it's important to acknowledge that pianists can only learn a limited number of works, we can only remember a limited number of stories, and so forth. With respect to dispositions to use a tool or beliefs, however, the possibilities for competition are more pronounced. Critics of the analogy between memes and genes have often asked enthusiasts to specify a cultural analogue for the notion of an allele. While it may be impossible to provide a context-independent specification of when two cultural items are rivals, I think we can make sense of the notion that with respect to particular psychological relations, within a particular population with a particular background of cultural items, there are families of cultural items whose members are in competition. That is, for a contextually specified population, there's a family F of cultural items such that (a) each member of the population is susceptible to standing in the relation Ψ to some member of the family, and (b) it is impossible that there should be two different members of the family c, c^* such that a person stood in the relation Ψ to both. We may all be susceptible to acquiring some belief or other about the existence of a deity, but it seems reasonable to suppose that such beliefs form a family such that each of us can have at most one. I'll consider this to be the basic circumstance of cultural competition, and the analysis of the next sections will presuppose it.

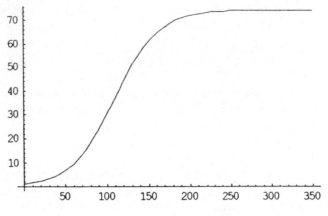

FIGURE 10.1

10. Competitive Infection with Indifference

Start with the simplest case. We have a two-member family of cultural items that are in competition with one another. Each person in the population is susceptible to infection by some member of the family. I adopt three assumptions.

Exclusivity. Each person in the population is infected at most once.

No recovery. Once infected, a person remains infected.

Indifference. Each person in the population is indifferent with respect to susceptibility to the two cultural items.

We now have two infected groups, indexed by I_1 and I_2. If the transmission rates are k_1 and k_2 respectively, the kinematics is governed by the equations:

(10.1) $$\frac{dI_1}{dt} = k_1 I_1 \frac{(N - I_1 - I_2)}{N}$$

(10.2) $$\frac{dI_2}{dt} = k_2 I_2 \frac{(N - I_1 - I_2)}{N}$$

Numerical solutions to these equations show that when the ratio k_1/k_2 departs only slightly from unity, there can be a pronounced bias toward the spread of the cultural item associated with the higher value. Suppose, for example, that we set $k_1 = 0.4$, $k_2 = 0.3$, $N = 100$. Then item one will eventually be about three times as prevalent as item two. (See figure 10.1, which shows the spread of item 1).

Even if the bias is slightly less, there can still be a significant difference between the long-term successes of the rival ideas. Suppose, for example, that $k_1 = 0.45$, $k_2 = 0.4$ ($k_1/k_2 = 9/8$). The long-term frequency of item one in the population turns out to be over 60 percent (see figure 10.2).

These results are readily comprehensible. For it's clear that

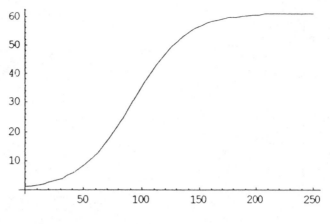

FIGURE 10.2

$$(10.3) \qquad \frac{dI_1}{dI_2} = \frac{k_1}{k_2} \frac{I_1}{I_2}$$

and, given that initially both cultural items are uncommon,

$$(10.4) \qquad I_1(0) = I_2(0) = 1$$

it follows that

$$(10.5) \qquad I_1 = I_2^{\frac{k_1}{k_2}}$$

The long-term values are obtained when

$$(10.6) \qquad I_1 + I_2 = N$$

and it's easy to see why these give rise to the kinds of asymmetries described above.

11. Competitive Infection with Differential Susceptibility

The next step is evidently to relax the idea of indifference. Let's divide the population into two classes, those who are more susceptible to the first cultural item and those who are more susceptible to the second. Thus we'll have

$$(11.1) \qquad N = N_1 + N_2, \text{ where } N_1 = sN$$

Let's suppose that people infected with idea number one engage in contact with others at a rate ρ_1; if the contact is with an uninfected person who is more susceptible to item number one, than the chance of transmission is τ_{11}; similarly, if they meet an uninfected person who is more susceptible to item number two, then the chance of transmission is τ_{12}. A similar notation applies to contacts involving someone infected with item number two. Plainly

$$(11.2) \qquad \tau_{11} > \tau_{12} \ \tau_{22} > \tau_{21}$$

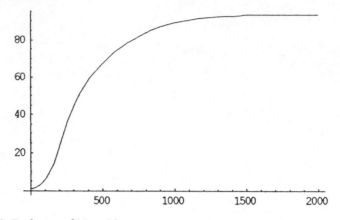

FIGURE 10.3 Evolution of $N_{11} + N_{21}$.

We have to look at the evolution of four populations: those more susceptible to item one and infected with item one, those more susceptible to item two and infected with item one, those more susceptible to item one and infected with item two, and those more susceptible to item two and infected with item one. Index these classes by N_{11}, N_{21}, N_{12}, N_{22} respectively. The kinematics is now governed by the simultaneous differential equations

$$(11.3) \qquad \frac{dN_{11}}{dt} = \rho_1(N_{11}+N_{21})\tau_{11}\frac{(sN-N_{11}-N_{12})}{N}$$

$$(11.4) \qquad \frac{dN_{21}}{dt} = \rho_1(N_{11}+N_{21})\tau_{21}\frac{((1-s)N-N_{21}-N_{22})}{N}$$

$$(11.5) \qquad \frac{dN_{12}}{dt} = \rho_2(N_{12}+N_{22})\tau_{12}\frac{(sN-N_{11}-N_{12})}{N}$$

$$(11.6) \qquad \frac{dN_{22}}{dt} = \rho_2(N_{12}+N_{22})\tau_{12}\frac{(1-s)N-N_{21}-N_{22})}{N}$$

These are not analytically solvable. But we can investigate special cases. Suppose, then, that the population size, N, = 100. Assume symmetry in transmission $\tau_{11} = \tau_{22}$ = 0.06, $\tau_{12} = \tau_{21} = 0.4$, and identity in rates of contact $\rho_1 = \rho_2$. Then the infection of the population will evidently be determined by the relative sizes of the susceptibility classes. If $s = 0.75$, then (unsurprisingly), cultural item number one quickly reaches a frequency of 75 percent. Although there are crossover effects (people more susceptible to item one who become infected with item two, and conversely), they cancel; interestingly, the population of those more susceptible to item two who are infected with that item is much the smallest (the final values of the four population sizes are roughly 62, 15, 15, 8). More interesting trajectories occur if there are opposite asymmetries in different sets of parameters. Suppose, for example, that we keep $N = 100$, and let $s = 0.4$. Abbreviate $\rho_i\tau_{ij}$ as k_{ij}. Then we can inquire into the values of $\langle k_{11}, k_{12}, k_{21}, k_{22}\rangle$ for which item number one becomes more prevalent. If

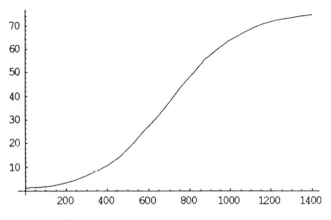

FIGURE 10.4 Evolution of $N_{11} + N_{21}$.

k_{11} is large in relation to the other values, then there can be pronounced bias towards item one; so, for example, if the vector is $\langle 0.04, 0.004, 0.003, 0.005 \rangle$, the final state will be one in which 93 of the 100 people in the population are infected with item one.

Even if we make the asymmetry less pronounced, by taking the values as $\langle 0.01, 0.004, 0.003, 0.005 \rangle$, it is still possible for over 75 percent of the population to become infected with item one.

Even when the vector of k_{ij} values is $\langle 0.01, 0.004, 0.003, 0.008 \rangle$, item one spreads to a majority of the population: this is because of a pronounced asymmetry between those who are more susceptible to item two but become infected with item one (over 20 percent) and those who are more susceptible to item one but become infected with item two (about 8 percent). Interestingly, the crossover asymmetry (that is the asymmetry between k_{21} and k_{12}) is sufficient to reverse the fact that a majority of the population are more susceptible to item two than to item one; the values $\langle 0.01, 0.007, 0.003, 0.01 \rangle$ yield a slight majority (53 percent) in favor of item one (even though, as with all of the foregoing instances, 60 percent of the population is more inclined to item two). Even a more extreme distribution of initial susceptibilities can be offset by appropriate relations among the k_{ij}. Thus if $s = 0.1$ (so that 90 percent of the population is more inclined to item two) and the k_{ij} vector is $\langle 0.01, 0.007, 0.001, 0.008 \rangle$, item one will attain a bare majority because it makes significant inroads into that part of the population more susceptible to item two (the value of N_{21} climbs to just above 40, about 44 percent of the people who are more susceptible to item two).

12. Competitive Infection: Trends and Resistance

The moral of this story is that, when we relax indifference, the kinematics of competition between cultural items involves complex trade-offs between the parameters that measure the relative sizes of the susceptibility classes and those that record

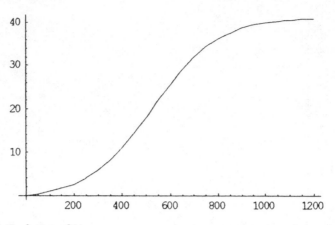

FIGURE 10.5 Evolution of N_{21}.

the intensity of the susceptibility. But this is only the beginning of further intricacies that await us. We need to take account of the possibility that susceptibilities are not constant but responsive to the prevalence of the items in the population. Thus let $\tau_{ij} = a_{ij} + b_{ij}(N_{1j} + N_{2j})$. Similarly, we can model trend-resistance by taking $\tau_{ij} = a_{ij} - b_{ij}(N_{1j} + N_{2j})$. Trend-following can greatly enhance the effects we saw in the previous section. Suppose that a majority is more susceptible to item two than to item one, but that their willingness to tolerate the less preferred item is slightly greater than that of the minority. Specifically, let s = 0.4 and the vector of infection probabilities be $c_{ij}(1 + 0.2(N_{1j} + N_{2j}))$, where the values of the c_{ij} are $c_{11} = 0.01$, $c_{21} = 0.007$, $c_{12} = 0.005$, $c_{22} = 0.01$. Despite the majority preference, item one becomes more prevalent in the population because N_{21} achieves a value of about 30. Indeed, the preponderance of crossing over to item one against the crossover to item two can be measured by the value of $N_{21} - N_{12}$, which reaches a value of around 20.

This crossover effect can even be sufficiently powerful to offset both majority higher susceptibility for item two and a high propensity for those more susceptible to item two to acquire it. Thus if the c_{ij} vector takes the values $\langle 0.01, 0.008, 0.003, 0.011 \rangle$ item one comes to infect over 80 percent of the population; if the values are $\langle 0.01, 0.008, 0.003, 0.013 \rangle$, item one still manages a bare majority; and the split between items one and two is almost equal when the values are $\langle 0.01, 0.008, 0.003, 0.0130385 \rangle$.

13. Abandonment and Switching

It's also important to take into account possibilities of abandonment and of switching. We can imagine four different types of scenario: abandonment at a fixed rate (see the discussion in section 7), abandonment as the result of transmissions of the alternative (thus if someone infected with item one encounters someone infected with item two there's a chance that each will abandon the pertinent item), spontaneous switching (perhaps in response to the prevalence of an item in the population or,

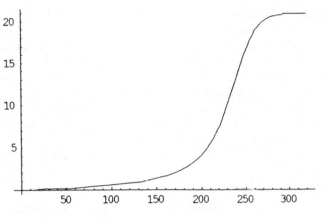

FIGURE 10.6 Evolution of $N_{21} - N_{12}$.

conversely, because the item is becoming too popular), and switching prompted by an encounter with someone infected with the alternative. Let's briefly consider the very simplest version of the second. Return to the condition of indifference described in section 10. The kenematics are now governed by the equations

$$(13.1) \qquad \frac{dI_1}{dt} = \tau_1 I_1 \frac{(N - I_1 - I_2)}{N} - \tau_2 I_2 \alpha_1 \frac{I_1}{N}$$

$$(13.2) \qquad \frac{dI_2}{dt} = \tau_2 I_2 \frac{(N - I_1 - I_2)}{N} - \tau_1 I_1 \alpha_2 \frac{I_2}{N}$$

where α_i is the rate at which item i is abandoned in contacts with its rival. Besides the obvious extreme equilibria, there's an internal steady-state when

$$(13.3) \qquad I_1 = \frac{\tau_2^2 \alpha_1 N}{\tau_1^2 \alpha_2 + \tau_1 \tau_2 \alpha_1 \alpha_2 + \tau_2^2 \alpha_1}$$

$$(13.4) \qquad I_2 = \frac{\tau_1^2 \alpha_2 N}{\tau_1^2 \alpha_2 + \tau_1 \tau_2 \alpha_2 + \tau_2^2 \alpha_1}$$

and this state contains an uncommitted population of size

$$\frac{\tau_1 \tau_2 \alpha_1 \alpha_2 N}{\tau_1^2 \alpha_2 + \tau_1 \tau_2 \alpha_1 \alpha_2 + \tau_2^2 \alpha_1}$$

(Of course, although the sizes of these populations remain constant, the people who belong to them change continually.)

14. The Influence of Ambiguity

In section 4 I noted the possibility of accommodating the possibility that the cultural item acquired may not exactly correspond to that held by the infecting person by deploying an analogue of mutation. On the face of it, cultural items that allow for significant variation are likely to follow a process of "sloppy spreading": if we consider the cluster of the original and its mutant offspring (grand-offspring and so forth) we discover that even though no particular member of the cluster may be common, the cluster as a whole dominates the population. Exploring the kinematics of such processes is difficult. Here's a relatively simple way to think about the issues. Imagine that, as mutations occur, the susceptibility of individuals to cultural items switches, so that some of those initially predisposed to prefer item two come to be just as disposed towards *some variant* of item one as those who were originally predisposed towards item one. Integrating this with the more exact treatment given in section 11 is difficult and we do better to work with an approximation. Instead of breaking the population into four groups, we can follow those infected with some version of item one (measured by I_1) and those infected by some version of item two (measured by I_2), and work with approximations to the probabilities that contact with an uninfected person will be with one of those more susceptible to some version of item one or with one of those more susceptible to item two. The kinematics can then be captured by

$$(14.1) \qquad \frac{dI_1}{dt} = \rho_1 I_1 \frac{[\tau_{11}(sN+(1-s)G(t)) + \tau_{21}(1-G(t))](N-I_1-I_2)}{N}$$

$$(14.2) \qquad \frac{dI_2}{dt} = \rho_2 I_2 \frac{[\tau_{12}(sN+(1-s)G(t)) + \tau_{22}(1-s) - (1-G(t))](N-I_1-I_2)}{N}$$

As with previous models, these equations aren't analytically solvable, and it's necessary to consider numerical examples. Suppose, then, that the items have identical transition parameters ($\rho_1 = \rho_2 = 1$, $\tau_{11} = \tau_{22} = 0.01$, $\tau_{12} = \tau_{21} = 0.005$). Let the initial susceptibility distribution be extremely skewed towards item two, $s = 0.1$, and let $N = 100$. A modest broadening of the class of people susceptible to item one (or, more exactly, to the variants that arise by mutation from the original item one) can be achieved by setting

$$(14.3) \qquad G(t) = \frac{0.001t}{1 + 0.001t}$$

The effect of allowing mutations to attract individuals who would otherwise have been more inclined to item two is to increase the frequency of I_1 in the population from 10 (see the discussion in section 11) to about 25. More dramatic results can be achieved if we imagine a higher rate of mutation, so that $G(t)$ increases more rapidly with t. Suppose

$$(14.4) \qquad G(t) = \frac{0.01t}{1 + 0.01t}$$

Under this regime, the value of I_1 goes to over 87.

The character of the process can be seen by considering the value of $I_1 - I_2$ Although this is negative at the very beginning of the spread of the cultural items (figure 10.8), as each attains a significant value, I_1 pulls strongly ahead (figure 10.9).

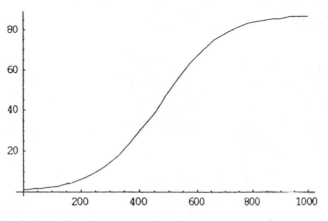

FIGURE 10.7 Evolution of I_1.

To appreciate the strength of these effects, consider two modifications of the example governed by (14.4). If we give item two a head start, letting its initial frequency be 10 (as opposed to 1 for item one), item one and its variants still attain a final value of 31. Alternatively, if we suppose that both variants are rare initially and that the transition parameters are skewed to favor item two, item one can still achieve a majority in the population. Keeping the parameters for item one fixed, if we increase τ_{12} to 0.006 and τ_{22} to 0.11, the long-term value of I_1 is 81; even if τ_{12} increases to 0.008 and τ_{22} to 0.02, the long-term value of I_1 is 57. This is striking support for the idea that cultural items that have a high rate of generating variations that appeal to broader groups can have a large competitive advantage. Texts containing ideas that mean lots of things to lots of people may exercise an enduring fascination, a phenomenon that might be called "the influence of ambiguity." As a paradigm, we might take the concept of a paradigm[8]—or even that of a meme.

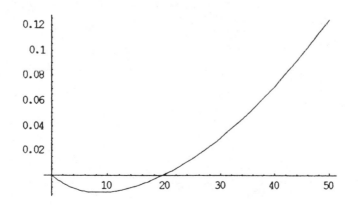

FIGURE 10.8 Early evolution of $I_1 - I_2$.

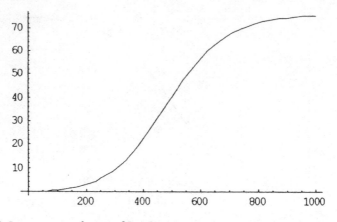

FIGURE 10.9 Long-term evolution of $I_1 - I_2$.

15. Cultural Infection: The "Three-Item" Problem

The last analysis I'll attempt will consider a version of the "three-item" problem. Imagine that we have three cultural items and three susceptibility classes. People in class one are more disposed to adopt item one than item three and item three than item two; people in class two are more disposed to adopt item two than item three and item three than item one; people in class three are more disposed to adopt item three than either item one or item two, and they're equally disposed with respect to item two. Class three is small relative to classes one and two. The interesting question is whether item three can exploit its ability to be second best for most people to become the most common item in the population. A full-dress treatment would extend the approach given in section 11 by considering nine variables (people in class i infected with item j, for three values of each). I'll use the

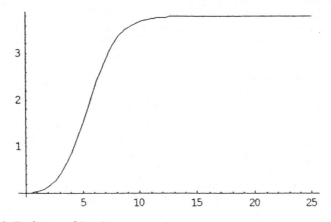

FIGURE 10.10 Evolution of $I_3 - I_2$.

simplifying approximative approach found in section 14 and take the kinematics to be governed by the equations:

$$(15.1) \qquad \frac{dI_1}{dt} = I_1(N - I_1 - I_2 - I_3)\frac{(\tau_{11}s_1 + \tau_{21}s_2 + \tau_{31}s_3)}{N}$$

$$(15.2) \qquad \frac{dI_2}{dt} = I_2(N - I_1 - I_2 - I_3)\frac{(\tau_{12}s_1 + \tau_{22}s_2 + \tau_{32}s_3)}{N}$$

$$(15.3) \qquad \frac{dI_3}{dt} = I_3(N - I_1 - I_2 - I_3)\frac{(\tau_{13}s_1 + \tau_{23}s_2 + \tau_{33}s_3)}{N}$$

where s_iN is the size of the ith susceptibility class. (For simplicity's sake, I've omitted mention of the contact frequency parameter ρ, which we can suppose to take the same value in all three cases.) The most obvious circumstance in which item three will predominate is when that item is a strong second choice for members of the first two susceptibility classes. Consider, for example, the following values for the τ_{ij} matrix:

$$(15.4) \qquad \begin{pmatrix} 0.01 & 0.001 & 0.009 \\ 0.001 & 0.01 & 0.009 \\ 0.001 & 0.001 & 0.01 \end{pmatrix}$$

Under this assignment, item three attains a long-run value of 73, leaving the other two items to share the remaining quarter of the population. Even if the secondary preferences for item three are less extreme, that item can still attain a majority.

$$(15.5) \qquad \begin{pmatrix} 0.01 & 0.001 & 0.007 \\ 0.001 & 0.01 & 0.007 \\ 0.001 & 0.001 & 0.01 \end{pmatrix}$$

Given (15.5), the long-run value of I_3 is 56. Even if we boost the tolerance of people in the first susceptibility class for item two (and of people in the second for item one), item three can still emerge as the most common. Consider:

$$(15.6) \qquad \begin{pmatrix} 0.01 & 0.004 & 0.007 \\ 0.004 & 0.01 & 0.007 \\ 0.001 & 0.001 & 0.01 \end{pmatrix}$$

Now item three spreads to 35 percent of the population. Its prevalence can be subverted if there's an asymmetry between items one and two. Thus

$$(15.7) \qquad \begin{pmatrix} 0.01 & 0.005 & 0.007 \\ 0.001 & 0.01 & 0.007 \\ 0.003 & 0.001 & 0.01 \end{pmatrix}$$

yields long-run values of 45 for I_2, 35 for I_1. Yet a slight adjustment restores the dominance of item three.

$$(15.8) \quad \begin{pmatrix} 0.01 & 0.005 & 0.008 \\ 0.001 & 0.01 & 0.007 \\ 0.003 & 0.001 & 0.01 \end{pmatrix}$$

Given (15.8), item three comes to infect 44 percent of the population, and is more prevalent than item two.

16. Distinct Generations

In the foregoing discussions, I've assumed that the time-scale on which the various processes occur is small enough to fit within the compass of a biological generation. Interesting variants can be obtained by relaxing this supposition and attempting to integrate the treatment here with the vertical transmission models offered by Cavalli-Sforza and Feldman.[9] Imagine, for example, that biological generations correspond to intervals during which the various logistic and quasi-logistic curves of the figures are unable to come close to their maxima. Suppose further that the character of the susceptibilities of the new generation is a function of the distribution of cultural items in the previous generation (corresponding to the intuitive idea of parental transfer of culture). Then it's quite possible for the spread of cultural items to take quite different forms, since there may be an advantage in attaining a slightly higher level of infection at an early stage. In this context, it's pertinent to note that figure 10.8 displays a case in which the early spread of cultural items doesn't correspond to their long-term prevalence.

17. Some Tentative Morals

Enough. Probably more than enough. Why have I lavished so much time and formalism on these models? For three reasons. First, as even the briefest glance at approaches that compare memes to genes will reveal, the kinematics of the processes I've been tracing is very different. People who want to use Dawkins's infectious idea to understand the spread of cultural items shouldn't use the genetic analogy and the infection analogy as if the two were equivalent. Second, despite the yen that many writers seem to have for Darwinian models of culture, it's important to think clearly and carefully about the kinds of parameters that affect the transmission of cultural items before building accounts of cultural advantages. My models, although primitive, begin to show how different kinds of features of cultural items—rates of contact involving infected individuals, probabilities of transmission in contact, different susceptibilities, mutation rates—interact with one another. It's premature to talk about Darwinian advantages in cultural competition without understanding exactly the ways in which such advantages accrue and how they might be offset because of other features of the kinematics. Contemporary evolutionary theory rightly recognizes that Darwinian accounts of how trait frequencies change are hostage to the details of population genetics. The same attitude ought to be standard in thinking about cultural change. If we want a slogan, it should be "No dynamics without a prior kinematics!" Third, if we're to give a positive answer to Dennett's question "Could there be a science of memetics?" it seems to me that the most promising approach is to explore the analogy with infection. For many of the complications that arise with respect to the explicit study of transmission—most notably keeping track of all the possible ways in which people can be influenced—can be bracketed if we assume that they underlie values of the parameters that occur in my models (thus, as noted in section 16, we might bundle a very disparate set of relations of cultural transmission between parents and offspring into a measure of the size of a susceptibility class or the value of a transmission rate). Nevertheless, it's quite apparent that the kinematics of cultural infection can be extremely complicated. Although I've pursued a few factors separately in relatively tractable contexts, it would be folly to claim that any of the models is likely to apply to the kinds of cultural evolution that most interest us. So I close with a conjectural answer to Dennett: Maybe. But I bet both that the infection approach is more promising than that which turns on the analogy with genes, and that implementing it will turn out to be *very* hard. The foregoing sections are intended to offer some support for my conjecture and, perhaps, to open up some avenues for exploring it further.

Notes

The numerical solutions in this essay were obtained using *Mathematica 4.0*, a tool for which I'm very grateful. I would also like to thank Dan Sperber for his patience, encouragement, and comments, and Erick Weinberg for advice about using *Mathematica* to solve differential equations, as well as for some valuable substantive suggestions.

1. Two outstanding examples are L. L. Cavalli-Sforza and M. Feldman, *Cultural Transmission and Evolution* (Princeton, N.J.: Princeton University Press, 1981), and R. Boyd and P. J. Richerson, *Culture and the Evolutionary Process* (Chicago: University of Chicago Press, 1985).

2. R. Dawkins, *The Selfish Gene* (Oxford: Oxford University Press, 1976), 206–207.

3. D. Dennett, *Darwin's Dangerous Idea* (New York: Simon and Schuster, 1995), chapter 12.

4. But see Cavalli-Sforza and Feldman, *Cultural Transmission and Evolution*, 29–53.

5. See, for example, S. J. Gould, "The Panda's Thumb of Technology," in Gould, *Bully for Brontosaurus* (New York: Norton, 1991).

6. D. Sperber, *Explaining Culture* (Oxford: Blackwell, 1996), chapter 4.

7. Reported in R. Stark, *The Rise of Christianity* (New York: Harper, 1996), 13.

8. T. S. Kuhn, *The Structure of Scientific Revolutions* (Chicago: University of Chicago Press, 1962).

9. Cavalli-Sforza and Feldman, *Cultural Transmission and Evolution*.

11

Race, Ethnicity, Biology, Culture (1999)

I

During recent decades, a number of prominent anthropologists have defended *eliminativism about race*, arguing that the notion of race, as applied to our own species, is of no biological significance.[1] One obvious motivation for discarding the concept of race is that it might provide the most effective way of undermining racism. Ironically, as I shall try to show later in this essay, important postracialist projects may require us to probe the connections between biology and race more deeply, to arrive at a clearer understanding of the concept of ethnicity, and to undertake empirical investigations of the connections between biological and social notions.[2]

However, whether or not eliminativism about race would achieve that goal, the first question concerns the truth of the thesis that races have no biological significance. Eliminativists have made two important points that should be recognized from the beginning. First, the phenotypic characters used to demarcate races—for example, the three "major races," Caucasian, African, and Asian—neither have any intrinsic significance nor have been shown to correlate with characteristics of intrinsic significance. Second, although generic and phenotypic studies have shown that certain alleles, dispositions to disease, and phenotypes occur at different frequencies in different racial groups, intraracial diversity is far more pronounced than interracial diversity. This latter point remains unchallenged. Painstaking research on human phenotypic variation has disclosed that, even with respect to the most evident marker of racial difference, skin color, there are profound differences within races.[3] Moreover, the growing mass of data on human genetic variation down to the minutest details of DNA sequence makes it plain that so-called races differ only in the frequencies with which various alleles are found, often in complicated and bewildering ways.[4] Jared Diamond[5] has made the point vivid by considering the ways in which various choices of genetic characteristics would subvert our standard racial classifications.

But if the facts of intraracial diversity are widely accepted, the idea that there are no correlations between familiar phenotypic differences and more significant traits remains controversial. Users of the notion of race have often maintained that the physical traits used to demarcate the different races are correlated with "mental" and "temperamental" differences. Nobody has been more forthright than the American champion of Mendelism, the early eugenicist Charles B. Davenport, who gave stark expression to the principal ideas in an essay, "The Mingling of Races."

> Not only physical traits, like eye color, skin color, body build and such characters as stature, color and form of the hair, proportions of facial features and many others are inherited in race-crosses but also mental traits. This is a matter which is often denied, but the application of methods of mental measuring seem to have produced indubitable proof that the general intelligence and specific mental capacities have a basis and vary in the different races of mankind. Thus it has been shown, by standard mental tests, that the negro adolescent gained lower scores than white adolescents and this when the test is made quite independent of special training or language differences and also when the children tested have a similar amount of schooling.[6]

What goes for brains goes for character. Davenport explained:

> Common observation shows that the emotional output of different peoples is very different. We note that the North American Indian is little given to emotional expression. On the other hand, the African negro expresses his emotions copiously. In Europe the Scotch Highlanders are characterized by a prevailingly somber tendency, while the South Italians are characterized by lightness of spirit.[7]

But there are even differences in instinct. Davenport continued:

> It is well known that most of the races of Europe are fairly stable and domestic, engaged in agriculture or industry. However, from eastern Europe and western Asia have come forth races of mankind with a strong tendency to wander over the face of the earth. Such are the Gypsies which have run through Europe and America and such are some nomadic peoples who are scattered across the face of Asia and Northern Africa and who even before the time of Livingstone had penetrated into the heart of Equatorial Africa. Now the instinct to wander, or nomadism, is one that has an hereditary basis. This has been worked out in some detail by the author and the results of his investigation have, so far, not been disproved.[8]

Plus ça change. Sixty years after these passages were written, we find contemporary authors adverting to the same themes as if the critiques of intermediate decades did not exist. Robert Herrnstein and Charles Murray confidently assert that IQ tests are free of cultural bias and that the 15-point gap between the means of Caucasians and African Americans points to genetic differences; J. Philippe Rushton suggests that the major races have different reproductive strategies that reflect temperamental differences.[9] True, Davenport's marvelously looney idea about genes for nomadism seems to have vanished, but it is remarkable how many of his claims are resurrected, more apologetically, by those who feel that the world should know the true facts about racial differences.

This is not the place to engage in a full critique of the recent revivals of Davenportism. Suffice it to say that many of the old charges have not been satisfactorily answered. Herrnstein and Murray make crucial assumptions in arriving at estimates of heritability, and they put the notion of heritability to work in ways that have been attacked as inappropriate for over two decades.[10] What we need to know about the genetic basis of intelligence are the shapes of the relevant norms of reaction,[11] and heritability estimates, even if correct, cannot enlighten us about these. Moreover, if we make some concessions, for the sake of argument, about the significance of IQ measures, there are interesting facts from the history of intelligence testing that point toward a quite different moral. One of the most noteworthy features of the data on which Davenport relied (the Army data from World War I) is the demonstration of a correlation between performance on the tests and quality of schooling (reflected in the differences between those educated in Northern and Southern schools).[12] Two groups once stigmatized for their "low" intelligence, the Jews and the Irish, currently perform better than members of other Caucasian groups. At the same time, data from Northern Ireland show that the mean score among Catholics is about 15 points below that of Protestants.[13] It is tempting to think that if differences in scores show anything at all, they reveal that people belonging to a group that is socially and economically disadvantaged often do significantly worse than the more fortunate members of the population. Rushton's work is equally insensitive to well-known criticisms. Since the mid-1980s, many scholars interested in the evolution of human behavior have learned to moderate their claims to avoid the excesses of what I have called "pop sociobiology."[14] Rushton writes as though there were no need for caution, investing anatomical and physiological differences with immense significance by spinning evolutionary scenarios that consistently ignore the possibility of alternative, more mundane, explanations.

So I begin from the position that the phenotypic characters used to pick out races neither have intrinsic significance nor are correlated with characteristics that are significant, and that intraracial variation is far greater than interracial variation. Does this mean that eliminativism is correct? I shall argue that it does not, and that, however admirably motivated, eliminativist approaches have failed to recognize more subtle ways in which divisions into races *might* have biological significance.[15] Further, in the light of this argument, I shall explore some of the ways in which concepts of race figure in social discussions, indicating questions that would have to be resolved if the practice of discarding racial divisions were to lead to desirable conclusions. We should all be worried by the thought that retaining concepts of race will foster racism—but my goal is to show that these should not be our only concerns.

II

It is helpful to begin an exploration of the biological significance of the concept of race by contrasting the uses that biologists make of this notion (and related notions) and those that figure in our social interactions. To fix ideas about the biological uses, we can turn to any of a number of standard examples that have been treated in contemporary neo-Darwinism. Dobzhansky's classic discussion[16] introduces three major illustrations: variant color patterns in the Asiatic beetle *Harmonia axyridis*, chromosomal races in *Drosophila pseudoobscura*, and shell coloring and patterning in the snail *Cepaea nemoralis*. Each of these instances involves a species with internal differentiation of groups. In the first and third examples, the groups are marked by readily identifiable phenotypic differences; in the second, the differences are solely at the chromosomal level. Underlying the phenotypic differences are differences in genes, while the chromosomal differences rest on heritable variations in the arrangement of genes. So, in all instances, the differences among members of the same species are heritable.

According to the neo-Darwinian synthesis, a species consists of a cluster of populations reproductively compatible with one another but reproductively isolated from other populations.[17] The notion of reproductive isolation, often misunderstood, rests on the idea that some organisms have a dispositional property: were they to be in the same place at the same time, they would not normally mate with one another. I shall explore the nuances of this complex idea shortly. First, however, let us note that the various groups of beetles, flies, and snails are not reproductively isolated from their conspecifics. Despite the heritable differences among the groups, they remain reproductively compatible. However, the genetic differences among the groups persist from generation to generation, so there are factors that prevent genetic homogenization. In some cases, there are selective pressures that tell against intergroup hybrids, in others geographical isolation. But whether the blurring of genetic differences is blocked by natural selection or by physical separation, the different groups appear to be taking the first steps toward speciation. They are *"species in statu nascendi."*[18]

There are three features of these examples that will be important in understanding a possible biological basis for racial concepts: the presence of phenotypic differences, the heritability of these differences, and the incipient reproductive isolation. All three deserve scrutiny, and will prove more problematic than might initially appear. First, however, it is worth contrasting, the biologist's demarcation of races with contexts in which the concept of race is employed in social discourse.

Some talk of race is overtly racist, and examples are too familiar to warrant recalling them explicitly. Yet there are other usages that might seem more benign, cases in which the concept of race fulfills a function in raising important problems. Consider discussions of the desirability of transracial adoption. In a society in which there is a practice of characterizing a majority race and a minority race, the adoption of a minority child by two majority parents might be opposed on the grounds that the child will be deprived of important parts of her racial identity.[19] The opposition recognizes, quite correctly, that in our species, genetic inheritance is one mode of transmission across the generations, accompanied by a different system in which items of culture are passed on. A particular style of cultural inheritance or, perhaps, a cluster of styles, regularly accompanies certain biological features; indeed, the division of the society into races on biological grounds maps onto a division into ethnic groups, ethnicities, marked out by alternative systems of cultural transmission. Because races are relatively broad categories, the mapping is hardly one–one.[20] Instead, the picture is of a cluster of related ethnicities, each more closely related to one another than with elements in the cluster associated with a different race. The picture reveals that at the basis of the opposition is the idea that the child will have an ethnicity that is at odds with her race. I shall later want to look at the notion of ethnicity introduced informally here, and at the assumption that it is desirable for ethnicity and race to be in harmony with one another.

For the moment, however, I simply want to place at the center of discussion the four elements whose interconnections I intend to explore: race, ethnicity, biology, culture. I want to review the ways in which a concept of race might be developed compatible with our present biological understanding, to explore the consequences of replacing the apparently biological concept of race with a social notion of ethnicity, and to ask if the social concept can play the role we intend for it without some biological notion lurking in the background. My strategy will be the inverse of one that is common in discussions of race. Rather than starting with our current conceptions of race, with all the baggage they carry, I want to ask how biologists employ the notion of race, and how we might regard our own species in similar fashion.[21] As I have already indicated, I believe that debates about the appropriate character of a postracialist society will be more sharply focused if we have information about the empirical issues which my probing of the notions of race and ethnicity will bring to the fore, specifically questions about the relationships between patterns of biological transmission and patterns of cultural transmission. It is also worth remarking, at the outset, that the notion of race I shall employ is minimalist: its ideas about racial division are far more modest than those to which defenders of race typically allude, and as I have been at pains to argue in Section 1, I concur in the eliminativist critique of the traditional views about the differences among races. Indeed, I am inclined to think that, if nothing corresponds to the notion of race I reconstruct, then eliminativists are quite right to maintain that no biological notion of race can be salvaged.

III

So much by way of introduction. Let me now begin more slowly and more carefully. If we propose to divide the human species into races, we offer a set of subsets, not necessarily exhaustive, that constitute the *pure races*. "Pure" here is shorthand, *and the usage of this term should carry no connotation of superiority.* "Pure races" might just as well be called "completely inbred lineages" (except that the phrase is cumbersome), for that is what they will turn out to be.

A necessary condition on any concept of race is the following:

(R1)   A racial division consists of a set of subsets of the species *Homo sapiens*. These subsets are the pure races. Individuals who do not belong to any pure race are of mixed race.

Now, there are all sorts of ways of dividing our species up that would by no means count as racial divisions. Suppose we considered subsets that marked out people according to income distribution, running speed, or average levels of ingestion of caffeine. One obvious reason why this kind of division is a nonstarter as a partition into races is that the characteristics that would identify the pure races are not heritable. Ruling out such proposals is very easy: we can simply impose a requirement of reproductive closure.

(R2)   With respect to any racial division, the pure races are closed under reproduction. That is, the offspring of parents both of whom are of race R are
also of race R.

Existing concepts of race honor (R2) but do not satisfy the converse principle

(R3)   With respect to any racial division, all ancestors of any member of a pure
race belong to that race. The parents of an individual of race R are of
race R.

Socially disadvantaged races consist of a pure core together with people any of
whose ancestors belongs to that core. Madison Grant's chillingly racist pronounce-
ment that one parent from an "inferior race" consigns the offspring to that race has
become a cornerstone of American notions of race, and Naomi Zack has insight-
fully explored the consequences of this presupposition.[22]

Racial divisions need not embody the idea that "inferior" races expand by
"tainting" their "superiors." It is possible to proceed symmetrically, honoring both
(R2) and (R3), and counting offspring of parents from different pure races as being
of mixed race.[23] However, even if both requirements are imposed, there are any
number of divisions of *Homo sapiens* that do not constitute what we would intuitively think of as racial divisions. Consider, for example, division by eye color. If
we were to partition people as blue-eyed or brown-eyed, this would fall afoul of
reproductive closure — brown-eyed heterozygotes can have blue-eyed children — but
this difficulty can easily be overcome. Let one pure race consist of people homozygous for the dominant (brown-eyed) allele, the other of people homozygous for the
recessive (blue-eyed) allele; heterozygotes will be of mixed race. (R2) is now satisfied, for, disregarding mutation, mating between two people both homozygous for
the same allele will only yield offspring also homozygous for that allele. However,
we have not yet secured satisfaction of (R3). To assure that, it is necessary to prune
the pure races, eliminating people who have any heterozygous ancestors. That can
readily be achieved if we proceed recursively, identifying *founder populations* and
the lineages to which they give rise.

Let us therefore fix a time in human prehistory, the *time of racial origination*.
The set of human beings existing at this time will be divided by identifying the
founder population of recessive homozygotes, the founder population of dominant
homozygotes, and the residue (the heterozygotes). The first generation of the blue-
eyed pure race is the founding population of recessive homozygotes; the $n + 1$st generation of the blue-eyed pure race consists of the offspring of matings between
parents each of whom belongs to the nth generation of the blue-eyed pure race (or
to some earlier generation). The pure races picked out in this way satisfy both (R2)
and (R3), but "racial divisions" of this kind are of little significance. Part of the reason
is surely that the overwhelming majority of the species would be counted as of mixed
race, and many of these people would be both genetically and phenotypically identical (as far as eye color is concerned) to members of one of the pure races.

So while (R1)–(R3) pick out important features of the concepts of race which
we employ, they are by no means sufficient to reveal what is distinctive about racial

divisions. Nevertheless, the construction that shows how to prune populations of homozygotes so as to satisfy (R3) is helpful, for it makes explicit the idea of a historical lineage within which inbreeding occurs. I take this to be essential to any biologically significant racial concept: instead of trying to draw racial divisions on the basis of traits of the contemporary population, it is necessary to consider patterns of descent. *The concept of race is a historical concept.*[24]

However, while a certain type of history is necessary for racial division, it is not sufficient. Whether or not we demand some special genetic feature in the founding population, it is possible to satisfy (R2) and (R3) by choosing a time of origination, splitting the temporal segment of the species at the time of origination into founding populations, and identifying the successor generation of a pure race as the offspring produced by matings between members of earlier generations of that race. We can pick times of origination as we please, gerrymandering founder populations as we fancy, but none of this will be of the slightest biological significance unless two further conditions are met: (1) the members of the pure races thus characterized have some distinctive phenotypic or genetic properties; (2) the residual mixed-race population is relatively small, at least during most of the generations between the time of origination and the present.

It is important to recognize, from the start, that the idea of a pure race is an idealization (and, once again, the notion of idealization should carry no connotations of special goodness). Just as meteorologists analyze the complexities of the weather by producing charts with lines marking "fronts," so it is possible to understand the messy facts of human reproduction and biological transmission by looking for approximations to historical lineages that are completely inbred. The descent of contemporary people might show any number of patterns. Our species might have been completely panmictic from a time in the distant past (*panmictic* populations are those in which each member of one sex has an equal probability of mating and reproducing with each member of the opposite sex). Or, at the other extreme, inbreeding might have been so tight that, for generations, brothers have only mated with sisters. The concept of a pure race that I have described will be a useful notion in charting human reproduction across the generations, if there are groups that persist for long periods during which they are *mostly* inbred. Such groups will contain a number of families, and, at any given time, most of the families in a group will be interbreeding with other families in the group, and, for each family in the group, most of its history will be one in which family members interbreed with other families in the group. This is the relevant sense in which the notion of a pure race might idealize (or approximate) actual mating patterns.

At this point it should be apparent how notions related to that of reproductive isolation enter the picture. For the residual mixed-race population to be small, interbreeding among the pure races has to be infrequent. Moreover, if this is the case, then the possibility of maintaining distinctive genetic properties for the pure races will be greatly enhanced.[25] Even if the initial differences between founder populations at the time of origination are small, if descendants of those populations face different selection pressures, and if they mate almost invariably with one another, it is possible that, after many generations, the pure races will have different distributions of genes and of allelic combinations.

At this point, we can begin to see how the racial concepts we actually employ might be generated.[26] Racial divisions start with the idea of a division of the species into founder populations (not necessarily contemporaneous), which generate pure races in the recursive way described. Through most subsequent generations, interbreeding between the pure races is low, initially, at least, because of geographical separation and limited dispersal. Thus we arrive at the idea that the phenotypic or genetic features taken to mark out particular races—skin color, physiognomy, distribution of blood types or of alleles conferring susceptibility to various diseases—gain their significance because lineages have differentiated in the absence of reproductive contact. But none of this would have the slightest importance, or interest, if geographical union produced a thoroughly panmictic population. The fact that lineages which have been geographically separated in the past have distinctive characteristics has no biological significance unless, when current populations in different lineages are brought together, there is an incipient form of reproductive isolation. If men and women with very different genealogies breed freely, then the separation of their ancestors is of no enduring biological significance.

The notion of reproductive isolation is frequently misunderstood. Clusters of populations are reproductively isolated from one another just in case, where populations in different clusters are in geographical contact, they interbreed only at low rates. The tendency in much nonspecialist thinking is to suppose that reproductive isolation requires the impossibility of mating under any conditions. But this is far too strong a demand: many species will interbreed when their natural environments are disrupted, as witnessed by the numerous instances of hybridization in captivity. Nor is it reasonable to demand that members of different species never mate in the wild. Naturalists know numerous instances of *hybrid zones,* regions within which two species meet and produce hybrids. In some cases, the hybrids are sterile, in others fertile; there are examples of hybrid species in frogs—and possibly even in chimpanzees.[27] What is crucial for preserving species distinctness is that the hybrid zones remain stable, so that genes from one species do not flow to the other. Stability of hybrid zones rests on the greater propensity of conspecifics to mate with one another than with a member of another species.

Hybrid zones typically occur at the edge of a species range. Here, members of the species seeking potential mates only encounter conspecifics at low density. If they are more likely to meet an organism from a closely related species, the lower propensity for mating with a member of the alien species may be overwhelmed by the greater frequency with which aliens appear. If we associate with each organism in a species a probability that that organism will mate with a conspecific, given that it mates at all, then that probability will vary from 1 (or a number infinitesimally close to 1) to a significantly lower number in those regions where conspecifics are rare.

Underlying this distribution of probabilities may be a species-wide propensity to favor conspecifics as mates. That propensity, in its turn, rests on the traits of the organisms that make them disinclined to interbreed, the so-called *isolating mechanisms.* Isolating mechanisms are of many types, ranging from incompatibility of genitalia, inability of sperm to fuse with ova, low survival probabilities for the embryo, through differences in time of activity or in microhabitat that keep the species separated, to complex behavioral differences. Some species of *Drosophila*, for example, are kept apart through subtle differences in the ritual behavior that precedes normal mating: males who perform a slightly deviant sequence of movements are only accepted as mates *in extremis*. Caribbean species of the lizard genus *Anolis* occupy the same area, but are differentiated in terms of habitat: one species is primarily found in the crowns of trees, another on the trunks, yet another on the ground around the trunks.

So far I have characterized reproductive isolation in terms of differences in mating probabilities, focusing on the probability that an organism will mate with a conspecific, given that it mates at all. However, it is also possible that mating within the species has a more fine-grained structure, so that the probabilities of mating with conspecifics with distinct phenotypic traits are different. So, for example, the species may divide into a number of groups with characteristic phenotypes, such that the probability of any group member mating with a member of the same group, if it mates at all, is very high, while the probability of mating with a member of another group, if it mates at all, is correspondingly low. If this occurs when the groups are in geographical contact with one another, then we can think of the groups as reproductively isolated *to some degree*, with the degree varying with (a) the probability of within-group mating and (b) the extent of the geographical contact. In the extreme case, in which the groups are thoroughly and completely geographically mixed within the range of the species, so that organisms are just as likely to encounter members of alien groups as they are to meet members of their own group, and in which the probability of mating out falls to the level that is usual for species within the interior of their range (i.e., little more than 0), then the groups have become distinct species. But long before the extremes are reached, the differences between inbreeding and outbreeding rates may be sufficient to preserve the genetic differences that underlie the distinct phenotypes—or, at least, substantially to retard the erosion of those differences.

If there is a workable biological conception of race, then it must, I believe, honor (R1)–(R3), employ the historical construction in terms of founder populations and inbred lineages, and finally, demand that, when the races are brought together, the differences in intraracial and interracial mating probabilities be sufficiently large to sustain the distinctive traits that mark the races (which must, presumably, lie, at least in part, in terms of phenotypes, since organisms have no direct access to one another's genes). Now, it is evidently possible for groups with distinctive phenotypic traits that have been geographically separated for many generations to form a completely panmictic population when they are reunited—so that the intergroup mating probability is exactly the intragroup probability. If this should occur, and there are m pure races occurring at frequencies n_i, at the time of geographic union, then, after k generations, the frequency of the ith pure race would be expected to be $n_i^{2^k}$. The significance of this point is that, if we contemplate an initial situation with two races in frequencies 0.9 and 0.1, then, after 10 generations the expected frequency of the majority pure race would be around 10^{-47}. If the distribution is less extreme, or if there are more races, pure races disappear even more rapidly.

IV

Let me now use the rather abstract and general approach I have been developing to consider the possible biological foundation for a division of our species into races. If my analysis is correct, then the core of any biological notion of race should be that phenotypic differences have been fashioned and sustained through the transmission of genes through lineages initiated by founding populations that were geographically separated, and that the distinct phenotypes are currently maintained when people from different races are brought together through the existence of incipient isolating mechanisms that have developed during the period of geographical separation. Part of this presupposition is probably correct. There surely were geographically separated populations that would serve as founder populations for making some racial divisions—although it is not clear to me that this can provide anything other than a coarse-grained division, picking Out the "major races."

In fact, the patterns of gene flow in the history of our species are complicated. Eliminativists insist on the connection of sub-Saharan African populations to northern African populations; these, in turn, to Middle-Eastern Arab populations, and so forth; much has been made of the flow of genes across central Europe. However, such linkages do not ensure that extreme populations are linked in ways that make them part of the same evolutionary unit at all levels. Studies of the history of marriage in southern England and in Italy testify to an amazing proximity of spouses, even comparatively recently.[28] It is not hard to show that if interbreeding is relatively tightly confined, then populations separated by large distances (at the opposite edges of a continent, say) are effectively independent with respect to the genetics of microevolutionary change. In effect, some populations—the Arabs of Mediterranean Egypt and the indigenous peoples of southern Africa, or Norwegians and Greeks—have not exchanged genes to any significant degree. The phenomenon is analogous to that of so-called ring species, illustrated in species of gulls around the north pole or snakes in Texas:[29] two species whose ranges join and which do not interbreed are connected by a chain of populations, each of which interbreeds with its neighbor. Just as biologists recognize two distinct species in such instances, so too they might view two populations that only interbreed to a very limited degree as constituting races, despite the fact that they occur at opposite ends of a transcontinental cline (a sequence of populations along which there is genetic variation in a particular direction, so that, while adjacent populations may be quite similar, differences in the extremes are quite pronounced).

So the first part of the presupposition—the commitment to a history of reproductive separation—strikes me as correct, at least for some ensembles of populations. In particular, the United States is currently home to many groups who represent the latest stages of lineages that have not exchanged genes for a very long time. What about the second part, the thesis that when the populations come together they still do not exchange genes at high frequency? Here, firm data are hard to find, and the picture that emerges from statistics and anecdotes is by no means uniform. Some groups, when reunited, interbreed more readily than do

TABLE 11.1 Black-white marriages in 1970

Race of husband	Race of wife	
	White	Black
White	99.7	0.7
Black	0.1	99.2

Source: U.S. Census, 1970.

others. However, if the incomplete studies I have managed to track down are reliable, they do show that rates of interbreeding between some groups are very low. In particular, some groups of people designated as "black" only mate infrequently with other groups designated as "white."

At this point it is worth being very explicit about what I am claiming. In reconstructing the notion of race, I have suggested that groups are racially separated if certain facts about reproduction obtain: this shows the *possibility* of a biological notion of race. Specifically, if the "blacks" and "whites" in a particular region at a particular time reproduce together at a relatively low rate, then we can say that there is an incipient racial division between those groups at that place and at that time; if the rate of interreproduction remains low across a period, then we can talk about two races in that region. Since I can only appeal to indications of relatively low rates of mating between American "blacks" and American "whites," not to firm data systematically collected over significant periods, I can only suggest, tentatively, that this division may answer to the notion I have reconstructed. I am, however, inclined to believe that this is likely to be one of the best (if not the best) examples of a racial division (although, here, as elsewhere, empirical research could prove me wrong).

Data on rates of interracial marriage are surprisingly hard to come by. I have not been able to obtain reliable recent figures. However, table 11.1, using data from the U.S. Census, shows the distribution for black-white marriages in 1970. Approval of interracial marriage apparently doubled between 1968 and 1978 (20 to 36 percent), although a recent poll (1994) has indicated that 20 percent of the American population still favor laws against miscegenation.[30]

Studies of other forms of intermarriage paint a different picture. It is reliably estimated that up to 50 percent of the marriages of Japanese people in the continental United States are with non-Japanese spouses (although by no means with non-Asian spouses[31]). The picture of interracial marriage in Hawaii is far more complex (see table 11.2).[32]

The "short version" of the recent survey of patterns of sexual behavior in the United States is very clear about the tendency to avoid interracial relationships.

> Almost as forbidden [as homosexuality] is interracial dating. The pressure to choose someone of your own race can begin as soon as teenagers start to date, and often sustains patterns of overt racism.
> That social pressure against interracial dating becomes greater the closer a couple comes to marrying.[33]

TABLE 11.2 Interracial marriage in Hawaii

Bride's ancestry	Groom's ancestry					
	Ca	Ha	Ch	Fi	Ja	Ot
Caucasian	517	230	36	86	79	52
Hawaiian and part-Hawaiian	177	515	20	121	94	72
Chinese	138	163	311	41	296	51
Filipino	114	159	26	584	69	48
Japanese	56	70	59	30	761	25
Other	201	18.5	.21	69	127	397

Interestingly, when the authors follow up these claims with several anecdotes about interracial couples who are cut off from their families and about the anger directed at people whose romantic friendships cross racial lines, the examples they choose all involve blacks and Caucasians.[34] The more technical version explores various preferences for kinds of similarities in sexual partners, suggesting that even casual relationships across racial lines occur at low rates.[35]

These sources clearly suggest that the second part of the presupposition for biologically significant racial divisions is partially satisfied. The United States consists of an ensemble of populations, some of which have been geographically separated before being brought into proximity with one another. Between some pairs of these populations, most notably between African Americans and Caucasians, the frequency of intermarriage is low, suggesting that these populations are behaving as separate units from an evolutionary and, perhaps, ecological standpoint. Emphatically, this does not mean that racial divisions can be drawn across the entire species, that the divisions into inbred populations that hold locally necessarily apply globally; my minimalist notion of race allows for the possibility that, within one geographic locale (say the United States, or even something narrower like the rural Midwest), two groups are racially divided, even though elsewhere they are not. The possibility of racial division that I am suggesting is specific to a broader group, an ensemble of populations that are present in a particular geographical region. Nor, even locally, need it honor all the traditional racial divisions. Although the evidence does appear to indicate a significant mating barrier between whites and blacks, the statistics about intermarriage between European Americans and Asian Americans (from at least some national backgrounds) tells a quite different story.

But why make such a fuss about intermarriage (or interbreeding)? If one grants, as I have done, that the phenotypic differences between groups are not significant and that intragroup variation swamps intergroup variation, why not let the race concept go? To answer these questions, it is helpful to adopt a conceit proposed by E. O. Wilson and recently taken up by Rushton.[36] Imagine a Martian naturalist visiting earth for the first time and observing our species. What infraspecific divisions, if any, would the Martian draw? Rushton announces confidently that they would spot three geographical "races" with quite different body types. But simply noticing the phenotypic variation in height, bone thickness, skin color, or whatever should not inspire the Martians to divide our species into races—Rushton's

Martians (and probably Rushton himself) make a mistake against which Ernst Mayr has inveighed for so long that it has become part of the standard equipment of any field naturalist concerned to identify the species in a particular area. Only the uninformed rush in and divide sexually reproducing organisms according to the differences that strike them, the outsiders, as salient. To repeat what is, perhaps, obvious: the notion of race I have been developing is not morphological, concerned with such features as skin color or physiognomy, but focused on patterns of reproduction; morphology plays a role only if morphological differences prove relevant to reproductive choices. In this, I am as much at odds with Rushton and others who deploy traditional notions of race as are the eliminativists who deny the biological significance of race entirely.

Taxonomic divisions should be grounded in distinctions that the organisms themselves make, in the propensities for mating and reproduction. Mayr named his conception of species "biological," both because it was founded on something of central importance to biology, the reproduction of organisms, and because patterns of reproduction reflect characteristics that matter to the organisms. So, a Mayrian Martian, looking at our species, would attend, above all, to the facets of our reproductive behavior, noting not simply the phenotypic differences but seeing that in some locales, like the United States, those phenotypic differences correlate quite strikingly with mating patterns. To return from our fantasy and state the moral more soberly, intermarriage statistics are crucial because those statistics (poor though they are) are proxies for what is biologically crucial in making taxonomic divisions.[37]

At this point it is important to confront an important objection. Many eliminativists have responded to the idea of articulating concepts of race along the lines I have proposed by suggesting that there are not significant intraspecific differences in gene flow, so that, despite the partial evidence from the incomplete statistics I have quoted, the presupposition for biologically important racial divisions is not satisfied. Two kinds of considerations prompt this line of response: (1) the familiar judgment that contemporary American "blacks" have some Caucasian ancestry, and, conversely, that many American "whites" have some African ancestry;[38] (2) the suggestion that, if there are indeed large differences in frequency between intraracial and interracial mating, this is a temporary phenomenon that is unlikely to produce biological effects.[39] I shall take up each point in turn.

In rough outline, what we know of the history of sexual relations in America between people of European descent and people of African descent suggests that there have been two main periods during which such relations were relatively common. First, in the early colonies, particularly in Virginia, indentured servants from Europe and Africans (either slaves or servants) flouted the strictures against sexual liaisons. Later, in the plantation South, there is no doubt that white men from slaveholding families often treated female slaves as sexual property. Since the offspring of these unions were counted as "black" (under the notorious "one-drop" rule), many "blacks" had one parent of European descent. The sexual relations between these "blacks" and others, some of whom also had European ancestors, spread genes from the "white" population into the "black" population. In similar fashion, those blacks with enough "European" features to pass as "white" sometimes married people of purely European ancestry, so that the genetic mixing went in both directions.[40]

If we now attempt to apply the concept of race I have developed to this history, there are two options: we can take the races to be ancient, setting the time of origination during the period of geographical separation, or we can suppose that the process of race formation begins at the time of Reconstruction. The first alternative appears to be blocked by the existence of two periods of substantial gene flow, and I think that it is the recognition of this fact that motivates sophisticated versions of eliminativism. In fact, however, matters are not so simple. For, in the first place, nobody has proposed that the probability that children born to people of African descent resulted from a union with a person of European descent was ever close to the probability that such children would result from a union with a person of African descent: neither the relations between indentured servants and Africans nor the exploitation of black women by white slaveowners ever came close to attaining the frequency of within-group unions. Second, from a purely biological point of view, it would be natural to redescribe the history by identifying two periods during which the proximity of people from two groups produced hybrid descendants, with the majority of these hybrids being assigned to one of the groups.[41] After these two periods were over, groups with somewhat modified gene pools (more extensively modified in the case of the blacks, only slightly modified in the case of the whites) once again engaged in cross-group unions, only at low rates. Even though the history does not strictly correspond to the requirements I have laid down for racial divisions, we might see it as an approximation to the idealized notion of separated, predominantly inbred lineages, disturbed only by two anomalous episodes in which the races are reshaped. From that perspective, the second episode, with its exploitation of black women, would not be viewed as the benign breaking down of interracial barriers, but as the coercive restructuring of the minority race.[42]

The second alternative would be to abandon the idea that the races are old and emphasize the low rates of interracial union during the past century.[43] This, of course, would be to invite the charge that such barriers are only temporary and thus of no significance for understanding human genetics and evolution. In response, it is worth noting two points. First, in introducing the biological species concept, Mayr insisted on a "nondimensional" version: populations at a given place at a given time belong to different species if they are not exchanging genes. In exactly parallel fashion, we could recognize "non-dimensional" races, groups at a particular place at a particular time that are not exchanging genes at substantial rates.[44] Second, and more important, I see no reason to conclude from the history that there has ever been a time at which people of African descent and people of European descent, with ample opportunities for mate choice, freely chose members of the other group at rates close to those with which they selected members of their own group. (I emphatically do not rejoice in this idea, but it does seem to represent our species' sexual past.) If that is so, then the incidents during which intergroup unions have been relatively common are the anomalies, and we should not think that the current tow rates are a temporary phenomenon that will lack biological implications.

I conclude, tentatively, that we can use the concept of race I have articulated to identify at least some divisions among contemporary Americans. This conclusion is tentative because further information about the history and current state of sexual unions in the American population (most pertinently those unions that produce children) might reveal a much greater rate of mixing than my account could allow.[45]

At this stage, there are a number of obvious questions both about the details of the approach I have adopted and about the division of our species into races. In the interests of making the position as clear as possible, it seems worth offering brief replies.

1. *Does this minimalist notion of race restore the status quo by yielding traditional racial divisions?* Although the evidence on patterns of reproduction is highly incomplete, it seems very likely that the view that there are three major races (Caucasian, African, Asian) will survive, if at all, only in highly qualified form. The statistics I have cited indicate that it is possible that there should be a division between Africans and Caucasians within the United States (although this might not hold elsewhere in the world), and that it is unlikely that there will be a division between Asians and Caucasians that will hold across the United States (although there might be more local divisions of this kind). I have given no grounds for even the most tentative opinion on the issue of whether there will be a division between Asians and Africans.

2. *How do divisions by race interact with divisions by social class?* There are two interesting issues about the interconnections of race and class. The first is whether the account I have given can always distinguish class divisions from racial divisions. In England immediately after the Norman conquest, for example, it seems possible that the population divided into two classes, an affluent class of landowners (often Norman) and a class of peasants (virtually all Saxon), and that these were reproductively disconnected. On the account I have given, these classes could be viewed as races, and we could describe the situation as one in which the English aristocracy was fashioned from the restructuring of a Norman population by the admixture of some (wealthy) Saxons. More generally, any situation in which there is limited intercourse (primarily sexual) among different classes could be viewed as one in which those classes function as different races (a judgment, interestingly enough, that members of the classes may express, albeit often with a different conception of race in mind). Interestingly, the institution of the droit de seigneur may have undermined any such racial division.

Second, just as a racial division may hold only in a particular locale, so too it may also obtain only within a particular social class. Consider the possibility that middle-class American "blacks" and "whites" are far more likely to reproduce together than are their working-class counterparts (a possibility that would invert the likely situation in the original colonies). Under these circumstances, there would be a class-relative racial division between Africans and Caucasians.

3. *Aren't the notions of reproductive disconnection and of the endurance of races both matters of degree?* Yes. I have talked, vaguely, of populations exchanging genes at relatively low rates and of divisions as enduring. Behind these vague remarks stand precise figures, as yet unknown, about the rates at which different groups interbreed over a number of generations. The same vagueness infects biological usage of subspecific (and even higher-order) taxonomic categories, and it is easy for there to be unclear cases. Surely, however, if we were to discover that the population

divided into As and Bs, that As interbred with Bs with probability .01, that Bs interbred with As with probability .03, and that these figures remained relatively stable
(showing some fluctuations but never rising far above the values given) for a century
and a half, then we could talk of a division into two races. Now, the actual data on
patterns of reproduction may be nowhere near so dramatic, and we may end up by
having to understand reproduction and biological descent by introducing explicitly
degreed concepts. This could be done, for example, by measuring the *strength* of
racial separation by the ratio of the probability of mating within to the probability
of mating out, and by measuring the *endurance* of a racial division by the number
of generations through which it persisted. Relations between groups could then be
indexed by their endurance at or above a given strength: so we might discover that
the African-Caucasian split relative to a geographic location (and perhaps to a class)
had endured at a strength of 20 for six generations. Development of such degreed
notions is straightforward, and I shall not pursue it here. It is sufficient to note that
some of the questions about the relationship between biological and cultural transmission could be raised by employing such concepts.

4. *What is the relationship between my position and eliminativism?* Even
though my approach and conclusions are at odds with eliminativism, I continue to
share the fundamental points that eliminativists have made against older, typol-
ogical, racial concepts: the characters that divide races (in my sense) are not significant, and the intraracial variation is greater than the interracial variation. What
I deny is the eliminativists' insistence that racial divisions correspond to nothing in
nature: I maintain that they correspond to patterns of mating, although I concede
that empirical facts about such patterns could show that they are adequately charted
only by using explicitly degreed concepts. However, even though I oppose the thesis
that races are purely social constructions, there is a deeper sense in which I want
to accept, and even to take further, this theme in eliminativism. When we look
behind the patterns of mating at the underlying causes, we see just the kinds of
factors that eliminativists emphasize. I shall explore this theme in the next section.

V

Given that members of some pairs of groups do not engage in sexual relations at a
very high rate, why does this occur? I can imagine all kinds of biological stories
about our greater propensity for mating with members of our own race than for
mating with members of different races. Perhaps our species has evolved "genes for
xenophobia," and the statistics represent the impact of these genes. Maybe, we
should take a cue from Patrick Bateson's beautiful experiments on mating preferences in Japanese quail, which show that the birds have a degree of attraction that
is low for very close relatives, low for their most unrelated conspecifics, and that
peaks at second cousins.[46] It is all too easy to lapse into pop sociobiology, either pos-
tulating genes and selective pressures to suit our fancy or extrapolating wildly from
meticulous animal studies.

But there is, I believe, a much more obvious explanation of the differences in
mating propensity. Isolating mechanisms may be very subtle, depending on the

nuances of an organism's responses to the behavior of others. *Drosophila*, recall, are very sensitive to the movements of potential mates. Furthermore, even when members of two species occur in the same region, they may be separated by differences in times of activity or in their microhabitats. Combining these points, it is not difficult to sketch an explanation of the reduced probability of mating between whites and blacks that accords with a host of familiar facts. Black people and white people may traverse the same terrain—the streets of the same city—every day without much significant contact. So long as whites and blacks live in different areas, work and pursue recreation in different places, geographical contact between the races is only superficial (recall the *Anolis* lizards of the Caribbean). Moreover, even when contact does occur, the people who meet may not provide one another with the right signals: from the tiniest gestures to ways of expressing ideas, expectations may easily be defeated.

In fact, a single dominant theme runs through the literature on the difficulties of interracial marriage. Successful relationships must surmount a barrier built up from local attitudes to the history of racial interactions. Oversimplifying enormously, that barrier is constructed in three stages. At the first stage is the history of colonialism, slavery, decades of injustice, and the perpetuation of economic and social inequalities in the present. This produces, at the second stage, attitudes of fear and resentment in families who see a relative contemplating an interracial marriage. The third stage consists in the recognition, by the protagonists, of the attitudes of their families, and their growing awareness that they may be cut off from those they love and that their children may grow up without any extended family whatsoever. Whether or not other forms of cultural signalling operate at earlier stages, so that people from different races are rarely initially drawn to one another, for those who find themselves attracted to members of different races, the barrier I have described is frequently acknowledged as the crucial obstacle to marriage. Interracial couples almost invariably mention this barrier and the ways in which they have overcome it.

The sources of the low rate of black-white mating lie ultimately, I suggest, in the history of slavery and colonialism, and, more proximally, in socioeconomic inequities. The *current* economic inequalities make significant contact between blacks and whites unlikely, and the past history of economic differences, with the social consequences of past exploitation and attempts at suppressing black culture, erect barriers that are hard to remove. The eliminativist emphasis on the role of social causes in the construction of race is thus not entirely misguided: at risk of solecism, we might say that races are *both* socially constructed and biologically real. Biological reality intrudes in the objective facts of patterns of reproduction, specifically in the greater propensity for mating with other "blacks" (or other "whites" respectively); the social construction lies in the fact that these propensities themselves have complex social causes.

To understand this apparently paradoxical view, we should recognize that there are three distinct views one might take about the biological significance of racial divisions. The two that have figured largely in twentieth-century debate are, on the one hand, that there are biologically significant divisions between races (e.g., between whites and blacks), and, on the other, that there are no such significant divisions and that the concept of race is an illegitimate social construct that should be discarded. In my judgment, this opposition intertwines a number of separate issues. First, if there is, as I have claimed, significant difference between the probability of intraacial mating and the probability of interracial mating, then the phenotypic and genetic characteristics that distinguish racial groups can be sustained, and, at a microevolutionary level, races are behaving as separate evolutionary units. Thus, if the empirical facts are as I have taken them to be, eliminativism with respect to the concept of race, while an attractive position, cannot be upheld — although it might be noted that traditional racial divisions might be no more biologically significant than other divisions with the structure I have identified. However, while, in the case of other species, the development of incipient isolating mechanisms during a period of geographical isolation might be conceived as a purely biological phenomenon, resulting from the increase in frequency of alleles that dispose organisms not to mate with members of the other group, I see no grounds for any such explanation for the different mating propensities in races of *Homo sapiens*. Here, the account of the separation of (say) blacks from whites seems to be purely cultural, a matter of the patterns of behavior that have been transmitted across the generations through modes of *nongenetic* inheritance, as well as the accidents (many of them tragic and disastrous) of the relations among the two groups. Hence, while the concept of human races may have biological significance, in the sense that there are differences in gene frequencies which can be preserved because of low probabilities of interracial mating, the explanation of the mating preferences may have no biological significance. Races may *quite literally* be social constructs, in that our patterns of acculturation maintain the genetic distinctiveness of different racial groups.

I do not have any definitive refutation of the hardline sociobiologist who insists that our propensities for mating within racial groups are caused by our genes and not by differences in culture and history. There is no evidence in favor of any such view and, as I have noted, plenty of familiar phenomena that suggest the third option I have sketched. In the remainder of this essay, I want to explore the implications of that "mixed" approach to concepts of human race. I shall start with a closely connected notion, that of ethnicity.

VI

The core of the view that there are ethnic groups is that distinct sets of cultural items, including lore, habits of interpersonal interaction, self-conceptions, and behavior, are transmitted across the generations by a process akin to biological inheritance. In recent years, careful studies of cultural transmission[47] have revealed both similarities and differences with the process in which genes are passed on. Plainly cultural inheritance can involve more than two "parents," and some of the "parents" may even belong to the same biological generation as their "offspring." Nonetheless, there are enough common features to enable us to pick out cultural lineages with the same formal structure previously discerned in races. Thus we can introduce a concept of *ethnicity* meeting the following conditions:

(E1)   An ethnic division consists in a division of *Homo sapiens* into nonoverlapping subsets. These subsets are the pure ethnicities. Individuals who do not belong to any of the subsets are of mixed ethnicity.

(E2)   Pure ethnicities are closed under cultural transmission. That is, the cultural "offspring" of "parents" all of whom are of ethnicity E are of ethnicity E.

(E3)   All cultural "ancestors" of any member of any pure ethnicity are of that ethnicity. If someone is of ethnicity E, then all their cultural "parents" are of ethnicity E.

However, if these conditions are to be realized in a world in which different cultures collide, it will be important to impose restrictions on cultural parentage. Liberal definitions of "parent" would allow anyone who transmitted any item to another person to count as a cultural parent—so that attendance in a classroom taught by someone of a different ethnic heritage would automatically disqualify a child from belonging to a pure ethnicity. I shall tolerate considerable vagueness in deciding how to resolve this problem, proposing that cultural "parents" be those who are responsible for the dominant items of the offspring's culture, where this should be taken to consist of those facets of lore, habits, conceptions, and behavior that are both central to the person's life and distinct from parallel items in the rival surrounding cultures. Roughly, the idea is that cultural parents transmit something that is important for the people they influence and play a role that could not have been filled by others from a different culture.

A second important modification that might be made is to recognize lines of cultural descent with respect to particular areas of human life: so we might focus on the transmission of religious beliefs and practices, musical tastes, food preferences, and so forth. Looking at the lines of descent generated in these various areas, we might discover that they were importantly different, that it was impossible to assign people to single "ethnicities," but that all of us belong to a variety of cultural lineages, some of which might match biological lineages while others were quite distinct. Alternatively, we might find that whatever field of human life we considered, the division into cultural lineages always produced the same divisions, in which case we would be justified in speaking about a single ethnicity to which a person belonged. We could then go on to ask the question of the relations between ethnicities so defined, and races.

Assuming that we obtained consistent lines of cultural transmission across different areas of human life, we could construct ethnicities by considering lines of cultural descent from founder populations, supposing, as before, that the $n + 1$st generation of the ethnicity consists of all those whose cultural parents belong either to the nth generation or are one another, and who have at least one cultural parent in the nth generation (these complications are needed to circumvent the problem of within generation cultural transmission). As in the case of races, if ethnicities are to be important they should be able to maintain themselves when they come into contact. So we should demand that genuine ethnicities have mechanisms of partial cultural isolation. Even in a multicultural society, the chief influences on new generations should not be thoroughly mixed. This criterion might be formulated by demanding that only a small percentage of offspring have cultural parents from different ethnicities, or by requiring that for almost all offspring the set of cultural parents has a very large majority from a single ethnicity.

The main thought behind the approach I have taken to the concept of race is that the two systems typically harmonize—indeed, that they reinforce one another. On the biological level, interracial mating is limited through the differences in the cultural items acquired by members of different races, that is, because different ethnicities belong to different races. On the cultural level, patterns of culture are preserved because culture is usually primarily transmitted by parents and other family members (who may also influence the receptivity to other potential cultural parents), who belong to the same race and share the same ethnicity. One particular consequence that I have emphasized above is that past racism shapes the attitudes of people today, in particular their attitudes to sexual union, and that this can maintain patterns of mating that are skewed toward one's own group.

The picture I have been assuming allows for the possibility that each race might correspond to many different ethnicities, although it suggests that the transmission of culture through any of these ethnicities serves to lower the probability that someone will marry a person of a different race. One way to question this assumption (already noted above) is to break down the notion that ethnicities are holistic entities that come one to a person. Just as eliminativists about race argue for the appreciation of human diversity without supposing a discrete system of divisions, so it might be suggested that cultural transmission affects all of us in slightly different—or very different—ways, and that ethnic boundaries are blurred. Further, following my oversimplified analysis of the causes of propensities for not mating out, it might seem possible to detach the general feature of many systems of cultural transmission that creates the incipient barrier to interracial marriage from the more specific characteristics of ethnicities. To put the point concretely, perhaps a society-wide readjustment of economic and social relationships among black people and white people would undercut both the fears and the resentments, leading to a situation in which, while certain distinctive cultures (religious traditions, styles of music and of literature) were retained within lines of cultural descent, the barriers to interracial marriage were substantially weakened. If the family of the white fiancée of the young black man no longer worries that she will be plunged into poverty, and if black women no longer see the black man as a scarce resource in a world in which few black men come to manhood with auspicious prospects, then whether or not differences in other forms of culture (ranging from tastes in food through styles of socializing to appreciation of forms of art and entertainment) are lessened, the pressures against interracial marriage may be substantially released. Hence it would not be necessary that cultural transmission *as a whole* become more mixed, but simply that certain background elements that affect part of every system of cultural transmission be changed.

We currently know too little about exactly how to reconstruct ethnicities and how to apply the reconstruction to understand their bearing on people's decisions and actions (for example, on their decisions about whom to marry). In proposing a fairly abstract account of ethnicities, I presuppose a particular apparatus which seems promising in coming to terms with systems of cultural transmission. It may turn out that this proposal fails at any number of levels: (1) the quasi-Mendelian formalism in terms of transmission of discrete items of culture is just inadequate to the phenomenon of cultural transmission; (2) when that formalism is applied it may reveal a parallel situation to that found in the case of conventional racial divisions, to wit that minority ethnicities are "mixed" whereas majority ethnicities are "pure"; (3) the mapping from ethnicity to race may not be many—one; and (4) detachable elements within the system of cultural transmission and/or common features which shape all such systems within the society (e.g., background economic and social inequalities) may play a dominant role in certain kinds of decisions and actions (e.g., decisions about marriage). All of these points need detailed exploration. Here I intend only to raise what I take to be important neglected questions about ethnicity and its connection with race, and to consider the consequence of *fallible* assumptions. Thus I do not wish to claim that it is plainly impossible to detach those features of cultural transmission that lower probabilities of interracial marriage from other parts of the system of cultural transmission.[48]

VII

I now want to explore the connections between race and ethnicity in a bit more detail, by considering how biological races and ethnic identities might both break down.

Suppose, first, that cultural transmission were to become much more heterogeneous, so that children became influenced by the ideas, habits, and lore of what now count as many distinct ethnicities. If my conjecture about the mechanisms underlying the differences in probabilities of intraracial and interracial mating is correct, then the more multicultural society might exhibit an increased frequency of interracial mating. This increased frequency would, in its turn, be likely to generate an increase in the proportion of children of mixed ethnicity. Perhaps that mixture, in its turn, would continue to erode the (partial) isolating mechanisms among races. We can envisage a spiral toward a point at which the divisions by race and by ethnicity both disappear.

All this is speculative, and the interface between biology and culture is a region in which speculations should be taken with great caution. Nonetheless, I think it is worth considering the consequences of this speculation, asking, in particular, whether it points to a constraint on our future social practices.

I began by considering what I characterized as a relatively benign social use of the concept of race, envisaging serious discussion of the desirability of trans-racial adoption. One important question to ask is whether the issues can properly be framed in terms of ethnicities—is the significant question whether trans-*ethnic* adoption is desirable? At first sight, this appears to be quite wrong. The child's ethnicity is not already defined at birth (or at the early age at which she is adopted): her ethnicity will be identified through the cultural influences that impinge upon her, and there is no issue of *violating* an ethnic identity she already has. Supposing we assume that the cultural milieu into which she will be pitched through adoption will be rich, whereas, if she is left where she is, she will have to struggle simply to survive, there would seem to be no reasons for opposing the adoption.

Now, there seem to me to be two important ways of undermining this argument, one that attends to the consequences at the level of the entire population (or species) and the other that makes explicit use of the concept of race. The most forceful way to express the first is to envisage a situation in which there are two ethnicities, E_1 and E_2. Suppose that children born to parents of one of these ethnicities, E_1 say, routinely experience various forms of deprivation, that they have little chance of benefitting from the richness of the culture, that, in many instances, biological parents simply lack resources to provide their children with access to significant parts of the culture, that with high frequency those children are simply left to rot. The alternative ethnicity, E_2 by contrast, is well-endowed, and children reared by people in that ethnicity are assured physical well-being and security as well as a rich cultural milieu. Acting in the interests of the children, well-meaning social planners allow the adoption of a large proportion of children from the economically disadvantaged ethnicity, so that the cultural traditions of that ethnicity are weakened and finally disappear. They reason, quite understandably, that issues of survival may swamp considerations of cultural transmission,

Conservationists are properly concerned about the extinction of biological species. We should probably be even more worried by the thought that major cultural traditions might vanish: diversity enriches our lives. However, for each individual, it may be better if that individual belongs to the dominant ethnicity. Hence a social policy directed toward individuals may bring about a situation in which valuable cultural traditions are lost.

An obvious remedy, roughly realized in contemporary treatments of Native American ethnicities, is to "protect" cultural traditions that are in danger of disappearing either by enhancing the benefits of remaining within that ethnicity or, more likely, by offering to people of a particular racial group only limited opportunities for transferring to the dominant ethnicity. What I want to note is that, once again, the concept of race, and the ideal of harmony between race and ethnicity, figures here. The strategy of preserving a culture threatened with extinction is not implemented by proceeding in race-blind fashion, so that biological ancestry is irrelevant to who undertakes to continue the ethnic traditions. Genealogy is felt to make a difference: people should preserve "their" culture.

The second way of questioning transracial adoption makes explicit use of the concept of race, and tries to defend the principle that ethnicity and race should be in harmony. As I pointed out at the beginning, the phenotypic and genetic differences among racial groups for which we have any evidence are trivial. Nevertheless, those differences, particularly the differences in skin pigmentation and physiognomy, have come to be taken as markers that signal membership in distinct clusters of ethnicities. In societies which make the biological mistake of overestimating the significance of variations in trivial aspects of the phenotype, and the moral mistake of showing at least sporadic intolerance toward the ethnicities associated with certain biological markers, a person's manifest biological traits will make a difference to the way in which she is treated. Thus, even if she comes to think of herself as part of the dominant ethnic group, if she has the phenotype associated with another ethnicity, it is likely that she will be treated, at least periodically, as if she were not a full member of the ethnic group with which she identifies. Nor will she have available to her the strategies for coping with the repudiation of the culture assigned to her which have been passed on in the ethnicity associated with her phenotype. So the simple argument for the harmony between race and ethnicity emphasizes the idea that the biological and moral mistakes of the past live on in the present, and that, in a society that has not completely freed itself from racism, mismatch between race and ethnicity will leave people rootless and defenseless.

This argument allows for the possibility of a future in which tolerance for alternative cultures is so widespread that insignificant phenotypic markers lose their significance in our social interactions. Harmony between race and ethnicity is valuable only because it serves instrumental purposes in societies with residues of racism. Yet it may well be thought that this does not uncover the deep motivation for insisting on the match. Other things being equal, we may feel that individuals should identify with the culture of their biological ancestors, that they should sympathize with the pains and struggles of great-great-grandparents whom they know only as dim figures in a shadowy past. Or, to put it more negatively, that failure to carry on the culture of one's genealogy is a kind of betrayal. I want to conclude by scrutinizing this idea.

On grounds of promoting cultural diversity, as I have remarked, it is important that some group of people should continue the lore and customs of each ethnicity—including the one of my biological ancestors. But why should it fall to me to continue those traditions? Why should I not pick and choose, identifying with bits and pieces of cultures that are quite alien to the practices of my forebears? After all, cultural inheritance, unlike biological inheritance, is multiparental, and it would be possible for each of us to make cultural linkages with all sorts of people and traditions, weaving their contributions together into idiosyncratic patterns.[49] We can envisage, and perhaps educational reformers are already envisaging, a multicultural society in which we are all ethnic hybrids. What exactly would that society have lost?

Moved by a biological analogy, we can appreciate the possibility that cultural mixing would quickly destroy the distinctive contributions of pure ethnicities, ultimately arriving at a state of relative cultural homogeneity. When populations that have been geographically, but not reproductively, isolated are brought together to form a thoroughly panmictic unit, the range of phenotypic variation may quite dramatically decrease. Setting on one side arguments from the intrinsic value of cultural diversity, there is a very different style of consideration that develops the thought that our biological ancestors should have a special role in our identification of who we are. Perhaps we have a natural tendency to identify with our biological parents, so that we take pleasure in developing a sense of values that accords with theirs and feel pain when we are at odds with their customs and ideals. A society which made a radical divorce between ethnicity and biological ancestry would thus rub against the grain of human nature.

Like most claims about the relationship between biology and culture, this seems to me to be pure speculation. It is possible that people are "hardwired" to feel this special cultural kinship with their biological ancestors. It is also possible that our sense of identification with our biological forebears expresses a pattern of socialization common to all, or most, societies, a pattern that may itself be part of the legacy of racism and xenophobia. We simply don't know if ethnic roots have to be biological roots to make us happy.

Of course, the consequences of the two assumptions are quite different. If the propensity to identify with our biological parents would develop in us across the entire range of social environments that we might contrive (or, more exactly, the entire range in which people would flourish), then we can expect ethnicities to remain relatively pure, to be in harmony with divisions into races, and for the practices of dividing people by race and by ethnicity to reinforce one another in the fashion I suggested earlier. It would not inevitably follow that we were committed to a racist society, for the appreciation of difference might not be associated with the idea that distinct groups have distinct worth. Nevertheless, there are surely grounds for concern that, either because of cognitive or moral limitations, people would, in practice, think of their own culture not as one among many but as the best.

By contrast, if our descendants could fashion their own eclectic mixes of culture without violating any sense of identification with ancestors, then we can envisage a future in which the concepts of race and ethnicity both become irrelevant. Cultural hybridization could be so promiscuous that we would simply recognize the different cultural identities of all individuals, and, as I suggested earlier, it is likely that the breakdown of ethnicities would promote mating between people now identified as belonging to different races, thus undercutting what I have exposed as the biological significance of racial divisions. Perhaps in this imaginary society the inability to demarcate clear groups would promote greater tolerance or even a celebration of human diversity

Something like this vision is what moves eliminativists. They worry that it is not enough to insist on the equality of races, and they propose that the most thorough way to combat racism is to discard the outworn concept of race. To this end, they contend that the concept of race lacks biological significance. I have been arguing that this is wrong, and that the interconnections between biological and cultural concepts are intricate. Those interconnections raise numerous empirical and moral questions that must be addressed if we are to decide if the vision of a society that abandons practices of racial division is either realizable or desirable.

Notes

I owe an enormous amount to many people who have helped me with this project. My colleagues Jim Moore and David Woodruff gave me excellent advice about issues in anthropology and evolutionary biology (they should not, of course, be blamed for my errors). The first version of this essay was written for a wonderful conference on concepts of race and racism, organized by Jorge Garcia at Rutgers University in the fall of 1994. The high standard of the discussion at that conference and the combination of incisiveness and open-mindedness among the people with whom I interacted were a vivid reminder of how productive philosophical exchange can be. I am grateful to Lorenzo Simpson for his extremely thoughtful comments which have helped me reshape much of my raw material. The comments of David Goldberg, Leonard Harris, John Ladd, Howard McGary, Michele Moody-Adams, Lucius Outlaw, Ken Taylor, and Gregory Trianosky-Stilwell have also been valuable. I am particularly indebted to conversations with Anthony Appiah, Amy Gutmann, Michael Hardimon, and Naomi Zack. Finally, I would like to thank Michael Hardimon for some wonderfully constructive comments on the penultimate draft and Leonard Harris for his encouragement, as well as for his invitation to contribute this essay to *Racism* (Amherst, N.Y.: Prometheus, 1999), the book in which it originally appeared.

1. A. Montagu, *The Concept of Race* (New York: Free Press, 1964); F. B. Livingstone, "On the Nonexistence of Human Races," in A. Montagu, ed., *The Concept of Race* (New York: Free Press, 1962); J. Marks, "Patterns of Human Biodiversity," unpublished manuscript, 1994. For the perspective of an evolutionary biologist, see J. Diamond, "Race without Color," *Discover* 15 (1994): 82–89.

2. This should become clear in section 7. See, in particular, the remarks about Anthony Appiah's postracialist project in note 49.

3. See R. C. Lewontin, *Human Diversity* (San Francisco: Freeman, 1982), 115.

4. See S. Molnar, *Human Variation* (Englewood Cliffs, N.J.: Prentice Hall, 1992), and L. L. Cavalli-Sforza, *The History and Geography of Human Genes* (Princeton, N.J.: Princeton University Press, 1994).

5. Diamond, "Race without Color."

6. C. B. Davenport, "The Mingling of Races," in E. V. Cowdry, ed., *Human Biology and Racial Welfare* (College Park, Md.: McGrath, 1930), 557.

7. Ibid., 558.

8. Ibid., 558–559.

9. R. Herrnstein and C. Murray, *The Bell Curve* (New York: Free Press, 1994); J-P. Rushton, *Race, Evolution, and Behavior* (New Brunswick, N.J.: Transaction Books, 1994).

10. See N. J. Block and G. Dworkin, "IQ, Heritability, and Inequality," *Philosophy and Public Affairs* 3 (1973): 331–409 and 4 (1974): 40–99; R. C. Lewontin, "The Analysis of Variance and the Analysis of Causes," reprinted in N. J. Block and G. Dworkin, eds., *The IQ Controversy* (New York: Pantheon, 1974), 179–193. The best diagnosis I have seen of the errors in Herrnstein and Murray's much-criticized work is offered by N. J. Block, "Review of *The Bell Curve* (Herrnstein and Murray, 1994)," *Cognition* 56 (1995): 99–128.

11. The norm of reaction of a trait is the function that, for given genotypes, maps constellations of environmental variables onto phenotype. It is most conveniently represented as a graph showing the variation of phenotype with environment. Fundamental points about the importance of norms of reaction are made in Lewontin, "The Analysis of Variance." I attempt to articulate those points and defend the notion of a norm of reaction against recent criticisms (including some from Lewontin himself) in Philip Kitcher, "Battling the Undead"; reprinted as chapter 13 in this volume.

12. See S. J. Gould, *The Mismeasure of Man* (New York: Norton, 1981).

13. See Block, "Review of *The Bell Curve*."

14. See Philip Kitcher, *Vaulting Ambition: Sociobiology and the Quest for Human Nature* (Cambridge, Mass.: MIT Press, 1985).

15. It is well to acknowledge, from the beginning, that the conceptual clarification I try to offer here should be viewed as a prelude to empirical investigations. Throughout the essay I shall appeal to the partial findings that are currently available. However, my main purpose is to pose precise questions whose answers ought to be employed in tackling the tricky moral and social issues that surround the notions of race and ethnicity.

16. T. Dobzhansky, *Genetics of the Evolutionary Process* (New York: Columbia University Press, 1970), 270ff.

17. See E. Mayr, *Systematics and the Origin of Species* (New York: Columbia University Press, 1942); Mayr, *Animal Species and Evolution* (Cambridge, Mass.: Harvard University Press, 1963); and many subsequent works by Mayr.

18. The phrase is Dobzhansky's.

19. Later in this essay (section 7) I shall indicate why I believe we need a biological notion to understand this example.

20. I shall elaborate this point at some length below, where I try to articulate the intuitive concept of ethnicity that is often employed in social discussions.

21. Caution is needed here. Many biologists are driven to avoid the notion of race because they are mindful of the harm that the concept has done in its application to Homo sapiens. Others are moved by the difficulties of any kind of intraspecific taxonomic category (lucidly pointed out in E. O. Wilson and W. I. Brown, "The Subspecies Concept and Its Taxonomic Application," *Systematic Zoology* 2 [1953]: 97–111). Yet, under many different names, the idea of intraspecific divisions lingers in ecological and evolutionary studies, where biologists recognize stocks, strains, breeds, evolutionarily stable units, geographical races, morphs, and so forth. Without commitment to the general applicability of an intraspecific category, it is possible to maintain that, with respect to particular species, a division of the species into biologically significant subunits is profitable. This kind of local pragmatism, which seems to me to permeate contemporary biological practice, provides the basis for my explorations here.

22. For the reference to Grant, see N. Zack, *Race and Mixed Race* (Philadelphia. Temple University Press, 1993).

23. The best argument for eliminativism is that if this requirement is honored, virtually the entire species will turn out to be of mixed race. I shall consider this argument in some detail below.

24. However, it is worth noting explicitly that the time of origination can be chosen to suit the purposes of the investigation, and there is no reason to think that this time must always be in the very distant past. As I shall point out below, the claim that racialization is now occurring in America might turn out to be defensible.

25. Or distinctive frequencies of particular alleles. Note that it does not matter whether or not the genetic (or phenotypic) differences are important or trivial. The concepts of race characterized here have no commitment to the racist doctrines repudiated in section 1 — just as Dobzhansky's use of a concept of race for nonhuman species made no claim that the variant characters maintained in the races of snails, flies, and beetles were particularly important.

26. Note that it is an empirical question whether the notion of race I am reconstructing will pick out any racial divisions at all, and, if it does, whether it will pick out (for example) the "big three." In the next section, I shall make some tentative claims about the racial divisions I think most likely.

27. For valuable discussions of hybrid zones, see M. J. Littlejohn and G. F. Watson, "Hybrid Zones and Homogamy in Australian Frogs," *Annual Review of Ecology and Systematics* 16 (1995): 85–112, and N. H. Barton and G. M. Hewitt, "Analysis of Hybrid Zones," *Annual Review of Ecology and Systematics* 16 (1985): 113–148.

28. J. K. Brierly, *A Natural History of Man* (Madison, N.J.: Fairleigh Dickinson University Press, 1970); L. L. Cavalli-Sforza, "Genetic Drift in an Italian Population," in *Readings from Scientific American: Biological Anthropology* (New York: Freeman, 1969).

29. E. Mayr, *Populations, Species, and Evolution* (Cambridge, Mass.: Harvard University Press, 1970).

30. Data from E. Porterfield, *Black and White Mixed Marriages* (Chicago: Nelson-Hall, 1984).

31. See J. N. Tinker, "Intermarriage and Assimilation in a Plural Society: Japanese-Americans in the United States," *Marriage and Family Review* 5 (1982): 61–74.

32. See M. M. Schwertfeger, "Interethnic Marriage and Divorce in Hawaii," *Marriage and Family Review* 5 (1982): 49–59.

33. R. T. Michael et al., *Sex in America* (Boston: Little, Brown, 1994), 57.

34. Ibid., 58–59.

35. See E. O. Laumann et al., *The Social Organization of Sexuality* (Chicago: University of Chicago Press, 1994), 254–266; the data on the infrequency of Asian-Hispanic sexual relationships are particularly striking.

36. E. O. Wilson, *Sociobiology: The New Synthesis* (Cambridge, Mass.: Harvard University Press, 1975); Rushton, *Race, Evolution, and Behavior*.

37. For present purposes I ignore all kinds of complexities about how to draw species divisions. I have argued elsewhere for the need for a plurality of species concepts (see Kitcher, "Species," reprinted as chapter 5 in this volume), but my pluralism is quite compatible with recognizing the role that reproductive patterns play in making taxonomic divisions among sexually reproducing organisms. The approach of the text is also consistent both with Mayr's more negative approach ("look at which organisms *don't* interbreed") and the more recent attention paid to positive cues ("mate recognition systems").

38. For elaboration of this point, see Zack, *Race and Mixed Race*.

39. In conversation Naomi Zack has made a forceful case for (1), and, in the original discussion at Rutgers, Michele Moody-Adams independently gave a lucid presentation of the same point. A number of people referred me to a seminal essay by Adrian Piper ("Two Kinds of Discrimination," *Yale Journal of Criticism* [1993]: 25–74), in which she summarizes the biological and anthropological sources for (1) (see her footnote 27). In public and private exchanges at Rutgers, both Anthony Appiah and Amy Gutmann argued for (2). I am grateful to all these people for pressing the case so well.

40. For a lucid account of this story, see F. J. Davis, *Who Is Black?* (University Park: Pennsylvania State University Press, 1991).

41. It is worth noting that the first episode, the intergroup unions in Virginia, looks very like the standard situation for hybridization among nonhuman animals. Individuals at the limits of their ranges sometimes have little chance of mating with conspecifics, and mate with a member of a closely related species. We could regard the two populations of servants in the colonies as cut off from the main body of the groups from which they came, partly by geographical, partly by social causes.

42. It is, I think, peculiar that eliminativists, who are surely moved by repugnance at the horrible things that have been done in the name of racial purity, should think that the breakdown of the notion of race is due to the fact that Southern white men were often prepared to rape black women or bribe them into sexual unions.

43. See Davis, *Who Is Black?*, 71–72.

44. Starting with the nondimensional notion, we could then build up the full notion of race, which I have explicitly taken to be a historical notion, by considering intervals of times at each of which the groups under study count as nondimensional races. This procedure raises the obvious question: How long must the reproductive disconnection last for the lineages to count as separate races? I shall come to terms with this, and other questions that turn on the difficulties of degreed notions, at the end of this section.

45. A recent report in the *New York Times* (July 4, 1996) suggests that rates of interracial marriage, especially those involving black women and white men, are on the increase. But the report has two interesting aspects. First, this seems to be a trend among middle-class people (and, as I shall suggest below, economic and class factors loom large in the maintenance of racial divisions). Second, a hypothesis about the cause suggests that young educated

black women greatly outnumber young educated black men, so that their choices of mates are limited unless they accept white men as potential partners. If this is correct, then the situation would again be analogous to that of members of nonhuman populations at the limits of the species range.

It is also worth noting that there is a significant amount of black-Hispanic mating as well as a significant amount of Hispanic-white mating, so that Hispanics could serve as a "bridge" population between the two groups (this point was made in the discussion at Rutgers by Gregory Trianosky-Stilwett). Only in the light of much more extensive knowledge will we be able to discover if the resulting population structure is more like that of an effectively panmictic population or the case of "ring races."

46. P. P. G. Bateson, "Optimal Outbreeding and the Development of Sexual Preferences in Japanese Quail," *Zeitschrift für Tierpsychologie* 53 (1980): 231–244; Bateson, "Preferences for Cousins in Japanese Quail," *Nature* 295 (1982): 236–237.

47. L. L. Cavalli-Sforza and M. Feldman, *Cultural Transmission: A Quantitative Approach* (Princeton, N.J.: Princeton University Press, 1982); R. Boyd and P. Richardson, *Culture and the Evolutionary Process* (Chicago: University of Chicago Press, 1985).

48. It is worth noting that evidence for (4) might be derived from study of patterns of mating among indentured servants, both black and white, in pre-revolutionary Virginia. If the population of servants showed a far higher rate of interbreeding than is now found between African Americans and Caucasians, an obvious explanation would be that economic disparities and the history of injustice have played a major role in the dramatic lowering of the rate of intermarriage.

49. This prospect is defended with great eloquence by Anthony Appiah in his contribution to A. Appiah and Amy Gutmann, *Color Conscious* (Princeton, N.J.: Princeton University Press, 1996). Appiah uses eliminativism about race as a steppingstone to recommending a future in which the contributions of all cultures are available to everyone. In my judgment, the points for which he wants to argue can be clarified by probing the connections between race and ethnicity in ways I have begun here. Indeed, many of the issues that divide Appiah from his critics seem to me to require empirical exploration of issues posed in section 6: we need to know just how lines of biological descent and cultural descent interact with one another. Even if Appiah were to view my earlier reconstruction of the concept of race as mistaken, I think he would still have to confront questions about patterns of mating and biological inheritance and how these affect and are affected by cultural transmission, for precisely those questions are pertinent to the kinds of futures he imagines. Those who oppose his "liberal cosmopolitanism" often seem to be making different assumptions about the answers to those questions, and I think it important to identify the empirical issues and try to resolve them.

12

Utopian Eugenics and Social Inequality (2000)

It is helpful to divide the philosophical issues surrounding the Human Genome Project into three main groups.[1] First, there are questions about the scientific significance of the project. Second, the project raises immediate practical problems, notably in decisions about genetic screening and in connection with the potential release of genetic information about individuals. Third, there are long-term concerns about the desirability of applying our new ability to identify the genotypes of the unborn, concerns often posed in the accusation that we are on the verge of a new eugenics. In this paper, I shall primarily be concerned with the third cluster of issues, for it is here, I believe, that the hardest *philosophical* problems lie. But, in order to frame my discussion, I want to begin by reviewing perspectives on the first two groups, perspectives that I have defended in some detail elsewhere.

Many critics of the HGP have offered what might be called the "boggle argument."[2] This consists in announcing that the end-product of the project will be a huge list of As, Cs, Gs, and Ts and asking, rhetorically, what possible point there could be in any such list. Although the "boggle argument" is perfectly appropriate as a corrective to some of the more grandiose statements made in defense of the HGP, it is quite irrelevant to the actual practice of the project. First, at present, virtually all of the research on *human* genomes is directed at constructing better maps (both genetic and physical), and this research is making ever more efficient the strategy of positional cloning. Second, the principal current sequencing efforts are focused on other organisms (the bacteria recently sequenced by Craig Venter, *E. coli, S. cerevisiae,* and *C. elegans*), and these efforts are revealing many new genes (including many that will be homologous to genes in our own species) and properties of the organization of genes. Third, while the problem of hunting for genes in pages of sequence data has no simple, elegant solution, and while the problem of understanding the conformation of proteins is unsolved in general, the construction of databases from results about genes in a variety of organisms and about the forms of well-known proteins enables ad hoc solutions that are likely to be able to recover an extremely high percentage of the genes and proteins from future bodies of human sequence data. Fourth, if sequencing becomes sufficiently easy during the next decade, sequencing the entire three billion base pairs of the human genome may be the most efficient way to identify (almost) all our genes; if large-scale sequencing remains problematic, then the smaller efforts on nonhuman organisms will be combined with "quick-and-dirty" techniques for finding human genes, to provide almost as much information about our own species. What is right about the "boggle argument" is that there is nothing sacred and wonderful about achieving the full list of As, Cs, Gs, and Ts that might stand for "the" human genome. But the conclusion ought to be simply that the emphasis on the *Human Genome Project* is faulty advertising for an exciting venture, the "Genomes Project," that will reveal masses of interesting and useful things about many organisms including human beings.[3]

When we reflect on the kind of information that will be achieved in the near future, we are bound to confront the second group of questions. The development of detailed maps has enabled gene hunters to clone and sequence genes that are implicated in a number of human diseases. These achievements can typically be translated, relatively quickly, into genetic tests, tests that might be useful in a variety of contexts: diagnosing a disease, identifying the particular form of a disease already diagnosed, identifying the risks of future disease or disability, exposing the genotype of a fetus, and discovering if a person carries a recessive allele. It is worth reminding ourselves that the first two contexts, although typically uncelebrated in the literature, are likely to provide real benefits without significant problems.[4] The principal difficulties with the power to engage in genetic testing concern the last three contexts, especially the possibility of identifying risks of future disability. As many authors have pointed out, there is little good to be done in telling people that they are at high risk for a disease when there is nothing that can be done to alleviate that risk (or nothing to be done that would not have been recommended whether or not the people had tested positive).[5] Testing is only useful when there is something special that those who test positive should do, something that would be inappropriate for the rest of the population.[6] Of course, medical lore does know of cases in which this necessary condition obtains: PKU is the obvious, much-cited, example.

From the medical point of view, PKU is a perfect case because there is a diet, necessary for normal functioning for people who have the abnormal genotype, and harmful to people who have the normal genotype. Yet when we expand our horizons, there are grounds for worry. Because the diet is unpleasant and expensive, many children who need it do not receive it for as long as they should. As Diane Paul has lucidly argued, we really don't know if PKU testing has done more good than harm.[7] We do know that, in the early stages, children who should not have received the PKU diet were given it (with severe disruptions of development), that some people have not stayed on the diet long enough to avoid the deleterious buildup of phenylalanine, that mothers who had come off the diet as young adults gave birth to gravely damaged babies—but we don't know how to balance these individual tragedies against the successes that have come from early intervention. At risk of committing myself to an untenable dualism, I suggest that those who appeal to PKU testing as a shining example are right *so long as we consider the problem as a purely biomedical one*. The trouble comes from the nature of the social context in which the applications of biomedical knowledge are made. Because we do not provide economic support for families in which a child is diagnosed with

PKU (as well as intensive counseling), we blunder away the possibilities of making a real difference to the quality of human lives.

The case of PKU testing has important implications for the future of genetic testing. Biomedical researchers are quite justified in thinking that some of their discoveries will have the potential to enable people to reduce their chances of future disease or disability. Actualizing that potential, however, may require significant changes in the ways in which medicine is practiced. So long as patients (and, very often, the doctors who advise them) are buffered by the forces of the market, so long as many segments of the population only have very limited access to advanced medical technology, so long as there is no attention to providing support (both economic and personal) for people who require expensive and uninviting procedures, what can be done in principle is unlikely to be achieved in practice.[8]

Just as the promise of new molecular knowledge must be evaluated by considering the social surroundings, so too the principal concerns about the flow of genetic information could be addressed by modifying features of contemporary medical practice. The idea that genetic information should be private appeals to us not because of the intrinsic value of keeping our genotypes to ourselves, but because, as things now stand, we could suffer serious harm if others came to know certain things about the alleles we carry. Envisaging the era of genetic testing simply reinforces straightforward arguments for ensuring that all members of our society, irrespective of genotype or economic condition, have access to affordable medical care. By the same token, genetic testing in the workplace should be restricted to those situations in which it is possible to argue that applicants with particular genotypes are less qualified for the jobs they seek. *Principled* solutions to problems about the flow of genetic information are not hard to come by. The real difficulties are not philosophical but practical. How can we ensure that the principles are embodied in practice? Or, if it seems impossible to achieve principled solutions, how can we broker some form of acceptable compromise?

During the years since the HGP began, there have been numerous, often repetitive, discussions about genetic screening, about the use of genetic information in insurance, and about genetic testing in the workplace.[9] I believe that philosophical analysis is important to these discussions because of its delineation of ideal solutions, and that, as philosophical problems go, the construction of such solutions is relatively straightforward. Debates go on only because political pressures seem to make the ideal solutions impossible. Once this is recognized, we face a choice. One can either expose with maximal clarity the rationale for the ideal solutions, hoping thereby to engender a change of view that will remove political constraints, or one can resign oneself to working within the current system and attempt to find a compromise that will not ride too harshly over the rights and aspirations of vulnerable people. Political scientists and lawyers naturally gravitate to the second option; philosophers, I hope, see their vocation in the first.

With these preliminaries, I turn to the main discussion of this paper. The gulf between an ideal solution and the practical realities is most apparent when we consider the use of genetic knowledge in prenatal decision-making. The deepest, most disturbing, debate about the HGP emerges from consideration of the possibility that we are committed to a revival of eugenics. My aim is to show that defenders of the project are committed to a benign enterprise, *Utopian Eugenics*, whose ultimate rationale stems from concern for the quality of future human lives. However, once that rationale is exposed, consideration of existing social inequalities and of the political pressures that support those inequalities yields two important lines of argument. According to the first, there is a deep inconsistency in promoting the HGP and the utopian eugenics it fosters, without attending to the social causes of human inequality. According to the second, we should recognize the fragility of utopian eugenics within any society that fails to eradicate widespread social inequalities. I shall suggest that the most important philosophical problems about the HGP arise from these arguments, and that the sources of skepticism about the project are really grounded in sensitivity to broad questions in social philosophy.

To tag an enterprise with a name that carries the burden of a terrible history sometimes short-circuits important discussions. So it is with that part of the application of human molecular generics that focuses on prenatal choices. Many authors have worried that the possibility of discovering the genetic characteristics of a fetus will spawn a new eugenics, and the stigma associated with 'eugenics' is powerful enough to end the conversation right there. There is no doubt that many past ventures in eugenics are truly appalling. However, if we are to condemn a future enterprise of choosing people, based in advanced knowledge of their genotypes, them we should try to identify the properties that have made previous forms of eugenics morally monstrous and see if these are shared by the envisaged practice of prenatal testing.

From the start, we need to we quite clear about what is in the offing. Philosophers who have considered the role of molecular genetics in designing future generations have often indulged in science fiction and ignored the more prosaic facts. Speculations about "perfect babies," "wonderwomen," and "supermen" are rife, as if there were realistic prospects of "engineering" people to our specifications.[10] Quite apart from the commonplace, but never-sufficiently-reemphasized, point that genotypes and environments work together to shape phenotypes, the technical difficulties of gene replacement therapy make *positive* programs of designing our descendants impossible. There are two major problems: delivering DNA to the right cells and regulating its expression within the cell. Even the most successful current procedures for DNA delivery only reach a fraction of the target cells (fortunately, in some cases, as with the most lucky of the Severe Combined Immune Deficiency (SCID) children, this is enough to restore roughly normal functioning), and the techniques used so far, which address only the crudest types of gene regulation (the alleles inserted are permanently expressed), show clear effects of interference with other cellular processes (often the percentage of cells expressing the inserted protein drops rapidly). Add to this the significant problem of delivering DNA to the brain, and the fact that nobody knows how to eliminate the mutant alleles that are present. The result is a picture, familiar to everyone who practises molecular medicine and to anyone who takes the time to look at the details of the practice: doctors inject patients with molecules they need, producing, when things go well, a cobbled-together approximation to normal functioning, usually rigged with all kinds of devices to ward off unwanted side-effects. That picture is a long way from the naive dream of precisely replacing, in all pertinent cells, the alleles one doesn't want with the alleles one does. The reality depicted is full of promise for ameliorating severe disease and disability by finding inefficient and cumbersome methods of managing tolerable levels of the right proteins.[11] It is no way to design a baby. *Positive* selection isn't the issue. What is possible on a small scale now, and will be possible on a large scale within a decade, will be *negative* selection. Provided that parents have the genotypic potential to yield the combination of alleles they want in a child, they can use prenatal testing to filter out unwanted progeny. From the very beginning of the HGP, it has been abundantly clear—although not the sort of thing champions broadcast in the halls of Congress—that one principal application of new knowledge in human molecular genetics would be a greatly increased number of abortions. Unless abortion is opposed on principle, most of these abortions would probably be viewed as benign and humane. As we learn more about the genetic bases of various diseases, not only the rare conditions that doom their bearers to permanent hospitalization but also more common cases that limit human functioning, prospective parents will have the ability to learn in advance whether the fetus one of them carries bears a relevant genotype, and on that basis, they will be able to choose to continue the pregnancy or to terminate it. There is no doubt that this will avert some human tragedies: doctors, counselors, and relatives of people who have had children with genetic diseases can recount numerous instances in which lives of promise were destroyed because a pregnancy proceeded to term. Nonetheless, as we envisage the many kinds of genetic tests that will become possible in the next decades, it is easy to fear that a benign policy of forestalling disease may become a program for enforcing social prejudice, a new eugenics.

Although there are some objections to *any* use of prenatal testing—even for diseases like Tay-Sachs or for the most disruptive trisomies and translocations—I shall take it for granted, for present purposes, that there are some instances in which identifying affected fetuses and aborting them would be morally justifiable. Doubts arise when we consider terminating pregnancies because the fetus has the "wrong" eye color or hair texture, when an allele "for" obesity or same-sex preference is present, when there are too few of the "alleles for intelligence." Now these doubts intertwine a number of issues which it is important to separate. But their great force comes from a resonance with the eugenic past, for these are the kinds of abuses that have figured largely in earlier attempts to apply human genetics.

A eugenic practice is an attempt, by some group of people, to shape the genetic composition of their descendants according to some ideal. Despite the large body of excellent literature on the history of eugenics, very little has been done to develop this characterization.[12] My own approach (which accords with that taken in a pioneering article by Diane Paul[13]) recognizes four dimensions of any eugenic practice. The first dimension identifies a subpopulation whose reproductive activity is to produce the desired results. The second dimension specifies the degree to which members of this subpopulation make their own reproductive decisions (or, conversely, the degree to which they are coerced by the ideals of others). The third dimension picks out the characteristics according to which the choices are made. The fourth appraises the quality of the genetic information that is used in making the reproductive decisions.

Nazi eugenics (and, to a lesser extent, the American eugenic ventures of the 1920s and 1930s) occupies a particularly loathsome region of the resultant four-dimensional space. Based on prejudices about desirable and undesirable human types, the Nazis selected particular subpopulations in which to promote breeding and particular subpopulations in which to reduce (or eliminate) reproduction. People in the latter groups experienced extreme forms of coercion, under the surgeon's knife or at the end of the storm-trooper's gun. The characteristics for which the future population was to be chosen (the glorification of the "pure Aryan" type) reflected a morally distorted and factually inaccurate account of human worth. Finally, the underlying genetic claims were frequently wildly wrong.

If the terrible evils of the past are in one corner of eugenic space, the envisaged future is at the opposite extreme. When human geneticists ponder the application of prenatal testing, they imagine that everyone in the population will have the same opportunities for shaping the descendant gene pool—there will be no prejudicial selection along the first dimension. They suppose that each couple will make their own free decisions—there will be no coercion along the second dimension. They envisage that those decisions will be made reflectively, that they will be educated and freed from superstition and prejudice. Finally, they assume that the underlying genetic information will be accurate. What they endorse is a vision of eugenics as different from Nazi eugenics as could be. I shall call it "Utopian Eugenics."

But, even if we grant that there is considerable difference between utopian eugenics and the horrors of the past, we should still ask why we should occupy any position in eugenic space. There is a straightforward answer. Once we know how to identify the genotypes of future people, eugenics is the only option. It is quite illusory to think that we have a noneugenic alternative, and the illusion can be exposed by considering exactly what it would be. Suppose, for example, that we said that nobody was to draw on any information from human molecular genetics in making reproductive decisions. This would be to institute a eugenic practice that has two highly problematic characteristics: first, its standard for the desirability of the descendant gene pool is that any such gene pool is preferable if it is brought about from ignorant decisions about the properties of offspring, if nature is left free to take its course; second, its position on the second dimension is highly coercive. But the fundamental point is that this is a *eugenic* practice. In a situation in which we have the option of bringing about future populations with various genetic characteristics, it commends one of those options, the one in which we act as we would have acted before the advent of detailed genetic knowledge. Once we lose our generic innocence, we have alternatives, and, because we have to elect one of the alternatives, we have to practise eugenics.

It should now be evident that labeling the HGP as opening the door to eugenics settles nothing. Everything depends on whether the form of eugenics that will result is better than the alternatives. Champions of the HGP claim two things: first, that utopian eugenics is the form that will be actualized; second, that this is superior to its rivals, and, specifically, superior to the coercive policy of pretending that we are genetically innocent.

As we shall discover, defending these claims is far more complicated than might initially appear. There is little doubt about the value of three of the dimensions of utopian eugenics. It is eminently reasonable to claim that a policy of nondiscrimination along the first dimension is morally justified, that leaving people free to make their own reproductive decisions is a good thing, and that the genetic information we can expect to apply will be accurate, so long as we avoid the pits into which behavioral genetics has so often fallen in the past. The major worries surround the specification of the third dimension, the delineation of the moral ideal that should inform free and educated reproductive decisions.

Once again, I shall abbreviate a story I have told in far more detail elsewhere.[14] A practice of prenatal testing should be oriented, I believe, toward a concern for the quality of future lives. In contemplating the quality of future lives, three kinds of factors must be considered. First, we should assess the chances that the person who would be brought into being would be able freely to develop a conception of what matters in his/her life, a sense of what is worth achieving and striving for. Granted that we can imagine the future person forming a life conception of this kind, we can identify the desires that are most centrally associated with it, and ask about the extent to which we could expect these desires to be satisfied. Third, and finally, we can evaluate the hedonic quality of the life, the balance between pleasure and pain that it would bring.

Now the most obvious worry about children with mutant alleles is that they will suffer, and in some instances, pain is untreatable. However, many of the principal uses of prenatal testing are not like this: palliatives are available, seizures can be brought under control. The real horror is the massive disruption of development, the neurodegeneration, the permanent confinement to a hospital bed, the need for daily grooming, assisted feeding, the absence of any cognitive life. What is missing in such tragic cases is the possibility that the person will ever form a sense of what matters to her/him, so that the envisaged life is viewed as having intolerably low quality when assessed according to the first factor. For other genetic conditions— such as the various kinds of muscular dystrophy—we recognize that the future person's choices of what matters in life will either be severely constrained or else that the resultant central desires will almost certainly be frustrated.

If decisions about the quality of future lives are genuinely to be free, then it is important that social prejudices, or lack of social support, should not coerce prospective parents into decisions to terminate pregnancies that might have issued in lives of significant potential. Societies that cramp the lives of people with particular conditions (for example, those who do not meet prevailing standards of bodily shape or those sexually attracted to members of their own sex) effectively undercut the freedom of choice of prospective parents who find that the fetus carries an allelic combination associated (I shall assume correctly) with the despised condition. In an even more obvious way, a society may impose its standards by denying people with particular conditions the opportunities to develop in ways that bring them much closer to normal functioning. Because of imaginative programs, many Down's syndrome children have achieved levels of development considered impossible thirty years ago. A society committed to providing such programs promotes the freedom of reproductive decision. Hence, I shall take it as a cornerstone of utopian eugenics that prospective parents can make their reproductive decisions in assurance that children born with debilitating conditions will receive the best support known, and that their children's lives will not be limited by social prejudices.

Utopian eugenics is obviously a program for a highly idealized world. However, since we are committed to some form of eugenics, it is worth being as clear as we can about the ideal and its demands. I think that utopian eugenics embodies a worthy ideal, and that the allure of this ideal is responsible for the view that human molecular genetics can be liberating.[15] The most important criticisms of the HGP do not stem from the faulty conflation of all forms of eugenics or from a failure to appreciate the worthiness of the utopian ideal. They result from advancing two theses: first, that the ideal commits us to much more than an enterprise in applied human genetics; second, that our failure to carry through with that commitment signals an acceptance of social inequalities that will reduce utopian eugenics, in practice, to a much darker and more morally problematic, program. In the rest of this essay, I shall be concerned with the merits of these two theses.

We are prepared, rightly I believe, to take steps to ensure that an unlucky legacy, passed on through the genes, does not doom a child to pain or to a pathetic life of restricted opportunities. But there are obvious questions: why should we rush to treat the unfortunate genetic inheritance of the few, while ignoring the unlucky social inheritance of the many? Shouldn't we commit ourselves to learning how to change the environments that break young lives as surely as defective proteins? While we should acknowledge our ignorance of the consequences of social interventions, it is worth remembering that the results of many forms of molecular therapy, are uncertain—and that, when the condition is sufficiently grave, we are prepared to take risks (as in gene replacement therapy for SCID). How bad must the plight of discarded children be to justify an analogous *social* experiment?

Critical rhetoric sometimes overstates the case, deriding the advertisement of future relief of human suffering. The anguish produced by genetic afflictions of all types, from Tay-Sachs and Lesch-Nyhan to the forms of cancer, is real and tragic. The potential of molecular medicine is equally real. It is right to celebrate the promise of human molecular genetics to enhance the quality of the lives that are led. Nonetheless, the questions I have posed are serious. If our moral venture rests— as I have suggested—on a concern for the quality of human lives, does consistency require us to undertake a more systematic assault on the pressures that shrink people's hopes and opportunities? Or are we properly conscious of our limitations, thankful for biological expertise and regretting the ineptness of social policymaking?

Our lives are the products of many lotteries, and only one of them shuffles and distributes pieces of DNA. Behind the often acrimonious controversy about the value of molecular genetics is a deep disagreement about the implications of this fact, a disagreement dividing people who may appropriately be called "pragmatists" from others whom I shall henceforth dub "idealists"[16] Idealists think that, when the underlying rationale for applying molecular genetics in prenatal testing is exposed, we should become aware of a commitment to the quality of nascent lives that ought to be reflected more broadly in social action. Pragmatists maintain that we should not hold the local good we can do (by developing and applying our biological knowledge) hostage to quixotic ventures (doomed to uselessness by our ineptness at social engineering). The dispute between these groups intertwines two large classes of questions. What is the extent of our obligation to aid people whose initial circumstances greatly reduce the quality of the lives they can expect to lead? What are the practical possibilities for meeting these obligations, specifically for combatting the environmental causes of pinched and painful lives?

I shall primarily be concerned with the first kind of question, the moral issue; I lack the expertise in social science that would be required to treat the factual questions in the detail they demand. My aim will be to scrutinize an analogy that moves idealists, to clarify an important debate, not to resolve it. The starting point for the analogy juxtaposes the fact that our lives are, in large measure, the products of lotteries, both genetic and environmental, which deal fortunes unequally, with an attractive social ideal, the ideal that people should have (in some sense that is to be specified) equal opportunities to live happy and rewarding lives. Mismatch between ideal and reality prompts citizens of affluent democracies to believe that justice demands some attempt to remedy the unequal accidents of birth and childhood. Some type of help is required. Help typically calls for money, demanding, in consequence, some form of redistribution of resources.

But which form should we aim for, and how far should we go? Because money can easily be counted and differences in financial resources readily compared, it is natural to frame discussions of social ideals in terms of economic assets.[17] Natural, but, I think, mistaken. As the deep rationale for utopian eugenics reveals, what is of primary importance to us is something far more nebulous, something with several dimensions, and something that is not easily compared across individuals—think of rating the qualities of the lives of your friends, or of famous people. In trying to clarify our intuitive conception of the demands on us, we do better to focus on what is fundamental, the quality of lives, acknowledging that comparisons will often have to be imprecise. Of course, it would be naive to neglect entirely the connection between economic status and the quality of life. Although decreasing differences in assets is not an end in itself, it may well be an indispensable means to realizing worthy goals.

If we could assign precise numbers that measure exactly the quality of lives, then it would be possible to identify the *expected quality of a person's life* by proceeding as gamblers do when they figure the odds. Sophisticated bettors calculate what they can expect to receive from various betting arrangements by multiplying the returns from an outcome by the probability of the outcome and adding terms corresponding to each eventuality. In analogous fashion, we could consider the person's current state, associate each of the possible lives that might ensue with the probability of its occurrence, multiply the probability by the measure of the quality, and sum across all the possibilities. Life is far too complex for any such calculations. We cannot quantify the various dimensions, we cannot weigh one dimension against another with any precision, and we are ignorant of the exact values of the pertinent probabilities. However, that does not interfere with our ability to make some comparisons of expected quality of life. Suppose that two children are born into roughly equivalent social environments, one bearing two copies of the common mutation for cystic fibrosis (delta 508), the other carrying two copies of the normal allele. There is little hesitation in declaring that the expected quality of life for the first child is lower than that of the second; the possible futures for the first child divide into two classes, those in which the phenotypic expression of the CF alleles is mild (or in which harmful effects are mitigated through treatment) and the life is roughly equivalent in quality to that of the second child, and those in which the phenotypic expression of the CF alleles restricts activities and curtails possibilities; if the child experiences a future of the first type, the quality of her life will not be higher than that of the second child; if she experiences a future of the second type, the quality of her life will be lower; because there is a significant (although un-known) chance that her future will fall into the second class, the overall expecta-tion is lower.

Nor do we have trouble in making some more complex comparisons. Imagine that the child bearing the two CF alleles is born into an affluent family, with two parents who are devoted to her and determined she shall thrive. By contrast, the child with the normal alleles is the sixth son of a single mother, who is unemployed and struggling to resist the pressures of a dangerous urban environment. Now there are clearly some possible futures for the girl in which her activities are so severely reduced, despite all her parents' efforts, that the quality of her life would fall below that we would expect for the boy. Nevertheless, with high probability, her range of opportunities and her ability to satisfy her central wishes will greatly exceed what is overwhelmingly likely to be available to the boy, so that there is a high proba-bility of a large positive difference, outweighing the lesser chances of negative outcomes. Overall, her expected quality of life is superior.

It is now possible to state the idealist argument more carefully. We have oblig-ations to improve the lives of the less fortunate, obligations that are honored in our commitments to attack genetic disease. By redistributing resources, taxing the well-to-do and using the revenues to provide compassionate care, as well as to invest in molecular research (both by seeking new ways of treating genetic disease and dis-ability, and by providing support so that the resulting treatments are available to those who need them), affluent societies conform to a principle of social justice. Idealists conclude that the principle requires more of us, that we are morally obli-gated to intervene in other ways to raise the expected quality of life of people who have been victims of social roulette, that assets currently used to enhance the quality of lives already at a high level might more justly be employed elsewhere. To eval-uate their case we need to have a clearer view of the principle on which they rely, and of its implications.

One way to think about social justice is to conceive society, highly abstractly, as a collection of individuals, each endowed with a particular amount of resources and each with a particular expected quality of life. Expected quality of life is fixed by many factors entirely beyond the person's control, the combination of genes that have been transmitted, the characteristics of the parents and the environment into which a child is born, and, of course, the resources that will be available during development. Redistributing resources would have an impact on the expected quality of the lives of all citizens. What ideals should guide us among the numer-ous options?

Three abstract principles have loomed large in democratic political theory, each of which seems to offer pure expression of an attractive ideal. One opposes schemes of redistribution in the name of individual autonomy: however beneficial the consequences may be, we cannot compel someone to give up assets which are his or hers to control. The second would seek to redistribute resources so as to make the total quality of life as high as possible: we are to imagine summing the measures of expected quality for the lives of all citizens, given all possible distributions of resources among them, and choosing that distribution that gives rise to the largest grand sum. Last is an explicitly egalitarian principle commending the distribution that comes closest to securing equal expected quality of life for all.

It is not hard to see that each of these principles is flawed. Although respect for individual liberty is a worthy ideal, that ideal cannot properly be expressed in a "hands off" attitude toward redistribution of assets. Once we realize that the freedom that counts—the freedom to choose that conception of what is significant in one's life—is a matter of degree, depending on the extent to which we have the opportunity to be guided by alternative possibilities, then redistribution might decrease the autonomy of the privileged by only a slight amount, while greatly enhancing the autonomy of the underprivileged.[18] The directive not to demand assets from the well-to-do, far from manifesting a concern for the autonomy of all, would be more accurately advertised as a maxim to respect the liberty of the winners in the lotteries that fix initial circumstances.

Maximizing total expected quality of life is heir to familiar troubles.[19] The grand sum is no respecter of the individual contributions. Everything comes out in the wash, and the highest total may easily be achieved within a society in which the majority have a very high expected quality of life, gained at the expense of a small number of people whose lot is doomed to be miserable. Perhaps we would increase the total expected quality of life within affluent democracies by withdrawing funds currently used to care for the most devastating genetic diseases, and spending them in quite different ways, possibly by building wonderful public sports facilities. It would surely be morally obscene to do so. Moreover, even if the arithmetic does favor our current allocations—maybe because families would be upset if the standard of care were dramatically reduced—that is surely not the correct moral basis for our determination to do what we can for those afflicted with genetic disease. The quality of their lives must be considered individually, not simply as some term in a colossal sum, and, despite the fact that expensive efforts do not increase expected quality by very much, both justice and compassion demand that we do what we can.

The third pure ideal is vulnerable because it insists on equality irrespective of total well-being. Antiegalitarians frequently voice the "dog in the manger" objection, claiming that appeals for equality are base on the envious suggestion that nobody shall have anything that somebody lacks. Despite the difficulties of specifying the pure form of egalitarianism—just how do we compare how well unequal distributions approximate perfect equality?—it would appear that there is always one way of securing *exactly* equal expected quality of life for all, to wit by reducing the expected quality of each person's life to some very low value. Since this may well be the *only* way to achieve exact equality in expected quality of life, and since it is plainly unacceptable, it is apparent that the ideal of equality must be compromised by attention to overall well-being.

Although we have to tolerate some inequalities, our respect for others might suggest that resources should be distributed to raise the expected quality of life of each person above a certain minimum level.[20] Depending on how the minimum is fixed, this requirement will either be toothless or unsatisfiable. Those afflicted with the most devastating genetic diseases will have an extremely low expected quality of life no matter what efforts we make, no matter what resources we assign to their care. If the minimum is set above the level they can attain, even when supplied with unlimited resources, then the requirement to redistribute so as to provide for everyone an expected quality of life above the minimum cannot be honored. If, however, the minimum is set lower, then the expected quality of life we demand for all with be so low that, except in the very gravest cases, redistribution will prove unnecessary. Even untreated, unsupported, children with cystic fibrosis have higher expected quality of life than those with Canavan's disease or Lesch-Nyhan syndrome, whatever resources we invest in them.

In the clearest cases of redistribution, greater equality and increased total expected quality of life go hand in hand. Guaranteeing public funds to provide the special diet for children with PKU would lower the expected quality of life of the affluent by a tiny amount, but would raise the expected quality of life of the PKU children enormously, producing both a more equal distribution and a larger total. However, it is sometimes just for the fortunate to make relatively large sacrifices to bring far smaller benefits to victims of circumstance. Perhaps the expected quality of the lives of people with the most serious genetic diseases would only be marginally reduced by providing a much cheaper form of care, and perhaps the savings could be used to offer much larger benefits to those who already live well. Even so, it would be morally indefensible to favor that distribution. Small gains for the unlucky count far more than larger improvements for the fortunate.

Reflection on these points suggests another way of formulating a social ideal. Our goal should be to redistribute resources so as to bring people's expected quality of life above a minimal level; when this cannot be achieved, it is just to transfer resources so that the expected quality of life of some people is decreased, provided that the result both raises the expected quality of life of people below the minimal level and does not depress beneath the minimum the quality of life of those who give. This formulation would sanction our support for the genetically disadvantaged, even though that support costs more, in expected quality of life, than it yields. (A pictorial presentation of the effects of the redistribution is given in figure 12.1.)[21]

However, the new formulation, like its predecessors, is inadequate. It is at least possible, even quite probable, that every increase we could make in funding for biomedical research would raise, if only infinitesimally, the expected quality of life of people who suffer the severest genetic diseases. Would it therefore be just to withdraw resources from other citizens, up to the point at which their expected quality of life approached the minimum level, in order to make these minute increases? Intuitively, it would not: the sacrifices made are so disproportionate to the gains that the redistribution appears quixotic rather than compassionate.[22]

It is a familiar point in general social philosophy that our ideals of justice need somehow to be combined, and my appeal to instances of genetic disease is only to highlight the fact that this point is plainly applicable to utopian eugenics. Respect

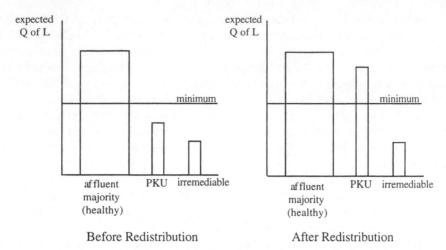

FIGURE 12.1 Possible effects of redistribution of resources on expected quality of life.

for liberty must be tempered with concern for equality, the yearning for equality must attend to considerations of average well-being.

Attempts to make further headway in clarifying the idealist's case can usefully being with the problem of deciding the level of support for those with genes that debar them from lives of any acceptable quality. We think it right to divert assets from people whose lives flourish to improve the lot of the genetically unfortunate, even though the positive effects are small in comparison with the benefits the affluent forego; yet it would not be right to proceed indefinitely, radically reducing the quality of life of the healthy to obtain vanishingly small expected gains for the genetically disadvantaged. How far, then, should we go? When is enough enough?

Here is an obvious thought. The expected quality of life of a person afflicted with a devastating genetic disease will inevitably fall beneath a particular value, the associated *quality of life ceiling*, however great the resources we make available for care and for research into the condition. More exactly, for a group of people who share a common genetic condition, I take the quality of life ceiling to be the maximum expected level of well-being they would attain, given any distribution of the resources available within the society.[23] Using this notion, we can explain the limits on a policy of giving aid to the genetically least fortunate. Moderate sacrifices on the part of those whose lives go well, which, while depressing the expected quality of their lives still retain it at a high level, can enable us to bring the genetically unfortunate close to their ceiling—we can make their lives go almost as well as it is possible for them to go. Large sacrifices appear quixotic because they generate a huge disparity between the expected quality of life of the fortunate and their corresponding quality of life ceiling. If those sacrifices are made, then some lives will be very much worse than they might have been. (See figure 12.2 for a graphical representation of these points.)

The farther people are below their quality of life ceiling, the more they deserve our support; the lower their expected quality of life, the more they deserve our

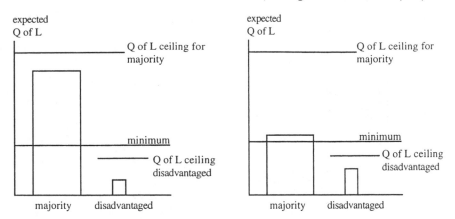

FIGURE 12.2 When large sacrifices can do little for the disadvantaged.

support. Each of these principles, I believe, represents an important moral impulse, corresponding to our urge to remedy lost potential (whatever the absolute level) and to our commitment to help those who are less fortunate (whatever their potential). Sometimes the principles are in harmony, and we would rightly divert resources from the affluent to provide special diets for children with PKU and compassionate care for those afflicted with Tay-Sachs. Sometimes they conflict, as when we envisage spending all societal resources in a massive effort to improve the lot of those who presently have the lowest expected quality of life.

A more complex illustration shows how the principles work together, and how they bear on utopian eugenics. Imagine that society consists of three groups of people: many are healthy, and all of them possess enough assets to raise their expected quality of life well above the acceptable minimum; a few suffer from a genetic condition, which, left untreated, will yield an expected quality of life well below the minimum, but which is susceptible to an expensive form of treatment that will raise their expected quality of life almost to the level of the healthy majority (think of a disease like PKU); most members of this second group lack the resources to purchase the treatment themselves; finally, there is a smaller group consisting of people who suffer from an even more severe genetic disease; even if unlimited resources were expended on them, their expected quality of life would always remain below the level of untreated members of the second group; they can be brought very close to their quality of life ceiling, if they are given extensive (and expensive) care; further investment of resources would increase their expected quality of life by minute amounts. The affluent members of the healthy majority are sufficiently numerous so that some of their assets could be diverted, with only small impact on their expected quality of life, providing both for the treatment of all members of the second group and for the extensive care of people in the third group. Redistributing resources in this fashion appears preferable to either of two alternatives: concentrating on the plight of those who are worst off, using all resources that can be spared without reducing the expected quality of life of the majority below the minimum until the expected quality of life of the third group

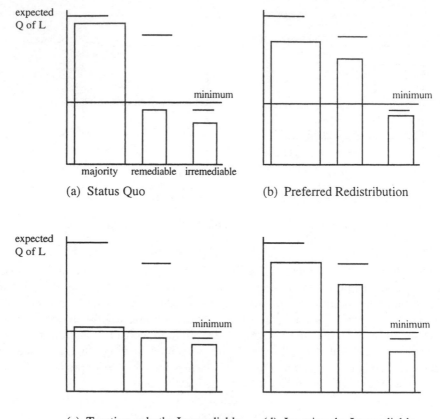

FIGURE 12.3 Resource allocation and quality of life under different redistribution models.

has been raised to that of the second (which is impossible); ignoring the third group entirely, and using some of the assets to the majority to treat members of the second group. It is wrong to focus on the most needy and neglect people whose actual prospects are far below their potential, people who might obtain significant benefits from aid. Equally, it is wrong to think solely in terms of lost potential that might be restored by redistributing resources, using the fact that some members of society may have a wretched existence, whatever is done, as an excuse for forgetting about them. (The options and their relations are presented graphically in figure 12.3.)

The example just considered represents, in schematic form, our predicament with respect to (some) victims of genetic disease. To a first approximation, the population of an affluent democracy divides into three groups: those with genetic conditions for which we can do nothing except to provide humane care, those with genetic conditions that we can reasonably hope to alleviate, and the large majority. Because the minorities are so tiny in comparison with the majority, because the majority has great resources, and because we already know (or have well-grounded hopes of knowing) how to intervene effectively, both in providing care and in ameliorating the conditions, it is possible to act in accordance with the principles without demanding large sacrifices. Indeed, many affluent democracies pursue appropriate policies, providing care and investing in biomedical research. Idealists think there is a direct lesson for social policy. With the basis of their analogy in clearer focus, we can now ask if they are right.

The most straightforward way of developing an argument by analogy is to envisage, as before, a society in which there are three groups. The first, once again, will be a large majority of healthy people, many of whom are affluent. The second will consist of children who, given the home environment they currently inhabit, have an expected quality of life below the acceptable minimum; each of the children in this group *could* achieve a life of far higher quality; if any of them were removed from the current brutalizing environment and provided with affection and stability, the expected quality of life would increase, perhaps, in some cases, to that typically attained in the first group. In the final group are children who, because of the environments into which they have been cast, have a quality of life ceiling that lies below the expected level of the second group, even if no resources are allocated to members of that group; with extensive care, these people can be brought close to their quality of life ceiling, and providing this care would not seriously affect the quality of life of the first group. To fix ideas, we can think of the third group as comprising those children whose cognitive and emotional functions have been irreparably damaged by environmental brutality. The second includes the children whose current environments severely limit their cognitive, emotional, and social development, but who could still be rescued if their environments were radically changed. There remains the first group, the relatively healthy and affluent remainder of the population.

In the genetic case considered previously, it appears right to redistribute resources from the affluent majority to provide care for those whose genes set a very low quality of life ceiling and treatment for children whose quality of life can be restored. Do we have similar obligations when social circumstances cause analogous relationships among the expected quality of life of three groups? Pragmatists will surely reply that the examples are not analogous. There is no form of social intervention (the analogue of the PKU diet) which we know how to carry out, whose cost is small, and which would restore the expected quality of life of members of the second group to the level enjoyed by members of the majority. Their objections intertwine a number of considerations. Most prominent is a pessimistic assessment of the effects of social intervention: many people, reflecting on the history of attempts to care for the disadvantaged, conclude that the problems are intractable, that well-meaning efforts to engineer solutions often fail to help those who are supposed to benefit, and, worse, have unforeseen consequences that generate further social disasters; the road to social catastrophe may be paved with the best of intentions. Additionally, pragmatists may stress the inefficiency of typical social interventions, claiming that, even where policies do bring benefits, they require expenditures much larger than the gains achieved. This thought alone would not prove telling—for, after all, we are willing to provide expensive care for victims of the most devastating genetic diseases, despite the fact that their quality of life is only raised by a small amount. However, where inefficiency on a small scale may be

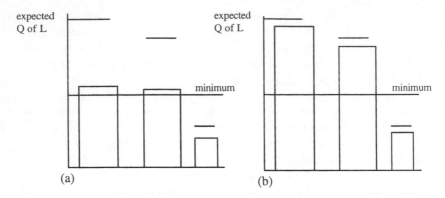

FIGURE 12.4 In (a) (the pragmatist's vision), large sacrifices are needed to raise the intermediate group above the minimum. In (b), the social analogue of the case of genetic diseases, small sacrifices produce dramatic gains for the intermediate group and bring the disadvantaged close to their ceiling.

tolerable, inefficiency on a far larger scale might prove ruinous. The size of the second group in the social case is far greater than the size of the second group in the genetic case: there are far more children who suffer because of desolate or threatening environments than children with the genes for PKU, or even children who are victims of genetic disease. Pragmatists thus view the idealist proposals for social policies based on redistribution of assets as quixotic ventures, liable to produce only uncertain benefits for members of the second group, in danger of generating damaging side-effects, and requiring, because of the scale of the problem and the inefficiency of social interventions, sacrifices by the majority that would lower their expected quality of life by an amount too large to tolerate. (See figure 12.4[a] for a representation of the outcome, as the pragmatist sees it; figure 12.4[b] depicts the outcome as it would be were the cases analogous.)

Idealists have a number of ways of trying to counter pragmatist pessimism. There is likely to be disagreement about the consequences of social interventions. Disclaiming the need for any sophisticated social science, idealists may contend that we know the sorts of things children need: better schools, safer streets and playgrounds, stable homes, and realistic prospects for jobs. Infusing money into job-training programs, teacher salaries, school renovation, group homes for neglected and abused children, and developing an effective police force, trusted by local residents, would make an enormous difference to the expected quality of lives. In response to the charge that there may be unanticipated harmful consequences, idealists may suggest that grave problems require us to take risks—just as we might consider risky interventions when confronted with a devastating disease. How many lives could be improved, and by how much, by the kinds of measures idealists recommend? Most of us have opinions—and perhaps some experts have answers. But prior to the wrangling over facts is a moral question. Because the effects of social programs are uncertain, and because, in the United States and, increasingly, in other affluent democracies, the scale of the problem is large, following idealist rec-

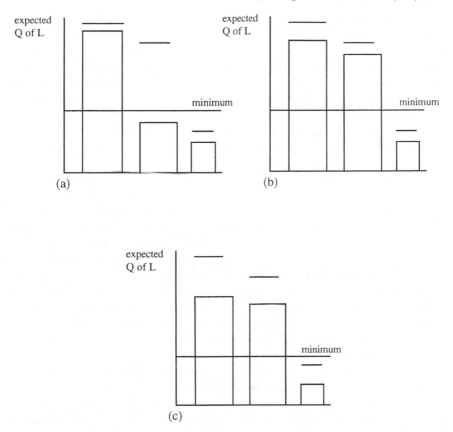

FIGURE 12.5 (a) shows very little redistribution of income, (b) displays redistribution with small sacrifices by the majority and large gains by the second group; (c) shows considerable sacrifices for more modest gains. Egalitarian commitments can be caused by judgments about which kinds of depicted states are preferable.

ommendations is unlikely to prove cheap. What is the extent of the sacrifice the affluent are morally obliged to make?

Depending on a host of factual details, reallocating resources might raise the expected quality of life of the second group by varying amounts, while demanding greater or lesser sacrifices by the majority. (See figure 12.5 a–c for three possibilities.) Reflection on the genetic cases teaches us that small losses in expected quality of life for the majority are legitimately required when we can raise the truly unfortunate close to their quality of life ceiling, and provide for the slightly less unfortunate lives of quality comparable to those of the majority. A whole spectrum of moral positions is consistent with this judgment. At one pole are those who place greatest emphasis on avoiding losses in expected quality of life: for them a redistribution that would cause the quality of lives of a large number of people to drop significantly below their quality of life ceiling is unjustifiable. At the opposite extreme are those who are most moved by the ideal of raising the expected quality of life of people whose prospects are currently bleak: even a large total loss in expected quality of life (computed by summing the changes in expected quality of life for each member of the society) should be tolerated in the interests of raising the expected quality of life of people currently below the acceptable minimum. Between the extremes are all the possibilities for balancing the two principles that emerged from the earlier discussion: the principle to avoid loss of potential quality of life (irrespective of actual level) and the principle to raise the quality of lives that currently have lowest expected quality (irrespective of potential). There is a *moral spectrum* along which different people, including pragmatists and idealists of various stripes, place themselves differently.

Molecular genetics does not confront us with the difficulty of positioning ourselves on this moral spectrum. The people who will be affected by the redistribution of assets—the children who receive care or whose lives are transformed by interventions like the PKU diet—are few enough in number to demand only small sacrifices. When we turn to evaluate the idealist analogy, however, disagreements about the pertinent social facts and about the proper position on the moral spectrum are interwoven. Many idealists, including the most vocal critics of the Genome Project, would surely offer an explanation for the apparent recalcitrance of social problems. Our failure, they believe, can be traced to the limitations of a moral perspective that embraces the redistribution of assets only timidly, withdrawing when our efforts encounter obstacles. Perhaps the remedy for our ignorance is to try a variety of approaches and learn which ones are beneficial. That can only be done, of course, if we are willing to commit substantial resources, prepared to position ourselves further toward the egalitarian end of the moral spectrum.

I have been exploring the most straightforward way of developing the idealist analogy—but the most straightforward way is not necessarily the right way, and the comparison so far considered may make it easier for pragmatists to dismiss idealists as starry-eyed optimists. In presenting the genetic example, I imagined the second group to consist of children whose quality of life could be significantly improved by applying our molecular knowledge, for example children carrying the genes for PKU. But PKU looms so largely in discussions of the fruits of human molecular genetics because it is a special case, something molecular medicine hopes to emulate repeatedly in coming decades. The promise of applications of molecular genetics is real, but it is likely to issue not in sweeping solutions to problems of major families of disease—total victory in the war against cancer, say—but to a motley of techniques, useful to varying degrees in a broad range of cases. The idealist analogy can be developed differently by combining the concern for the quality of the lives of the least fortunate with a counterpart for this therapeutic pragmatism.

If we were to reconceive the second group of people in the genetic example, thinking of them as sufferers from genetic disease who may benefit, to varying extents, from future molecular medicine, then the original pragmatist response would have to be slightly modified. No longer would it be possible to claim that there is a known way of raising the quality of life of members of this group, effective across the entire class. Public commitment to funding programs of molecular research would be grounded in the prospect of an assortment of more or less useful therapies. Pragmatists would have to contend that nothing similar can be expected in the social case, that we cannot anticipate any such range of successful local interventions.

Statistics on rates of violent crime, unemployment, drug abuse, and education levels in some regions of urban America are sufficiently horrifying to inspire idealists to call for sweeping plans of social action—or, perhaps, sufficiently numbing to cause pragmatists to resign themselves to the impossibility of tackling problems of such vast scope. Yet it is worth taking a look behind the statistics. I shall draw, very briefly, on two recent discussions of the consequences of social inequality in the lives of American children.

In a moving book on the lives of some young boys in the Chicago housing projects, a book aptly titled *There Are No Children Here*, Alex Kotlowitz exposes quite specific needs that could be addressed, very directly, by a commitment of public funds.[24] The public defender who represents one boy does not have time to listen to his story because of the overload of cases; the energetic and inspiring teacher who encourages the boy's younger brother has inadequate funds for books and supplies; parts of the projects are always dark because the city has given up trying to devise a system of lighting that would be proof against vandalism; and, perhaps most importantly, the resourceful administrator who successfully sweeps a few buildings for drugs and guns (earning the respect and gratitude of the inhabitants) is unable to extend his operation to twenty other complexes—"money from the Department of Housing and Urban Development has not been forthcoming." It is hard to maintain that these difficulties are any more insuperable than those biologists face in trying to repair the damage done by defective, or missing, enzymes—and hard not to believe that solutions to these mundane problems would bring real improvements in the quality of human lives.

Jonathan Kozol's *Savage Inequalities* is an equally scathing indictment of the conditions in which citizens and their elected representatives allow some American children to grow up. Kozol's book begins with the children of East St. Louis, a town in Illinois that has served as the home to chemical plants and to companies specializing in the burning of hazardous wastes. He quotes a St. Louis health official, who compares the conditions in which East St. Louis children live to the Third World.[25] The incidence of lead poisoning is high, but "[t]he budget of the city's department of lead-poison control, however, has been slashed, and one person now does the work once done by six."[26] Despite the fact that the schools stand on grounds that have been heavily polluted by chemical spills (from the plants that no longer employ the parents of the schoolchildren), and despite the recurrent overflows and flooding by raw sewage, the state government has refused money to help clean up the mess. (The state government blames the local administration for its inefficiency. Finger-pointing, of course, fails to remedy the plight of children who become sick from the poisonous fumes or who repeatedly find their schools closed.)

Nobody can read these books without recognizing clear possibilities for concrete solutions to some of the problems that reduce the expected quality of life for poor children in the United States, solutions that money can buy just as easily as a special diet for PKU or the new apparatus that may advance the identification of genetic diseases. Resolute efforts could greatly reduce the ill-effects of lead poisoning in American cities, and, quite possible, do more to enhance the quality of the lives of American children than the attacks on all the single-locus genetic diseases combined. If pragmatists are to resist the idealist argument, they cannot do so simply by reiterating their general distrust of social interventions. They must produce economic and social analyses that reveal that the apparent benefits of specific programs are unreal, or are offset by excessive costs. In short, they must engage in the same kinds of detailed data-collection and calculation that is needed to assess the levels of social support that applications of molecular genetics will require.

Although idealists and pragmatists may find some common ground in the thought of redistributing resources for parallel programs that seek specific, local ways of reversing the depressing effects both genes and environments have on the quality of lives, significant moral differences are likely to linger. Because of the scale of the social problems, there will be disagreements about how far societies are obliged to go, about how to balance concern for those who are worst off with avoidance of society-wide losses in quality of life. The debate between idealists and pragmatists is multifaceted—perhaps more many-sided than is commonly supposed. But I believe the idealist analogy cannot simply be dismissed; and, if the analogy does nothing else, it should force each of us, privileged citizens of affluent societies, to reflect on the extent of our obligations to promote the quality of lives that are currently bleak, to think about our own chosen place on the moral spectrum.

Let me now turn to the second part of the idealist critique. Champions of molecular medicine commonly believe they can achieve its benefits without attending to the broader social inequalities. The difficulties of avoiding serious harm, of preventing the reinforcement of prejudice and inequality, already surfaced in my discussion of the short-term problems of the HGP. Some of those difficulties can be resolved—at least in principle: we could ensure adequate counseling for all, provide universal health coverage, enact strict regulations to block discrimination in employment.

Yet are these important social programs likely to be implemented and sustained in a society whose prevailing concerns are with efficiency rather than equality, whose members position themselves on the moral spectrum at a far remove from the egalitarian pole? We live at a time in which the dominant political rhetoric opposes the idea that affluent members of our society owe anything to children currently born into poverty. Utopian eugenics requires the construction of expensive forms of social support. Consider the important conditions that secured it against the charge of repeating the moral blunders of the past.

1. All members of the society must have access to information that will enable them to make free and informed reproductive decisions.
2. Personal choices must not be limited by background social prejudices.
3. Those born with genetic conditions that can be ameliorated by social support must be guaranteed those forms of support.

Translating these ideals into practice requires bringing millions of Americans, whose connections with the system of medical care are currently tenuous, into a position in which they can draw on extensive genetic counseling and receive costly forms of support (special diets, wheelchairs, special classes for developmentally disabled children, and, quite possibly, an increasing number of drugs patented by biotechnology firms anxious to recoup initial investments). Will a society that grudges the children of East St. Louis unpolluted schools take on these obligations?

As I was writing the current draft of this essay, this question took on renewed force. The governor of California (Pete Wilson) and California legislative leaders agreed on a proposed budget, providing more funds for public schools and for the state university system. At the same time, welfare funding has been cut. Some salient points are made clear in a newspaper report:

> Welfare for families will be cut $141 million statewide, pushing the monthly grant for a mother with two children down 4.9% from $595 to $566 in high-rent areas, including Los Angeles, Orange and Ventura Counties. In lower-cost areas . . . the grant level will drop to about $538 for a mother with two children.
>
> Aid to the poor who are elderly, disabled and blind will be cut by similar percentages—pushing the grant for an individual from $614 a month to $584 in high-rent areas, and $566 a month in low-rent areas.
>
> As legislative leaders tried to sell their colleagues on the deal, several disabled people in wheelchairs blocked the front door to Wilson's outer office, vowing to stay until they were arrested. Twenty-three demonstrators were cited for disturbing the peace and released, but none was jailed.
>
> "I'm having trouble making ends meet now," said Daryl Wisdom, 47, who is in a wheelchair because of polio. Wisdom's pharmacy bill is $200 a month, he said.[27]

The apparent cruelty of these budgetary cuts might be understood as a necessary measure. In hard economic times, perhaps California must sacrifice the interests of some to attain other worthy ends—pumping money into an ailing educational system by drawing from the impoverished. But a later paragraph in the same report belies the idea of economic necessity: "Although there is no general tax cut, high earners stand to benefit. The two upper-income tax brackets for people whose annual taxable income exceeds $100,000 will expire at the end of the year. As a result, the state will lose about $300 million in revenues."[28]

It is not hard to do the arithmetic. If the tax brackets had been renewed, then, instead of cutting funds to the poor, the elderly, the disabled, and the blind, those people could actually have benefited from a modest increase (at least 3%). The 1994 Tax Schedule locates the top bracket at $429,858 (for couples filing jointly) and $214,929 (for a single person); the second bracket begins at $214,928 (for couples filing jointly) and $107,464 (for a single person). Under the new measure couples whose joint taxable income exceeds $61,000 will all pay at the same top rate. Mr. Wisdom, on the other hand, will receive $384 each month after he has paid his pharmacy bill. We should ponder the high ideals of utopian eugenics in the light of these political realities.

Utopian eugenics is a fragile enterprise. Critics of the promise of molecular biology are wrong to fear that we shall *inevitably* be swept into repeating past eugenic excesses—there are theoretical possibilities, exemplified by utopian eugenics, for a far more benign program. It would be small consolation if that were simply a theoretical point, if the possibilities could not be actualized. Ironically, we can now see that *both* idealist critiques of applications of molecular medicine and pragmatist defenses mix optimism and pessimism. Idealists believe that we can take steps to eradicate the inequalities in affluent democracies, but are skeptical about the possibility of preventing new injustices, spawned by applications of molecular knowledge, if the broader social issues are not addressed. Pragmatists maintain that piecemeal efforts are sufficient to ward off the danger, and that we can sustain a benign form of eugenics, but they doubt our ability to remedy more general social ills. What if both the pessimistic assessments are correct?

We don't need the HGP to inform us that there are important social problems that face our society, and that concern for the welfare of others cries out for a more egalitarian (and compassionate) approach to those who are born into poverty. Reflection on the HGP doesn't somehow, suddenly, make an argument in social philosophy that wasn't available before. Rather the HGP is a lightning rod, a program that draws fire because of the presence of background instability. Its critics have, I believe, muddied the main issues by using rhetorically powerful devices: they have charged that the HGP overestimates the idea that genes are powerful, or that they are destiny, and that the program is a return to eugenics. Intelligent supporters of molecular medicine recognize that these charges are quite unfounded. We don't have to become genetic determinists or revive the eugenic errors of our predecessors to appreciate the real benefits that advances in human molecular genetics may bring.

I have been trying to identify what I think is most important in the critics' case. Because of urgent social problems that arose before the HGP was proposed, problems that we have had a moral obligation to address quite independently of the possibilities opened up by molecular genetics, the beautiful world that the enthusiasts envisage may be a long way from reality. The main argument of this essay is that discussions of the HGP ought to lead into a much broader, and more difficult, set of issues—that leave genetics far behind—because these issues are deeply implicated in the social setting that will surround the new medicine.[29] We need, in short, explicit discussions about the moral spectrum and appropriate regions along it, supplemented with careful analyses of the potential costs and benefits of various programs, both social and medical. If the HGP leads citizens and policymakers, as well as philosophers, to engage in serious discussion of these broad questions, then, whatever else it achieves, it will have done us an important, and unexpected, service.

Notes

1. This essay draws on my previous efforts to organize the issues around the HGP. See, in particular, Philip Kitcher, "Who's Afraid of the Human Genome Project?" in *PSA 1994: Proceedings of the Biennial Meeting of the Philosophy of Science Association* (East Lansing, Mich.: Philosophy of Science Association, 1995), and Kitcher, *The Lives to Come: The Genetic Revolution and Human Possibilities* (New York: Simon and Schuster, 1996).

2. Alex Rosenberg offers an entertaining version of this argument in "Subversive Reflections on the Human Genome Project," in *PSA 1994: Proceedings of the Biennial Meeting of the Philosophy of Science Association* (East Lansing, Mich.: Philosophy of Science Association, 1995), 329–335.

3. These points are developed and defended in chapter 4 of Kitcher, *The Lives to Come*.

4. This is a point that Eric Lander has emphasized in public lectures.

5. One might reply that knowledge increases the autonomy of the subject. However, this is too stern an ideal—for many people could easily be crushed by news that they will inevitably suffer from a late-onset disease. This is most clearly shown in the studies of people's reactions to testing for Huntington's disease. See Lori Andrews et al., *Assessing Genetic Risks: Implications for Health and Social Policy* (Report of the Institute of Medicine) (Washington, D.C.: National Academy Press, 1994), and Nancy Wexler, "Clairvoyance and Caution: Repercussions from the Human Genome Project," in Daniel Kevles and Leroy Hood, *The Code of Codes* (Cambridge, Mass.: Harvard University Press, 1992), 211–243.

6. A point stressed by Ruth Hubbard and Elijah Wald, *Exploding the Gene Myth* (Boston: Beacon Press, 1993).

7. Diane Paul, "Toward a Realistic Assessment of PKU Screening," in *PSA 1994: Proceedings of the Biennial Meeting of the Philosophy of Science Association* (East Lansing, Mich.: Philosophy of Science Association, 1995), 322–328.

8. This was dramatized at a recent meeting in Bethesda, Maryland, organized jointly by the NIH/DOE working group on the Ethical, Legal, and Social Implications of the HGP and the National Action Plan on Breast Cancer. At that meeting, several women who had had positive diagnoses for familial early-onset breast and ovarian cancer recounted their struggles with insurance companies and employers. Insurers refused to cover prophylactic mastectomies that the women, and their doctors, saw as measures required to avoid grave risks to life. Some women lost their jobs. All have had great difficulty obtaining subsequent health coverage. Indeed, in order to protect anonymity, they spoke at the meeting under assumed names, and video recording of their presentations was not allowed.

9. For some of the best discussions, see N. A. Holtzman, *Proceed with Caution* (Baltimore: Johns Hopkins University Press, 1989), and Dorothy Nelkin and Laurence Tancredi, *Dangerous Diagnostics* (New York: Basic Books, 1989).

10. See for example, John Harris, *Wonderwoman and Superman* (Oxford: Oxford University Press, 1993), which discusses the ethical issues in a sensitive fashion but seems at a far remove from the scientific realities.

11. Kitcher, *The Lives to Come*, chapter 5.

12. Particularly outstanding is Daniel Kevles, *In the Name of Eugenics* (London: Penguin, 1985).

13. Diane Paul, "Eugenic Anxieties, Social Realities and Political Choices," *Social Research* 59 (1992): 663–683.

14. Kitcher, *The Lives to Come*, chapter 13. My discussion there draws on ideas of Ronald Dworkin, *Life's Dominion* (New York: Knopf, 1993), and James Griffin, *Well-Being* (Oxford: Oxford University Press, 1985).

15. Negative discussions of the HGP often overlook its potential to avoid real tragedies. At a symposium in Washington D.C. in the spring of 1994, a man from the audience responded to the worries of three speakers (Patricia King, Eric Lander, and me) by telling us of his daughter, who had given birth to two children afflicted with neurofibromatosis. After describing, in a very restrained way, how her life, and that of her husband, had been blighted, he asked why discussions of the HGP so quickly gravitate to the potentials for danger, and never dwell on the real promise. That reminder seems to be highly salutary.

16. Paradigm idealists are Richard Lewontin (*Biology as Ideology* [New York: Harper-Collins, 1993]) and Ruth Hubbard (Hubbard and Wald, *Exploding the Gene Myth*). Because the debate has not been explicitly presented in the form I adopt here, it is hard to identify pragmatists—but I suspect that many defenders of the HGP would embrace pragmatism.

17. Or perhaps with "primary goods," as with John Rawls's rightly influential *A Theory of Justice* (Cambridge, Mass.: Harvard University Press, 1971). My own approach is much

closer to that of Amartya Sen and his emphasis on equalizing capabilities. See in particular Sen, *Inequality Reexamined* (Cambridge, Mass.: Harvard University Press, 1992). Comparisons between genetic and environmental differences provide the basis for many interesting defenses of Sen's points, but I shall not try to articulate them here.

18. This argument is a highly compressed version of a line of reasoning that I develop at much greater length in the last three chapters of Kitcher, *The Lives to Come*.

19. There are any number of good formulations of the point. See, for example, Bernard Williams's contribution to J. J. C. Smart and B. Williams, *Utilitarianism: For and Against* (Cambridge: Cambridge University Press, 1973).

20. The discussion of this paragraph is influenced by Derek Parfit's illuminating exploration of related problems in part 4 of *Reasons and Persons* (Oxford: Oxford University Press, 1984).

21. Figure 12.1 and those that follow adapt to the present context the style of representatin used by Parfit in part 4 of *Reasons and Persons*. See also Larry Temkin, *Inequality* (New York: Oxford University Press, 1993).

22. It might naively be thought that this is at odds with Rawls's "difference principle." However, trying to apply the difference principle to the present context would be quite misguided, for several reasons—not least because the principle is intended to govern "basic social institutions" and thus to function at a much more abstract level. It is an interesting question whether the genetic example of the text subverts an intuition on which Rawls's treatment of justice draws.

23. After I had reached this way of approaching the issue, I discovered, through Sen's discussion (89–93), that a formally similar idea has been developed by welfare economists. See, in particular, A. B. Atkinson, "On the Measurement of Inequality," *Journal of Economic Theory* 2 (1970): 244–263.

24. A. Kotlowitz, *There Are No Children Here* (New York: Doubleday, 1991).

25. J. Kozol, *Savage Inequalities* (New York: Harper, 1992), 10.

26. Ibid., 11.

27. *Los Angeles Times*, 27 July 1995, A18.

28. Ibid.

29. In short, the big ethical and social issues surrounding the HGP are not *implications* of the HGP. They are huge social problems that American society has shamefully neglected for a decade and a half.

13

Battling the Undead

How (and How Not) to Resist
Genetic Determinism (2000)

"But wait," the exasperated reader cries, "everyone nowadays knows that development is a matter of interaction. You're beating a dead horse."

I reply, "I would like nothing better than to stop beating him, but every time I think I am free of him he kicks me and does rude things to the intellectual and political environment. He seems to be a phantom horse with a thousand incarnations, and he gets more and more subtle each time around. . . . What we need here, to switch metaphors in midstream, is the stake-in-the-heart move, and the heart is the notion that some influences are more equal than others."

—S. Oyama, *The Ontogeny of Information*

Nobody has done more to combat genetic determinism than Richard Lewontin, whose writings, from the original IQ controversy to present debates about the implications of human molecular genetics, diagnose errors that have seduced influential scholars and their readers into believing vulgar slogans about genes and destiny.[1] Lewontin's reward for his decades of effort has often been the irritated response that what he claims is uncontroversial: once the intellectual poverty of a version of genetic determinism has been exposed, there is a rush to denial ("That is not what we meant; that is not what we meant at all"). Yet, within months or years, some new version of the view that human behavior is largely shaped by the genes returns, inspiring a new rash of popular discussions and, in some instances, framing debates about social policy. It is small wonder then, that people appalled by the sloppy thinking Lewontin has exposed yearn for the "stake-in-the-heart move."

Lewontin's own response to the continued reemergence of genetic determinism has been to deny the correctness of the interactionist credo, the conventional wisdom to which purveyors of determinist claims retreat in the face of criticism.[2] Although many of his best arguments consist in demonstrating how determinists have ignored interactions among genetic and nongenetic factors, Lewontin seems to believe that acceptance of these arguments is not enough, that we need to free ourselves from the grip of the Cartesian picture of the world as a machine, that we should recognize the interdependence between organism and environment, and that we should formulate a "dialectical biology." He is convinced that there has to be a fundamental error—an error that can be corrected only by reconceptualizing some parts of biology.

In my judgment, no such reconceptualization is needed, and Lewontin's positive proposals are in constant danger of relapsing into the obscurity that he rightly sees as affecting traditional forms of biological holism.[3] Genetic determinism persists not because of some subtle error in conventional ideas about the general character of biological causation but because biologists who are studying complicated traits in complex organisms are prone to misapply correct general views. Ironically, the existence of this tendency to error testifies to the social pressures on biological practice—pressures that Lewontin has been at some pains to point out. The search for the stake-in-the-heart rests on a misunderstanding of the problem and may even undermine the effectiveness of the more limited measures that Lewontin and others have crafted.

It is high time to back up assertion with argument. Let us begin more slowly by asking what the thesis of genetic determinism claims.

Here is a first version. To suppose that a particular trait in an organism is genetically determined is to maintain that there is some gene, or group of genes, such that any organism of that species developing from a zygote that possessed a certain form (set of forms) of that gene (or a certain set of forms of those genes) would come to have the trait in question, whatever the other properties of the zygote and whatever the sequence of environments through which the developing organism passed. Although this is a relatively simple way to articulate the idea that genetic causes take priority, it is of little use for reconstructing the debates about genetic determinism. Perhaps, with sufficient ingenuity, one can discover traits that are genetically determined in this sense, but any such traits will be causally "close" to the immediate biochemistry in which DNA is involved—they will not be the characteristics for which we wonder about the rival contributions of nature and nurture. Even if we apply the definition to a relatively uncontroversial exemplar, investigating whether it counts Huntington's disease (HD) as genetically determined, we encounter trouble. True enough, human beings who carry abnormally long CAG repeats at a particular locus near the tip of chromosome 4 undergo neural degeneration, typically between the ages of 30 and 50, and doctors know of no preventative treatment. Does this mean that no way is known of contriving an environment in which the terrible decay does not occur? Not really. Huntington's disease could be forestalled by giving those with the long repeats the opportunity to end their lives before the onset of the disease, and it is overwhelmingly probable that some people with such repeats have suffered accidental death early in life. Hence, strictly speaking, there are environments in which people who have abnormally long CAG repeats at the HD locus do not develop HD, and thus, according to the definition, HD would not count as genetically determined.

Of course, the existence of environments in which the expression of the HD phenotype is forestalled by early death is hardly comforting, and it would be rea-

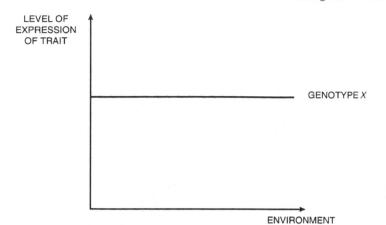

FIGURE 13.1 A graphical representation of the simplest type of genetic determinism. The level of expression of the phenotypic trait of interest in individuals with the focal genotype ("Genotype X") remains constant no matter how the environment varies.

sonable to suggest that the account of genetic determinism ought to be refined in one of two obvious ways: (a) by replacing the demand that the trait be acquired in *all* environments with something weaker ("almost all") and (b) by restricting attention to complexes of causal factors that enable the organism to develop to the age at which the trait would normally first appear. But it seems more illuminating to make explicit the strategy that underlies the proposed definition. That strategy begins by isolating certain properties of organisms for exploration of their causal impact, regarding the phenotype as the product of contributions from particular kinds of DNA sequences, on the one hand, and from *everything else*, on the other. It goes on to inquire how the phenotype varies as the DNA sequences are held constant and as other factors (the cytoplasmic constitution of the zygote, the molecules passed across cell membranes, etc.) change. The graphical representation of this, the *norm of reaction* of the genotype, is a familiar concept in genetics, and the crudest sort of genetic determinism consists in claiming that the norm of reaction for the trait of interest is flat (see figure 13.1). Just the kinds of difficulties that appeared in the HD example make doctrines of so simple a form implausible, but the pictorial style of representation suggests plenty of ways in which the genetic factors can be seen as playing important causal roles—perhaps the norm of reaction will be flat almost everywhere, perhaps it will vary only slightly, perhaps the norms of reactions for different genotypes will show a universal relation, perhaps the norm of reaction will be flat if we restrict ourselves to those complexes of other factors that we think of as healthy for the organism. (See figure 13.2.) We might thus see genetic determination as a matter of degree and, instead of quibbling about the proper definition of genetic determinism, investigate the shapes of the norms of reaction in the cases of interest to us.

One of the great insights of Lewontin's early discussions of these questions was his recognition of this as the real issue to which claims of genetic determination

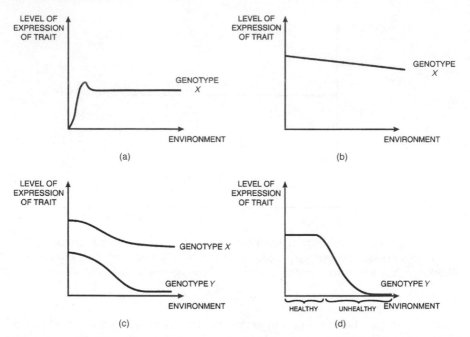

FIGURE 13.2 Some determinist themes. In (a), the level of expression of the trait is constant (for genotype X) in almost all environments; in (b), the level of expression is almost constant across all environments; in (c), despite variation in levels associated with genotypes X and Y, the level of expression for individuals with X is always greater than that for Y, no matter what the environment; in (d), there is considerable variation in the level of expression but only in environments that are unhealthy. These themes admit further refinements, combinations, and variations.

were directed.[4] Moreover, Lewontin explained with admirable lucidity why the methods employed to establish those claims could not deliver such conclusions. Estimates of heritability do not reveal the contours of norms of reaction; cross-cultural surveys are only likely to do better if one can be confident that the entire space of nongenetic causal variables is covered.[5] If, as I believe, Lewontin was right in his diagnosis of the errors of popular behavior genetics (most evident in doctrines about the determination of IQ) and popular human sociobiology (manifested in conclusions about the ineradicability of sexual differences in behavior), then the besetting sin was the tendency to draw certain kinds of pictures on the basis of woefully inadequate evidence.

It should now be obvious how a weary critic of hasty generalizations about norms of reaction might go further. Perhaps the tendency to suppose that the relative invariance of a phenotypic trait, given a particular genotype across a manageable range of environments, indicates a flat norm of reaction might be scotched by denying the legitimacy of any such representation altogether. During the 1980s and 1990s, Lewontin and others (most prominently Susan Oyama, Paul Griffiths, and Russell Gray) began to argue that our entire view of genotype–phenotype relations needed to be changed, and that the framework within which I have been posing issues about genetic determination ought to be rebuilt.

Is the notion of a norm of reaction well defined? The writings of those who demand a new conception of nature and nurture—a "dialectical biology" (Lewontin) or "developmental systems theory" (Oyama, Griffiths, and Gray)— suggest several worries about the notion and its relatives (such as the standard genetic idiom of a gene "for" such-and-such a trait). Organism and environment, it is said, are interdependent; there is "developmental noise" in the production of phenotypes; the singling out of genes as causal factors is an unwarranted abstraction from a complex causal situation wrongly giving priority to some determinants of the phenotype; the notion of a gene "for" a trait cannot be coherently reconstructed. These are important concerns, and I will take them up in order.

Lewontin has argued that an organism's environment should not be thought of as identifiable prior to the organism and its distinctive forms of behavior:

> Are the stones and the grass in my garden part of the environment of a bird? The grass is certainly part of the environment of a phoebe that gathers dry grass to make a nest. But the stone around which the grass is growing means nothing to the phoebe. On the other hand, the stone is part of the environment of a thrush that may come along with a garden snail and break the shell of the snail against the stone. Neither the grass nor the stone are part of the environment of a woodpecker that is living in a hole in a tree. That is, bits and pieces of the world outside of these organisms are made relevant to them by their own life activities.[6]

The facts reported here are uncontroversial, and the last sentence strikes me as completely correct. What exactly follows?

Lewontin uses these observations to oppose both the idea that we can think of organisms adapting to environments that are independent of them and the idea that we can think of the phenotype as dependent on causal interactions between genotype and environment. The latter conception is the principal concern here, although similar remarks apply to both types of criticism. Lewontin is moved by a principle about causes and causal dependence: C cannot be a causal factor in the production of P if C is dependent on P. Applying his conclusions about the dependence of environment on organism, he maintains that we cannot see the environment as a causal factor in the production of the phenotype, and thus the idea of a norm of reaction, with its partitioning of causal variables along different axes, is confused.

There are two related points to be made about this line of reasoning: first, it is not obvious what notion of dependence figures in the causal principle, and, second, it is not clear that just one notion of environment is pertinent here. Consider the notion of dependence. In one very obvious sense, the stone in Lewontin's garden is independent of the presence of phoebe, thrush, and woodpecker—it sits there before the arrival of the birds, before the eruption of fledglings from the nest. So, if we understand "dependence" to mean that the existence of one thing is an effect of the presence of the other, then Lewontin's principle, although plausible, does not apply to the case at hand: there is no reason to think that the contents of the garden cannot play roles in the formation of phenotypes. On the other hand, if we understand "dependence" to mean that the causal relevance of one thing varies with the properties of the other, then the principle does apply to the relations between the birds and the garden. Whether grass, stones, or holes in trees are causally relevant to the development of the birds varies with the properties of the birds, as Lewontin's illustrations show. But now there is no great plausibility in the causal principle itself, for, eleaborated, it says that if the causal relevance of C to P varies with the properties of the bearers of P, then C cannot be a causal factor in the production of P in *any* case, and this claim seems to verge on paradox.

The point can be clarified further by focusing on the other murky term in the argument, "environment." Biologists typically think of environments as those parts of the world outside the organism that are causally pertinent, and in this, the *functional* environment, great tracts of nature are not part of the organism's environment. Lewontin's observations reveal very clearly that an organism's functional environment can depend on what the organism does. However, when we think about the development of an organism, we can pick out some potential causal factors—say the organism's DNA—and take the environment, the *total* environment, to be everything else. In Lewontin's phrase, the total environment is all "the bits and pieces of the world outside the organism" plus some more "bits and pieces"—to wit those inside the organism but not the DNA.[7] The phenotype the organism acquires is determined together by the genotype (the DNA sequences) and the total environment, and of course, a large part of the total environment will be causally irrelevant. Furthermore, it is quite correct to note that the functional environment, the bits and pieces that are pertinent, depends on (in the sense of varying with) the properties of the developing organism. But this is quite compatible with the causal analysis of phenotypes in terms of genotypes and total environments and with the attempt to draw norms of reactions that identify the causal contributions.

Yet there is an important point behind Lewontin's argument, one that becomes misfocused because of his eagerness to drive a stake into the heart of genetic determinism. To produce a picture indicating the shape of a norm of reaction is to advertise oneself as understanding how to order environments along the axis, and that is typically false advertising. In most instances, we only have the most rudimentary knowledge of how to identify the functional environment, and our ignorance affects the pictures and the conclusions drawn from them.[8] Typically, we can divide the factors outside the DNA into three categories: those we can identify and know to be causally relevant, those we can identify and know not to be causally relevant, and those we either cannot pick out or whose relevance we do not know. (It is, of course, quite possible for us to realize that there is much about which we are likely to be ignorant.) Confronted with a claim about the genetic determination of human propensities to violent behavior (for example), modesty should urge us to think that the last category is quite large, and thus a demonstration that the norm of reaction for a genotype remains flat over a wide range of the nongenetic variables known to be relevant ought not to inspire much confidence that the result would survive a more detailed and fine-grained partitioning. Thus, the right point to make is that we should not leap to premature conclusions about the character of the functional environment, that we should recall the fragility of our representations of the non-genetic causal factors, and that, in consequence, even though the notion of a norm of reaction is perfectly well defined, even though norms of reaction are just what we are trying to discover, knowledge of such norms is very hard to come by. Lewontin has miscast the important methodological point about the difficulty of settling the questions of concern (the shapes of norms of reaction) as an incorrect conceptual point about the incoherence of the notion of a norm of reaction.

The second concern about interactionism focuses on the possibility of "developmental noise." Lewontin argues that even knowledge "of the genes of a developing organism and the complete sequence of its environments"[9] would not allow prediction of the phenotype. In support of this claim, he notes that fruitflies typically have different numbers of bristles at the left and right sides of their thorax, that the difference cannot be explained by a difference in genotype and is not traceable to differences in environment.

> Moreover, the tiny size of a developing fruitfly and the place it develops guarantee that both left and right sides have had the same humidity, the same oxygen, the same temperature. The differences between left and right side are caused neither by genetic nor by environmental differences but by random variation in growth and division of cells during development: *developmental noise.*[10]

Once again, it is important to ask what is being counted as part of the environment and what standards are being used to assess identity of environment.

There are three main types of answer to the question Why do fruitflies have different numbers of bristles at the left and right sides of their thorax? One is to suggest that Lewontin has just counted environments as the same in too coarse a fashion. Perhaps the temperature on the left is the same as that on the right so long as we measure to two or three significant figures, but there are minute differences from side to side, and at crucial stages of cell division, these differences make a difference. This answer would broadly accept Lewontin's conception of the environment but would eliminate the notion of developmental noise in terms of a more precise understanding of the environmental variables.

The second response would take advantage of the fact that, when interactionists undertake causal analysis of phenotypes in terms of the contributions of DNA and other factors, some of these other factors might be internal to the organism. One of the principal achievements of developmental biology in recent years has been the demonstration of how initial asymmetries in the cytoplasm interact with the DNA in the first stages of ontogeny to produce patterns in early embryos (worked out in greatest detail so far for *Drosophila*). It is quite possible that the differences in rates of cell division do account for the difference in bristle number and that these rate differences are remote effects of the inhomogeneity of the zygote. Although we could reasonably describe them as "random" in the sense that there is no uniform process that determines the distribution of molecules in the cytoplasm of the ovum—so that the initial state of the zygote is the result of contingencies of the formation of a particular egg—they are not *irreducibly* random. A fine-grained specification of the total environment of the DNA would provide a causal explanation of the asymmetry in bristle number. Thus, once again, the form of the phenotype can be viewed as fixed by the genotype and the environment provided that we conceive of the environment in the proper (total) fashion. There is no need to invoke developmental noise or to think that the notion of a norm of reaction breaks down here.

The last possibility is that even the initial distribution of molecules throughout the zygote together with the fine-grained structure of the sequence of environments through which the fly develops does not determine the bristle number. Perhaps the asymmetry is irreducibly random in that no further introduction of causal factors will account for it. I do not know if Lewontin has this possibility in mind, but the existence of fundamental indeterminacies in quantum physics makes it necessary to consider it. There are no well-established instances of quantum events playing a significant role in ontogeny, and many biologists and philosophers seem convinced that subatomic indeterminacies will wash out because of the enormous numbers of molecules that play a role in the development of an organism (the law of large numbers is often thought to be suggestive here). If irreducible randomness does not "percolate up" from the quantum level, then, of course, there is no challenge to the notion of a norm of reaction and no reason to think that subatomic indeterminacies are a source of developmental noise. But, even if some differences in phenotypes ultimately trace to random subatomic events, a simple revision would save the concept of a norm of reaction. Instead of thinking in terms of a single phenotype, fixed by the genotype and (total) environment, we would have to suppose that this congeries of factors determines a probability distribution of phenotypes: pictorial representations would thus illustrate expected values of phenotypes, and given the elusiveness of quantum effects at the phenotypic level, it would be entirely reasonable to suppose that the spread around the mean was very small.

I turn now to the third worry, the idea that singling out the genotype and considering its effects against background environmental conditions is misguided abstraction from a complex causal situation. No interactionist denies that many causal factors are involved in development (that, after all, is the point of interactionism). However, interactionists defend the legitimacy of a general strategy of causal analysis—the strategy of isolating some of the causal factors, holding them constant, and investigating how the effect varies when other factors are altered. Interactionists ought to support a principle of causal democracy: if the effect E is the product of factors in set S, then, for any $C \in S$, it is legitimate to investigate the dependence of E on C when the other factors in S are allowed to vary. Taking E to be a phenotypic trait, C to be a particular genotype, and S to be a large (probably mostly unknown) set of factors in the total environment (that is factors in the rest of nature outside the genotype), the democracy principle endorses the legitimacy of seeking norms of reaction for phenotypic traits. But it should already be clear that the democracy principle endorses lots of other ways of undertaking causal analysis. For example, we might consider a particular environmental factor and investigate what happens to the phenotype when we vary the genotype and other parts of the environment or we might pick out some mix of genotypic and (total) environmental factors, investigating how the phenotype varies with respect to the rest of the causal factors. The democracy principle accords no special privilege to the representations that foreground the role of genes.

But why, then, do we always end up discussing whether genotypes are all-powerful in development? Why does democracy in principle always translate into elitism in practice? As we shall see, the answers turn out to be complicated, but, for the present, the interactionist's claim is simply that we should not suppose that efforts to investigate the effects of some factors, while others are allowed to vary, are incoherent or illegitimate. Complex causal situations do not demand that we perform the impossible feat of considering everything at once; rather they challenge us to find ways of making these factors manageable.[11] One defense of the prevalence of efforts to chart genotype–phenotype relations against the background of other variables would cite the epistemic benefits of such investigations: this is something we know how to do and that we can expect to prove informative. I will argue below that this cannot be the whole story.

For the moment, we can move on from the blanket charge that any kind of separation out of causal factors does violence to the causal complexities of development and turn to the last line of objection. Russell Gray (both writing on his own and in collaboration with Paul Griffiths) has provided the sharpest version of the charge that thinking in terms of genes "for" traits is a confusion. Alluding to an earlier attempt to suggest that talk of genes "for" traits always presupposes a relativization to "standard" genetic backgrounds and "standard" environments, Griffiths and Gray offer the following counter:

> Consider the DNA in an acorn. If this codes for anything, it is for an oak tree. But the vast majority of acorns simply rot. So "standard environment" cannot be interpreted statistically. The only interpretation of "standard" that will work is "such as to produce evolved developmental outcomes" or "of the sort possessed by successful ancestors." With this interpretation of "standard environment," however, we can talk with equal legitimacy of cytoplasmic or landscape features coding for traits in standard genetic backgrounds. No basis has been provided for privileging the genes over other developmental resources.[12]

There is much here with which I agree, although the last sentence contains an ambiguity that enables Griffiths and Gray to arrive at more exciting conclusions than those to which they are entitled.

Kim Sterelny and I have proposed to reconstruct the everyday talk of genes "for" traits by developing the intuitive idea that "we can speak of genes for X if substitutions on a chromosome would lead, in the relevant environments, to a difference in the X-ishness of the phenotype."[13] The notion of environment we appealed to was that of *total* environment conceived as everything outside the locus (or loci) of interest, and we sketched accounts of standardness for the genetic background and for the part of the environment that does not consist of other parts of the DNA. With respect to the extraorganismal environment, we offered three theses: (1) there are alternative ways of explicating the notion of "standard conditions," (2) one of these ways is to count as standard those environments frequently encountered by organisms of the species under study, and (3) another is to count as standard only those environments that do not substantially reduce population mean fitness.[14]

Although (1) remains untouched, Griffiths and Gray have shown that (2) and (3) are problematic if the aim is to reconstruct standard genetic discourse. Botanists

studying the oak genome want to identify some loci as affecting particular structures in the mature tree, but for most acorns, genetic substitutions at the pertinent
loci do not affect the form of the related structures because those acorns rot. In
the accounts of standard environment offered in both (2) and (3), individuals with
genetic differences at the loci do not manifest any phenotypic differences in the
trait that is supposed to be influenced when they grow in standard environments
because the most frequent environments, which also happen to be environments
that do not reduce population mean fitness, are environments in which no mature
tree grows.

Consider a locus "for" root proliferation. A botanist declares that the allele A
is "for" root proliferation, meaning thereby that AB trees generate more roots than
BB trees given standard complements of genes at other loci, standard distributions
of molecules in the zygotes, and standard sequences of environments. Suppose now
that we interpret "standard" in the fashion of (2) or (3). We have to acknowledge
that, in most standard environments, the number of roots generated by an organism growing from an AB zygote is no greater than that generated by an organism
growing from a BB zygote (both numbers are zero). However, the botanist could
still claim that, for any standard environment, the number of roots generated by the
organisms developing from AB zygotes is never less than the corresponding number
for BB zygotes, and in some standard environments, it is greater. So let the allele
A be "for" root proliferation just in case in all standard (total) environments the
number of roots generated by AB individuals is greater than or equal to the number
of roots generated by BB individuals with the inequalty holding strictly in some
cases. Obvious challenge: surely, by luck, the sole oak tree growing in one environment might be BB whereas thousands of acorns around (some BB, some AB)
rot; thus, in that environment, the inequality would be reversed. Response: once
again, we have to be careful to individuate environments; at the fine-grained level,
the environment encountered by the lucky acorn is different from that encountered
by the unlucky ones, and, if an AB acorn had found itself in precisely that
fortunate environment, then it would have generated more roots than its BB
counterpart.

This strategy for reconstructing the "gene for X" locution allows us to retain
the interpretation of "standard" as "statistically normal" by weakening the demand
that genes "for" X promote X-ishness in every standard environment. Alternatively,
we could decide that a standard environment is one that allows for the development of those features required for the manifestation of the general (determinable)
property of which the trait on which we are focusing is a particular (determinate)
instance. So, in the case at hand, to talk about genes "for" root proliferation is to
suggest that there are differences among individuals with various genotypes—
specifically individuals that have the capacity for producing roots (that is, trees).
Environments that prevent the maturing organisms from manifesting the general
property (exhibiting any form of the trait) are thus ruled out as nonstandard, but,
in accordance with the pluralistic line offered in our thesis (1), that demarcation
will vary with the kind of trait in which we are interested.

I conclude that talk of genes "for" traits can be coherently reconstructed
(indeed along the lines that Sterelny and I originally suggested). However, Griffiths

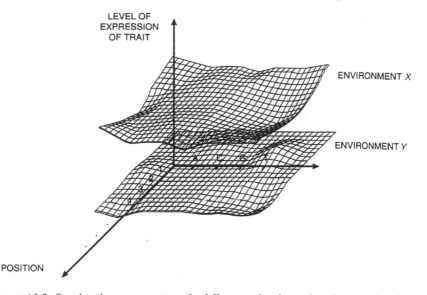

FIGURE 13.3 Graphical representation of a different style of causal analysis. In the plane of the two horizontal axes, we code genotypes at the locus of interest by specifying the nucleotide at each position. For a fixed environment, the variation of the level of expression of the trait, as the genotype varies, is represented by a surface in the space. (This can be thought of as a dual to the notion of norm of reaction.) For the example shown, the level of the trait for environment X is always greater than the level for environment Y.

and Gray are right to note that a similar form of reconstruction would enable us to speak of "cytoplasmic or landscape features" for traits (here I drop their reference to "coding" since it is a rhetorical flourish irrelevant to the discussion). Indeed, the molecular developmental genetics of *Drosophila* has already begun to emphasize the causal role of proteins deposited by the mother in the cytoplasm of the ovum: to say that the *Bicoid* protein is "for" head–tail polarity is to note that variations in the forms or concentrations of that protein will lead zygotes with standard complements of genes, given environments standard in other respects, to develop variation with respect to the anterior and posterior structures. Moreover, we can speak of some environments as "stunting" the growth of plants of particular taxa, meaning that plants with standard complements of genes, grown in those environments, will be shorter than those grown in different environments. Far from being a reductio of the interactionist view, this point simply testifies to the democracy principle introduced above. Interactionists want to allow for various ways of analyzing the complex processes of development, *one* among which is the identification of norms of reaction for genotypes, or the discovery of genes "for" traits. (See figure 13.3.)

There is a standard temptation to think that all scientific disputes can be readily resolved into differences of principle. Finding that people who advance genetic determinist claims assent to interactionism, critics of genetic determinism want to find some substantive thesis that separates the two camps, and this accounts, I believe, for the repudiation of interactionism. At bottom, however, this dispute, like other significant debates in contemporary biology, is not quite like this.[15] Instead of thinking of two groups of biologists who differ on general principles, we should view biological practice as supplying a toolkit that different people draw from in different ways. Faced with the complexities of ontogeny, biologists have some techniques of causal analysis—of the many forms sanctioned by the democracy principle. For reasons that will be probed shortly, the model of causal analysis that looks at the effects of a single genotype across varying environments is attractive when people are trying to fathom the causes of human behavior, but working out rigorous conclusions about the pertinent forms of reaction proves very difficult, and it is easy to leap to conclusions. Many of Lewontin's most pointed critiques expose the ease with which scholars have leaped to conclusions.

One moral we might draw is that we have a defective instrument, but that, I have been urging, is incorrect. There is nothing the matter with the type of model that has been applied. Rather, the trouble lies in the difficulty of the task and the tendency for the impetuous to bungle. Of course, we might do better if we had different tools—so maybe, after all, there is a case for moving beyond interactionism (not now dismissed as false or incoherent doctrine, but as a source of models too primitive for the important tasks of fathoming human ontogeny) toward "dialectical biology" or "developmental systems theory."

A different set of models for analyzing human development would be welcome, especially if they could be used to achieve insights into the causes of complex capacities and disabilities. Unfortunately, neither Lewontin's "dialectical biology" nor the "developmental systems theory" pioneered by Oyama offer anything that aspiring researchers can put to work.[16] If we want to understand why people become addicted or resist addiction, have the sexual orientations they do, give way to violence or live peacefully (and I will consider, shortly, why we might want insight into these issues), then both versions of the transinteractionist approach to nature and nurture leave us helpless. In effect, they are primarily critiques of the past misuses of old tools and at best blueprints for new tools that we might develop. When problems of analyzing human behavior seem socially urgent, and when investigators believe that new advances in molecular genetics have given new scope to the old models, pleas for "dialectical biology" or "developmental systems theory" are likely to fall on deaf ears.

There is a profound irony here. Nobody has been more sensitive than Lewontin to the social pressures that shape biological research—especially in attempts to evaluate the contributions of nature and nurture. Oyama, too, clearly recognizes these pressures. Unless there are cogent reasons for thinking that past methods of analysis are fundamentally flawed (and I have argued that there are not) rather than simply misapplied in the episodes that Lewontin and Oyama view (rightly) as politically mischievous, then the social pressures to find answers will make fledgling ventures in transinteractionism seem vague and underdeveloped rivals to well-articulated techniques that promise resolution of important questions. Furthermore, the critics of conclusions about the important effects of genotype on phenotype will be seen as taking refuge in nebulous appeals for a new general view of the causation of behavior and as driven to this predicament solely by their sense of outrage at the determinist claims.

Contemporary human genetics, including human behavior genetics, is full of promises largely because of the possibilities of using sequencing techniques to identify shared alleles (combinations of alleles) in different people.[17] Instead of the dubious passages from heritability to conclusions about causation, genetic research can hope to discover norms of reaction more directly by finding large numbers of individuals who share a genotype and tracking the variation in phenotype across environment.[18] Of course, our pervasive ignorance of the causally relevant features of the extraorganismic environment, to which I alluded earlier, should lead us to be tentative in evaluating the results, for we may well be overlooking some crucial environmental variable. Yet this is precisely the point on which the critique of genetic determinism should focus, and it would be unfortunate, perhaps even tragic, if we were to overlook it because the only way of opposing determinist theses was seen as the acceptance of some underdeveloped transinteractionist biology.

The confident behavior geneticist believes that new molecular techniques will enhance our understanding of socially important facets of human behavior. Sometimes the motivation for applying those techniques is impeccable. Researchers into addiction or alcoholism want to understand the causal pathways so that they can prevent human misery: in these areas, many investigations are continuous with attempts to fathom mechanisms behind diseases.[19] They begin with genetic causes not because they are convinced that these are the most important (that the norms of reaction for certain "addictive" genotypes are virtually flat) but because they want to unravel the neurochemistry, and they see the investigation of genotypes as a thread that will lead them into the tangle. For they know how to sequence DNA, and, by finding allelic sequences that correlate with addiction, they may be able to see how abnormal proteins make a difference to certain reactions in the brain and thus understand the molecular details of the interactions between organism and environment that go differently in addicts and in others. There is no question of "privileging" the genes in this kind of inquiry but rather a pragmatic criterion for using a particular type of model and a readily comprehensible, even admirable, medical motivation.[20]

At its best, research in behavior genetics is driven by a morally defensible motivation (that of alleviating human suffering). The investigator tries to understand the plight of the unfortunate by beginning with particular alleles and tracing how the associated phenotypes vary across environments because this is a readily applicable strategy. Yet the goal is to move from singling out certain loci as playing a causal role to identifying differences in the chemical reactions that occur in the formation of healthy and unhealthy phenotypes and from there to discovering what kinds of contributions the environment makes. For, at the end of the day, the goal is to bring relief by adjusting the input from the environment.

However, the reasons for entering on a program of genetic research are not always so easily defensible. Consider the much-disputed example of the genetics of violent behavior. There are good reasons to suspect that the environmental factors causally relevant to eruptions of violence are complex and varied, that there are fine-grained differences in environments that can have large effects, and that, in consequence, our attempts to construct the norms of reaction for "violence"

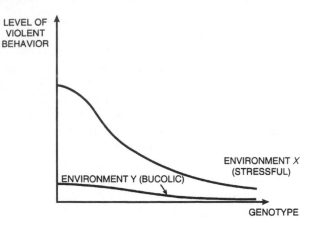

FIGURE 13.4 Representation of a pronounced environmental effect on tendencies to violence. The two axes for representing genotypes of figure 13.3 have been condensed into one
(surely more plausible than representing environmental variation on a single axis!), and the
graphs show the variation in levels of violence for fixed environments as the genotype varies.
The representation is purely hypothetical, but it is worth noting that the pronounced role of
the environment is compatible with discoveries of "violence alleles"; individuals with genotypes near the origin who encounter stressful environments show much greater levels of violence than others — even those who share similar environments. This does not detract from
the obvious fact that there is a very marked effect of environment.

genotypes will be highly fallible. Further, unless we are profoundly deceived, there
are some readily identifiable features of the physical and social environment that
have major impact: rates of crime are much higher in decaying inner cities, but I
doubt that there is a "violence" allele that has the pleiotropic effect of sending its
bearers into grim urban environments. Thus, there is an obvious form of causal
analysis that could harness the techniques of molecular genetics and that would be
sanctioned by the democracy principle enunciated above. Perhaps students of the
causes of violent behavior should show how immersion in hostile environments
generates greater levels of violence when genotypes vary, compared with seques-
tration in the leafy suburbs (see figure 13.4). These students could expect to show
something important about the causes of violence and to support their conclusions
with greater rigor than the hunters of "violence alleles." Yet, there has not been any
notable impetus to do the work.

    And, of course, we know why. In a society that consistently and callously turns
its back on programs that might aid the unfortunate and that sees taxation as a form
of robbery rather than a necessary means to social cooperation, the investigation I
have outlined has no obvious point. (It might, after all, lead to campaigns for expensive new social programs.) Better, then, to take a different tack, to find out who the
people are who are likely to become violent and do something about them in
advance. Thus, a politically palatable solution would be to discover genotypes
whose norms of reaction show a high propensity for violent behavior virtually invariant across environment. Perhaps there are a few such rare genotypes (the possibil-

ity should not be excluded), but the overwhelming likelihood is that we will mistakenly come to believe that they are far more frequent than they are (because of our massive ignorance about how to partition environments) and that these conclusions will reinforce the prevailing sense that social solutions are hopeless.[21]

In fact, the motivations for the study are doubly illicit because they are blind both to the serious dangers of reaching erroneous conclusions and, when articulated, the practical policies are morally disreputable. What precisely is the "something" that is to be done to those who bear the "violent" genotypes? Are they to be branded as criminals, labeled from childhood up, even before they have done anything? Should they be forcibly restrained or treated with tranquilizing chemicals? It is precisely because the motivations for the investigation of the genetics of criminality are economic—after all, we could spend money and invest in jobs for inner city youths, clean up their environments, and make hope possible—that we know in advance that the solution has to be cheap. Hence, we cannot anticipate that great moral niceties (always expensive) are likely to bulk large in the application of "discoveries" about "violence genotypes." Add the difficulty of discovering such genotypes (or, at least, common genotypes), and the potential for injustice is obvious.

Thus, there are two major questions that we ought to ask about proposals to unearth genes "for" complex human traits (including dispositions to forms of behavior that prove either personally or socially disruptive). First, is the investigation informed by the interactionist's commitment to explore the impact of some factors, while others vary, in a way that recognizes our ignorance about environmental causes and that pragmatically deploys the genetic techniques to remedy that ignorance? Second, does the information to be acquired lead to a social policy that is both applicable and morally defensible? As my pair of examples indicates, the answers will be quite different in different instances, and there is no shortcut for considering cases individually.[22]

Some scientists bridle at the thought that my second question should ever figure in the evaluation of a program of scientific research, insisting that the business of science is to uncover the truth, however unpalatable, and that inquiry cannot be subordinated to moral critique. Lewontin has often been criticized for introducing extraneous "political" considerations into discussions of biological investigations,[23] but, in my judgment, his recognition of the wider framework in which science is done is profoundly correct. Researchers cannot hide from themselves the fact that their findings will be applied, often by people who do not grasp the nuances of their positions, nor can they take refuge in the division of labor proposed by Tom Lehrer's brilliant song:

> "When the rockets go up, who cares where they come down?
>   That's not my department," says Werner von Braun.

Many workers in contemporary human genetics, including the genetics of behavioral traits, are convinced that their inquiries will promote human well-being, although critical discussion of the ways in which genetical information can affect people's lives may sometimes undermine their confidence. Unless we have a scientifically informed and ethically sophisticated public discourse about possible programs of genetic research, we are likely either to lose important benefits or, more likely, by accepting the most extravagant promises at face value, mix in significant social harms with the improvements we seek.

Because he sees the latter possibility so clearly, Lewontin has come to advocate a "dialectical biology" that will move beyond interactionism. I have tried to argue that the critiques of interactionism are flawed, that they do not respond to the genuine problems of using biology to promote human good, and that there is no substitute for a detailed examination of the merits of individual cases. It is appropriate to close by noting that, in carrying out the much-needed piecemeal critique, there is no better paradigm than the writings of Richard Lewontin.

Notes

It's an honor and a pleasure to dedicate this essay to Dick Lewontin, who has inspired so many people in so many ways.

I am extremely grateful to Peter Godfrey-Smith for valuable discussion, to Paul Griffiths and Richard Lewontin for illuminating correspondence, and to the editors for their suggestions for improvement. I have sometimes followed my readers' advice but surely not as resolutely or as frequently as they would have wished. E-mail exchanges with Lewontin have convinced me that the position ascribed to him here is not always his own, but he and I agree that discussions of genetic determinism and of the notion of a norm of reaction have been marked by important confusion. So, while I must apologize for my misreading, I hope that this essay clarifies some of the underlying issues.

1. For prominent examples, see R. C. Lewontin, "The Analysis of Variance and the Analysis of Causes," in R. Levins and R. C. Lewontin, *The Dialectical Biologist* (Cambridge, Mass.: Harvard University Press, 1985), chapter 4 (originally published in 1974); Lewontin, S. E. Rose, and L. Kamin, *Not in Our Genes* (New York: Pantheon, 1984), esp. chapters 5 and 9; and Lewontin, *Biology as Ideology* (New York: Harper, 1991).

2. See Levins and Lewontin, *Dialectical Biologist*, chapter 3 and conclusion; Lewontin et al., *Not in Our Genes*, chapter 10; and Lewontin, *Biology as Ideology*, esp. 3–37.

3. Lewontin et al., *Not in Our Genes*, 279; Lewontin, *Biology as Ideology*, 14–15.

4. Levins and Lewontin, *Dialectical Biologist*, 114.

5. Lewontin et al., *Not in Our Genes*, 245–251.

6. Lewontin, *Biology as Ideology*, 109–110.

7. The importance of this point is clear from the development studies in *Drosophila* pioneered by Christiane Nüsslein-Volhard and her co-workers (see P. Lawrence, *The Making of a Fly* [Oxford: Blackwell, 1992]); the general point has been made very forcefully by Evelyn Fox Keller in *The Century of the Gene* (Cambridge, Mass.: Harvard University Press, 2000).

8. There are a few biological instances in which we might be entitled to some confidence that we have identified important environmental determinants—in studies of the growth of corn plants or of the development of particular structures and behaviors in fruitflies. But it is easy to be misled by these simple cases, concluding that we have a more general ability to map the environmental axis in representing norms of reaction. (Of course, with respect to many of the traits in which we are interested, it is probably grossly inaccurate to think in terms of a single environmental dimension: to distinguish environments adequately would require coding them by some vector in a space of large dimension.)

9. Lewontin, *Biology as Ideology*, 26.

10. Ibid., 27.

11. "Dialectical biology," "developmental systems theory," or both might come to do this, providing better ways of abstracting from the mix of causal factors. But until they do so, we are stuck with *different* schemes of abstraction. Our predicament is discussed in more detail below.

12. P. E. Griffiths and R. D. Gray, "Developmental Systems and Evolutionary Explanation, *Journal of Philosophy* 91 (1994): 277–304; quote on p. 283.

13. K. Sterelny and P. S. Kitcher, "The Return of the Gene," *Journal of Philosophy* 85 (1988): 339–361; reprinted as chapter 4 of this volume. We avoided the idea of "coding for" that Griffiths and Gray attribute to us.

Paul Griffiths has acknowledged, in correspondence, that the addition of the "coding" idea goes beyond what Sterelny and I actually said. Conversations with Peter Godfrey-Smith have, however, convinced me that there is a rationale for the Griffiths–Gray attribution, for the "standard view" current in *contemporary* biology does seem to make the addition. Hence, according to Godfrey-Smith, the reconstruction offered here does not defend the "standard view" but presents a position intermediate between that view and the approach of the critics—a position to which (perhaps) the "standard view" ought to modulate. I leave it to readers to judge if this is right.

14. Sterelny and Kitcher, "Return of the Gene," 350. As a reader of an earlier version noted, there is a fourth possibility—namely, that "standard" is defined by an arbitrary convention. I will not consider this possibility here.

15. In particular, three of the most heated disputes in recent theoretical biology seem to swirl around issues of which kinds of models are to be taken as defaults, which to be seen as exceptional. I have in mind the controversy about punctuated equilibrium, the adaptationist controversy, and the debate over human sociobiology. Stephen Jay Gould has been very clear in seeing that the issue about punctuated equilibrium is a matter of frequency, that it concerns what paleontologists take to be the "standard" situation. Similarly, Gould and Lewontin ("The Spandrels of San Marco and the Panglossian Paradigm: A Critique of the Adaptationist Programme," *Proceedings of the Royal Society of London B* 205 [1994]: 581–598) opposed a tendency to think that any trait that strikes the evolutionary biologists' eye should be assumed to be an adaptation until reasons are supplied to the contrary. I argued for a similar approach to sociobiology, urging that problems come not because of the flaws of the theoretical tools for building models but because of the ways in which these tools are used (P. S. Kitcher, *Vaulting Ambition: Sociobiology and the Quest for Human Nature* [Cambridge, Mass.: MIT Press, 1985], 117–121).

16. Here I give little detail. But it may help to imagine a sympathetic biologist reading Lewontin, Oyama, or both. The obvious question that would arise for this biologist would be, How do I put these ideas to work in concrete situations?, and to that, neither Lewontin nor Oyama supplies much by way of answer. This does not mean that "dialectical biology" and "developmental systems theory" should be abandoned but that the kinds of work needed to make them viable pieces of biological theory are specific models for tackling interesting problems (rather than philosophical diagnoses of previous errors). It would be *very* interesting, for example, to see a developmental systems analysis of early development in *Drosophila* or a dialectical biology substitute for some part of population genetics.

17. For investigations of traits in which a single locus (or a small number of loci) is assumed to have major effects, this may make monozygotic twins redundant. Perhaps twin studies will only continue to be useful when the trait is assumed to be highly polygenic.

18. Despite the arrival of more direct methods that make the announcement of heritability measures irrelevant, some behavioral geneticists continue to include such measures in reporting their investigations. This seems to be an unfortunate tic from which they cannot

free themselves. At a behavioral genetics workshop, organized by the National Academy of Sciences, David Goldman gave a presentation in which he accompanied interesting findings by using molecular approaches with heritability estimates. When asked by Marcus Feldman why he had included the heritability measures, and whether he would be inclined to use heritability estimates as a prelude to molecular investigations of, say, the genetics of religious belief, Goldman replied that he would take this to be suggestive. Feldman clearly viewed this as a reductio of the continued deployment of heritability measures. In my opinion, he was quite right to do so.

19. It is certainly wrong to conjure up a stereotypical behavior geneticist—socially insensitive and methodologically crass. Irving Gottesman (to single out one example) has been formulating standards for careful analysis for decades, and his own work has been sensitive to the social uses to which behavior genetics might be put.

20. Of course, Lewontin has expressed skepticism about the role that molecular genetics can be expected to play in future therapies (see *Biology as Ideology*, 67–68). I agree with his assessment that there is no *automatic* translation of molecular insights into practical treatments but resist the idea that molecular genetics is either a universal panacea or else useless. We can expect varying degrees of therapeutic success in different cases, and it is impossible to predict in advance where molecular insights will be fruitful. (See P. S. Kitcher, *The Lives to Come: The Genetic Revolution and Human Possibilities* [New York: Simon and Schuster, 1996], 105–112.)

21. One recurrent problem is the propensity for lumping, supposing that the causes of all cases of conditions that share a name are the same. Thus, discovering a "violence" allele in a family with a history of antisocial behavior not only induces the conviction that any environmental factors can be ignored but also inclines people to think that there are lots of other "violence" alleles waiting to be discovered.

22. The case of sexual orientation seems intermediate in character because the human consequences of the research are so uncertain. S. LeVay, *Queer Science* (Cambridge, Mass.: MIT Press, 1966), gives a sensitive response to the moral questions surrounding the search for biological causes of sexual orientation, although, as I argue (Kitcher, "Review of LeVay, *Queer Science*," *The Sciences* [November–December 1996]), it is not clear that his defense of research in this area is successful.

23. See the comments by David Botstein quoted in C. Burr, *A Separate Creation* (New York: Hyperion, 1996), 274.

14

Developmental Decomposition and the Future of Human Behavioral Ecology (1990)

1. Introduction

Every gardener knows the problem. Simply removing the weeds from your soil is useless, for the bare space that results will surely be invaded anew unless you take more active steps, introducing and nurturing the varieties of plants that you want. Similarly, a critique of human sociobiology that simply uproots the current occupants of this part of the intellectual landscape will not suffice. Effective criticism should supply the seeds for a new enterprise.

In 1985 I surveyed the main efforts of sociobiologists to provide conclusions about human social behavior.[1] On the view I developed (see especially *Vaulting Ambition*, chapter 4), the field of sociobiology is generated through the application of evolutionary ideas to the study of social behavior. Nobody can protest that application in principle, even when the species under study is our own. However, there are deep confusions in attempting to use evolutionary analyses to resolve the nature-nurture problem or to erect a grand theory of human nature.[2] Thus, part of the sociobiological enterprise as it has sometimes and most popularly been conceived should be simply abandoned. What remains should be thoroughly cleansed. For, as I endeavored to show,[3] the applications of evolutionary ideas to human social behavior are notably less rigorous and careful than the best work in the evolutionary study of the behavior of nonhuman animals.

Homo sapiens has an evolutionary history and it is reasonable to think that there must be *some* connection of *some* kind between that history and our social behavior. The challenge is to specify the connection. In three previous articles,[4] I have offered a blueprint for the transformation of human sociobiology, insisting on the disavowal, once and for all, of inferences about genetic determination of traits, on the need for precise models and detailed data, and on the importance of recognizing the role of cultural transmission in the history of human social practices. However, this does not complete my task, for as I have argued, especially in chapter 9 of *Vaulting Ambition*, evolutionary analysis simplistically applied to overt behavior, rather than to the proximate and developmental mechanisms that underlie behavior, can easily generate mistakes and confusions. The blueprint presented in my three essays of 1987 urges integrating evolutionary studies with inquiries into proximate and developmental mechanisms. My goal here is to elaborate this suggestion, which I (like Gould)[5] regard as the most important step in the replacement of the "cardboard Darwinism" of traditional human sociobiology with a more adequate way of introducing biology into the social sciences.

2. Behavioral Ecology and Functional Description

Following Niko Tinbergen's famous taxonomy of questions for ethology,[6] we can distinguish four different inquiries. The first concentrates on the *proximate mechanisms* of the behavior, seeking a physiological or psychological description of the immediate causes. The second looks further back in the causal history, searching for the factors in ontogeny that bring about the proximate mechanisms. In addition to such *developmental* investigations, there are also efforts to identify the *function* of behavior. Here the investigator tries to offer an analysis of the fitness differences among various forms of behavior, with the aim of showing how the form actually found might be maintained by selection. Finally, there are *historical* questions directed at eliciting the causes that figured in the origination of the behavior in the species (or some ancestral species). When the species lacks any significant system of cultural transmission, historical questions are likely to become evolutionary questions, and the answers may well involve offering an account of the selection processes through which the behavior originated. Nonetheless, as sophisticated students of behavior are well aware, the selection pressure (if any) that figures in the origination of behavior may be quite different from the selection pressure (if any) that maintains it. Functional and historical inquiries are related but distinct.

Some sociobiologists—most prominently those who prefer to call themselves "behavioral ecologists"—have illuminated aspects of nonhuman animal behavior by constructing ecological models that identify the sources of the fitnesses in alternative forms of animal behavior. In the best work (see, for two examples, Woolfenden and Fitzpatrick and the essays in the collection edited by Krebs and Davies),[7] we find precise models allied to detailed observations, the articulation and testing of rival hypotheses, and even some attention to possibilities of nongenetic transmission. There is no misguided commitment to the idea of solving nature-nurture problems or of undertaking the impossible task of deriving conclusions about the norms of reaction of (typically unknown) genotypes from premises about the selection pressures that operate in maintaining behavior.

These studies advance our understanding of behavior by supplying a description of what the animals are doing, a description that may replace an incorrect, possibly anthropomorphic, conceptualization of the situation that has been adopted almost unconsciously, or that may introduce order into a situation that we find confusing. A male baboon picks up a juvenile baboon at the approach of a larger heavier male, and with the juvenile in his arms, the smaller adult male turns to face the newcomer. Is he protecting the young baboon? Appeasing a dominant male? Trying to impress an onlooking female? Male speckled wood butterflies descend from the canopy and gyrate in the patches of sunlight below. What are they about? In cases like these, behavioral ecologists attempt to find a functional description of the behavior by relating the observed movements to hypothetical selection pressures that maintain the disposition to perform such movements.

3. Human Behavioral Ecology: Promises and Pitfalls

Human sociobiology—or, as I shall henceforth call it, in hopes of producing an exercise in imitation, human behavioral ecology—might prove as enlightening. There's no reason why we should not correct our naive views about the significance of various forms of human social behavior in a way that is exactly parallel to the tutoring of naive observations in nonhuman behavioral ecology. If the behavioral ecologist continues to be rigorous and careful, what's the difference between studying butterflies and focusing on human beings?

The most obvious dissimilarity consists in the fact that there are systems of cultural transmission that operate in human societies that *might* maintain a behavioral trait in opposition to natural selection, so that in using natural selection as a guide to forming evolutionary expectations, we might be misled. Specifically, we can ask if there are possible systems of cultural transmission that can divert a population from an equilibrium it would have attained under natural selection *and whether such systems can themselves be maintained in the face of selection*. Thanks to the efforts of Boyd and Richerson,[8] we know that the answer to the last question is "Yes." Hence, the task of the human behavioral ecologist should apparently be to construct analyses of the maintenance of human traits that take account both of the selection pressures and of the effects of cultural transmission.

Boyd and Richerson's work holds two different morals for human behavioral ecology. Traditional human sociobiology has sought to explain two different kinds of things: distributions of individual behavior and social institutions. So, to anticipate the example I'll consider below, we can try to explain the frequency with which incest occurs among siblings, why it occurs just where it does, or on the other hand, we can attempt to explain the presence of incest taboos in different societies. Only the former explanatory task is strictly analogous to the endeavors pursued by behavioral ecologists studying nonhuman animals, and my aim will be to understand how this kind of task should be undertaken in the case of our own species. The first moral is that explanations of individual behavior need to take into account systems of cultural transmission. The second is that, in pursuing the more ambitious enterprise (not considered in this essay) of accounting for the origin and maintenance of social institutions, the models of transmission adduced by Boyd and Richerson may be profoundly relevant.[9]

Cultural transmission affects us at many stages of our development, and the multiplicity of the influences raises a further complication for human behavioral ecology. The human case *is* different from that of butterflies—although it may not be different from that of the higher primates and the social carnivores—in that serious attributions of function are heavily dependent on presuppositions about proximate and developmental mechanisms. Evolutionarily and developmental studies need to be pursued in tandem.

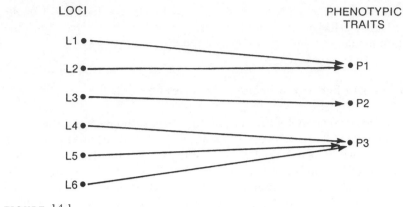

FIGURE 14.1

4. Developmental Constraints and Evolutionary Simplicity

I'll proceed by offering an abstract account of the difficulty, followed by a relatively detailed example. The latter will be intended to show not only how human behavioral ecology might go astray, but also how it might overcome the problem posed by the existence of developmental constraints.

Behavioral ecology will work straightforwardly when the proximo-developmental connection between genes and behavior is *evolutionarily simple*. Consider the very simplest type of organism, one for which the *internal* causes of behavior are as depicted in figure 14.1. If an animal exemplifies this simple kind of causal picture, we can enhance our understanding of what it is doing by thinking in terms of what would contribute to reproductive success. Tacitly, we suppose that there are alternative alleles available at one of the loci, that these correlate with the forms of the behavioral trait and that selection maintains alleles generating superior forms of the trait. Matters do not change if allelic substitutions do modify characteristics besides the target trait provided that these ancillary substitutions are selectively neutral (see figure 14.2). So I'll say that a causal net encompassing the internal factors that affect a behavioral trait is *evolutionarily simple* just in case the difference in expected reproductive success given an allelic substitution is due to the difference in just one facet of the behavioral phenotype.

Suppose however, that the animal we are studying instantiates something like figure 14.3. We imagine that there are probabilities associated with the connections among the nodes, and these probabilities are set by the external environment. Assume for the moment that all members of the populations under study face a common external environment so that the probabilistic connections are the same in each case. Unless we are very lucky, and the situation reverts to the type of evolutionary simplicity depicted in figure 14.2, there will be obvious limits on the power of selection to shape an individual phenotypic trait. What is fixed or maintained by selection will be a *behavioral spectrum* or some underlying disposition that is implicated in each of the elements of a behavioral spectrum.

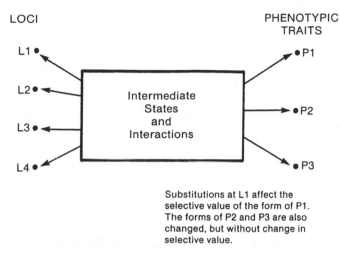

LOCI

PHENOTYPIC
TRAITS

L1 ●

L2 ●

L3 ●

L4 ●

Intermediate
States
and
Interactions

● P1

● P2

● P3

Substitutions at L1 affect the
selective value of the form of P1.
The forms of P2 and P3 are also
changed, but without change in
selective value.

FIGURE 14.2

Matters are likely to be far more intricate than I have represented them as being. Some of the critical allelic variation available in a population may not only affect the forms of the behavioral traits in which we are interested but also modify the structure of the causal trees that underlie them. This is especially obvious if one considers the relatively large ways in which substitution of alleles might affect behavior. Imagine that an allelic substitution changes the form of a particular neurotransmitter. The consequence may be that a young child responds in abnormal ways to certain types of stimuli. That early effect is likely to have consequences of its own: parents and others who interact with the child may do so in atypical ways,

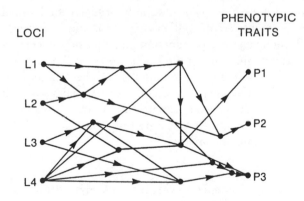

LOCI

PHENOTYPIC
TRAITS

L1 ●

L2 ●

L3 ●

L4 ●

● P1

● P2

● P3

Substitutions at L1 affect
the selective value of the
forms of P1, P2, P3.

FIGURE 14.3

thus provoking an abnormal sequence of events in individual ontogeny. An early deviance may be magnified in a whole variety of ways as the child comes to encounter an unorthodox developmental environment. More abstractly, the causal connections depicted in figures like 14.2 and 14.3 may vary with the substitutions at certain loci because underlying allelic combinations affect the probabilistic connections among nodes in the causal tree.

While it is hardly controversial that *some* complex causal structure underlies human behavior, there is room for much debate on how to interpret the kinds of diagrams I have drawn. What exactly are these connected nodes and what do the connecting lines represent? There are as many competing answers as there are views about the study of the proximate mechanisms and the ontogeny of behavior. Because none of these views has won anything like universal acceptance, there has been hope in some quarters that human behavioral ecology might enlighten us.

To succeed, we need to know the ways in which various forms of behavior are bound together by connections among their proximate mechanisms. We need what Darwin called "the laws of correlation and balance." The task does not require that we fathom the full details of the ontogeny of behavior, but that we delineate the structure of the causal trees that link the genes to aspects of the phenotype (possible facets of morphology and physiology as well as behavioral traits). If we are entirely ignorant about this structure, then there is a very real danger that we shall focus on the wrong *explananda*.

Developmental and proximate studies cannot furnish ready-made constraints for investigations of the functions of human behavior. Our ignorance of the relevant facts about development and about how to characterize the proximate mechanisms is usually, as Darwin would say, profound. But, lacking any information about how patterns of behavior are interconnected and bound together into spectra that selection can either maintain or modify, specifications of the selective advantage of a particular piece of behavior are, at best, stabs in the dark. How then, can the investigation of human behavior possibly proceed? I think only by a tentative and exploratory attempt to combine the two types of study. We must free ourselves both from the illusion that some nascent human behavioral ecology will clarify mysteries about proximate mechanisms and behavioral ontogeny and from trying to foster some proximo-developmental inquiry that will beg no questions about function.[10] We need to fuse the enterprise of evolutionary analysis with that of *developmental decomposition*, an integration that must be tentative since we start from nothing that is firm, but that may prove usefully exploratory if the two investigations are used to constrain one another.

5. Incest Avoidance: The Westermarck Hypothesis

So much for the abstract and general part of my story. How does it translate into concrete advice for a concrete situation? I shall try to dispel the fear that my ecumenicism is too nebulous by showing how the perspective I recommend can be elaborated in a particular example.

One of the favorite cases of human sociobiology is human incest-avoidance. The orthodox story consists in a revival of the Westermarck hypothesis.[11] For our purposes, the hypothesis may be divided into three parts:

a. In the course of normal ontogeny, people acquire a disposition to avoid copulating with those with whom they have been reared, and because this disposition is almost universally acquired the frequency of incest (at least among siblings) is low.
b. The disposition to avoid copulating with those with whom one has been reared is selectively advantageous because it will typically lead its bearers to refrain from mating with close kin and producing offspring that combine deleterious recessive alleles.
c. The taboos on incest that are prominent in many societies are the expression of a *prior* hostility toward copulations with kin; they are not major proximate factors in preventing incest.

Quite evidently, this cluster of claims involves ideas about the mechanisms of behavior and about the evolutionary forces that have shaped behavior. What I want to expose for scrutiny is the notion that there is a unitary phenomenon, incest-avoidance, that is susceptible to evolutionary analysis.

Let us briefly rehearse some of the evidence that stands behind the Westermarckian revival. Studies of mate choice in a number of species yielded some striking results: Patrick Bateson has shown that Japanese quail prefer to mate with conspecifics who are different, but not too different, from themselves; Ann Pusey has documented a change in patterns of association among chimpanzees at the times that females first become estrous—newly mature females forsake their formerly favorite male associates (typically, though not exclusively, kin).[12] In a study of marriages and romantic liaisons among Israelis who had been brought up from infancy in the same kibbutz, Shepher offered the initially compelling finding that *nonsiblings* (who were thus under no pressure to avoid sexual relationships) brought up together did not become sexually involved with one another. A similar example, the institution of arranged "minor marriages" once found in Taiwan, seems to support Shepher's conclusions: the marriages of women who had been adopted as the designated spouse of an adoptive brother had a higher rate of divorce and a lower incidence of children than ordinary marriages.[13] Finally, a much-cited study comparing the frequencies of birth defects in two Czech populations, one consisting of the offspring of incestuous unions and one of the non-incestuous offspring of the mothers involved, seems to show that copulations between close kin are more likely to produce defective children than other matings.[14]

Some of this recent evidence for the Westermarck hypothesis has been criticized. Shepher's conclusions from the kibbutz are vulnerable to the charge that the young people whose careers he followed were reared at a time when the educational policy emphasized ideological commitment and an almost puritanical attitude on sexual matters.[15] John Hartung has also charged that Shepher's statistics are worthless because he has no adequately formulated null hypothesis.[16] Careful reading of the Wolf and Huang study of minor marriages reveals that there are many potentially interfering factors that might be adduced to offer alternative explanations of the low success rate of minor marriages—most prominently the contempt that is often expressed for the institution, contempt that frequently surfaces in childhood and adolescent teasing. The significance of the Seemanova study can be disputed on grounds that the children of incest were often conceived and born under conditions that promote a higher incidence of birth defects independently of inbreeding (one obvious factor here is the age of the mother). Finally, as becomes clear as soon as one thinks about serious modeling of mating behavior, the costs of inbreeding might be outweighed by costs associated with exogamy (for example, dangers associated with migration).[17]

Besides these specific objections, there's a major worry about part (c) of the Westermarck hypothesis. Why should there be a taboo on behavior that people are not disposed to perform?[18] Interestingly, some anthropologists, especially those inspired by Freud, contend that the strength of a taboo reflects not so much the extent to which there is a disposition to refrain from the banned act, but the strength of the repressed wish to *engage* in that act.[19] On this account, incest taboos are viewed as the public expression of a psychodynamic conflict, and they can be expected to play an important role in the etiology of individual behavior. Moreover, if the psychodynamic view is even partially correct, then it appears that we ought to scrutinize closely the notion that those who are reared together do not become sexually attracted to one another after they reach puberty—whether or not the attraction is actually expressed in overt sexual behavior.

Part (c) of the hypothesis should be abandoned. There's no reason to think that the presence of a taboo is simply the generalized expression of individual horror at the idea of incest. As explicitly noted above, my concern will be with explaining individual behavior in order to understand the frequencies and conditions of incest and of incest-avoidance. Can we do better than (a) and (b) by integrating the functional considerations on which Westermarck and his successors draw with attention to proximate mechanisms and their ontogeny?

6. Incest-Avoidance: Preliminary Clarifications

The newly trimmed version of the Westermarck hypothesis potentially explains the frequency of three different kinds of behavior: sexual activity among relatives reared together (expected to have low frequency), sexual activity among nonrelatives reared together (also expected to have low frequency), and sexual activity among relatives not reared together who meet for the first time as adults (not necessarily expected to be low). Now there are at least three areas of important vagueness both in the hypothesis and in the phenomena that it is supposed to cover. First, what kinds of relationships are to be included in the scope of the hypothesis? Second, what exactly do we require for people to be reared together? Third, and most important, what is to count as sexual activity?

The most obvious ways to answer the first two questions are to say that the Westermarck hypothesis is concerned with *sibling* incest and that two individuals are reared together just in case they are brought up in the same household and interact with the same family members (parents, siblings, and other caregivers) from birth on. But now it is worth posing two questions.

First, if we narrow the scope of the Westermarckian account so that it only deals with siblings, how is the account to be integrated into a more general view of causes of incest-avoidance? Consider the father-daughter relationship. Should we suppose that daughters acquire a disposition to avoid sexual relations with fathers just as they do with respect to brothers, but that fathers do not develop any such disposition with respect to their daughters? If so, then some new factor—perhaps fear of being punished for engaging in a forbidden form of behavior?—would have to be adduced to explain why fathers do not more frequently coax or coerce their daughters into sexual activity.[20] It would then have to be shown why the newly invoked factor does not play a primary role in avoidance of incest among siblings. Furthermore, notice that we cannot now mount an argument parallel to (c) in our formulation of the Westermarck hypothesis to explain why the sexually coercive behavior of fathers is forbidden.

Second, except in the case of twins, there will always be an asymmetry in siblings' rearing environments. One child, the elder, will have undergone a period of his/her development without any contact with the younger, and throughout their development, each will have *different* interactions with the other. Presumably, the perfect case for the Westermarck hypothesis is that of twins in whom the disposition to avoid postpubertal sexual contact can be expected to develop as completely as it ever does. It is interesting to note that this is not only a prominent case for sibling incest in mythology, but that there is one case in the clinical literature of an incestuous relationship between male identical twins, reared together from birth, that persisted (intermittently) for twenty-three years from the age of six.[21]

The most fundamental issue to be resolved in clarifying the Westermarck hypothesis concerns what is meant by "sexual activity." Any attempt to review the vast literature on incest will turn up numerous definitions. In some earlier anthropological writings, incest is closely linked to marriage, and rules forbidding marriages among close kin are regarded as prohibitions on incest.[22] This linkage is surely inappropriate, for as Robin Fox lucidly points out,[23] incest is about sex, not marriage. More recent anthropologists, expecially those who have drawn inspiration from sociobiology (Shepher, van den Berghe), favor a narrow definition: at the narrowest extreme, we may say that incest only occurs when there is full sexual intercourse between close relatives (coefficient of relationship no less than 0.25) who are of opposite sex and who have been continuously together since the birth of the younger. Clinicians who are concerned with the plight of incest victims typically adopt a much more inclusive definition, allowing cases in which participants engage in mutual masturbation, fondling, intrafemoral intercourse, homosexual intercourse, oral-genital contact, and even exhibition of the genitals to count as cases of consummated incest.[24] The most liberal conception is that of Judith Herman who takes the sexual relationship in incest to be any physical contact that has "to be kept a secret."[25] This definition obviously allows for the confusion of sexual contacts with other types of physical abuse; but for a clinician who is concerned with the causes of abuse, with the prevention of abuse, and with the treatment of those who have been victimized, that is unlikely to be problematic.

For our purposes, I suggest that we take an *incestuous advance* to be any action that expresses sexual feelings toward a family member. Family members are characterized disjunctively, either as relatives with a coefficient of relationship of 0.125 or greater or as people who share the same household. (There are biological family members, sociological family members, and of course, family members who satisfy both criteria.) An *incestuous act* occurs when the agent's sexual feelings receive their full expression: crudely, the agent gets what he (or she, but it is usually he) wants. The act is *reciprocal* when both (or all) participants have sexual feelings that are fully expressed in it.

My suggested definition allows most of the clinicians' cases to be included. Why is it an appropriate way to think about incest? The significant point for our purposes is that incestuous sexual contact is the terminus of a causal process involving at least several and probably many actions and reactions on the parts of two individuals, and that incest-avoidance will come about if this causal process is blocked at any point. The Westermarck hypothesis proposes that among a subclass of cases—consisting of those dyads who are of the same generation and who are *social* family members (who have been reared together)—the causal chain will not begin. In moving from traditional human sociobiology to the human behavioral ecology that I envisage, I suggest that we understand why sibling incest occurs when it does by focusing on the proximate causal history of acts of sibling incest, identifying the psychological mechanisms involved, and trying to fathom the functional significance (if any) of the corresponding mechanisms that operate in people who do not engage in incest. Conceiving incestuous advances in terms of their expression of sexual desire is plainly congruent with this shift away from overt behavior and towards underlying mechanisms.

Let's begin with a point that has been implicit in earlier remarks. Consummate incest may either be consensual or coercive.[26] As the prototype of the causal history of a coercive act of consummated incest we may take the following: first, X comes to feel sexual desire for Y; second, in consequence of the sexual desire, X makes a sexual advance toward Y; third, Y either does not respond or rebuffs X's sexual advance; fourth, X has the power to coerce Y and X has no inhibitions against using that power to force Y to engage in the sexual contact. By contrast, the prototype of the causal history of a consensual act of consummated incest will be: first, X comes to feel sexual desire for Y; second, in consequence of the sexual desire, X makes a sexual advance toward Y; third, either as a standing disposition or in consequence of X's advance, Y feels sexual desire for X; fourth, in consequence of Y's sexual desire, Y signals willingness to engage in sexual contact, which thereupon takes place.

Now these prototypes are obviously much too simple and schematic. (Related points and similar acknowledged simplifications, are also made by those clinical psychologists who have thought most seriously about incest and incest-avoidance.)[27] The actual etiology of human sexual behavior is much more complex, and there may be elaborate signaling and sexual negotiation that takes place over a lengthy period (days, weeks, or months). Incest-avoidance isn't a unitary phenomenon, and it's highly unlikely hat a propensity to incest-avoidance is a trait that has been shaped or maintained by natural selection. For consider the various ways in which our prototypical causal chains could be broken. X may want to consummate incest and may have no scruples about forcing Y, but X may be either powerless or afraid of reprisals from others. X may be sexually attracted to Y and have the power to coerce Y, but X may have a strong disposition against forcing another into sexual activity. X may be sexually attracted to Y and Y may be sexually attracted to X, but either (or both) may be unwilling to translate their desires into action because they are afraid. Notice that all of these varieties of unconsummated incest are at odds with the Westermarck hypothesis, which takes for granted a point that ought to be investigated in detail: to wit, the thesis that a low frequency of sibling incest reflects a low incidence of incestuous desires.[28]

But the main point that I want to emphasize is that sexual activity between siblings—or the absence of such activity—cannot be understood apart from consideration of a range of dispositions that are implicated in other areas of sexual and nonsexual behavior. If we are searching for traits that have functional significance, then we ought to be looking at more general properties such as the dispositions to be sexually stimulated by the presence of certain kinds of individuals, the dispositions to translate sexual desire into action, more general dispositions for taking risks, and so forth. In an attempt to illustrate the position I am recommending, I shall conclude by offering a highly oversimplified model of one aspect of the oversimplified scenarios that have so far been considered.

7. A Preliminary Model

Suppose that the proximate mechanisms of sexual advances work in something like the following way. We can imagine that there is a standing evaluation function that rates the sexual attractiveness of individuals according to a number of variables, including indicators of sex and age (or possibly, relative age[29]), size, and, I'll suppose, familiarity. The reason for including this last variable is to enable us to co-opt some of the virtues of the Westermarck hypothesis. Assume that the evaluation function is relatively constant over time, at least for postpubertal adolescents, but that there is a threshold function that assigns values to combinations of the same variables and that fluctuates with time.

Let us now adopt a very simple view of the making and receiving of sexual advances. Suppose a person X encounters a person Y, that the evaluation and threshold functions for X and Y are e_x, e_y, t_x, and t_y respectively, and that the vectors representing the values of the variables corresponding to X and Y are x and y. Then X makes a sexual advance to Y just in case

(i) $e_x(y) > t_x(y)$

and Y reciprocates a sexual advance from X just in case

(ii) $e_y(x) > t_y(x)$.

(Here I assume, without loss of generality, that X is the person to make the first move.) I suppose that sexual contact takes place just in case the advance is made and reciprocated, that is, just in case both (i) and (ii) hold.

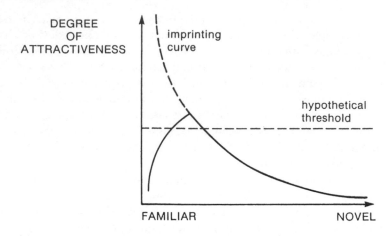

FIGURE 14.4

Within this very simple framework we can articulate a number of hypotheses about proximate mechanisms, development, and function. Consider first, a way of refining a central idea of the Westermarck hypothesis. In his pioneering studies on mating preference among Japanese quail, Patrick Bateson has shown that birds exhibit a preference for mates that are familiar but not too familiar. Assuming (and it's a very large assumption) that something similar holds for our own species, we can suppose that the projection of the evaluation function along the axis representing familiarity/novelty takes the form shown in figure 14.4. Bateson hypothesizes that the evaluation function takes this form in a mature individual as a result of the interaction of two processes. In ontogeny, there is an initial tendency to be attracted to the familiar but the process of habituation decreases the attractiveness of those with whom one has frequent contact. Using the notation so far introduced, we can suggest that there is a mapping h taking the evaluation function at the nth developmental stage onto the evaluation function at the $n + 1$st developmental stage, with the property that, for values of x at the familiar end of the familiarity/novelty axis $h(e)(x)$ is always less than $e(x)$ with the difference being greater the closer x is to the familiarity/novelty origin. Notice that the realization of h might involve the kinds of psychodynamic variables stressed by Freudians. For example, the transformation may be effected as a result of advances toward people who are initially attractive, advances that are not met in ways that satisfy the child.[30]

Focus next on the threshold function. This will surely vary with time in obvious ways. For example, a prolonged period of sexual abstinence may lower the values of the threshold function across all arguments with the result that the class of potential sexual partners is greatly increased. One interesting issue is the question whether public incest taboos or parental commands about relationships among family members affect the form of the function. Those who believe that there are social pressures that play an important role in diminishing the frequency of incest can frame their hypothesis as the claim that certain kinds of teachings or warnings cause a relative increase in the values of t for those arguments x that correspond to family members. The relative increase would be assumed to be maintained as the absolute values of t fluctuate.[31]

There's a different way to think about the springs of sexual behavior and about the translation of sexual desire into action. Suppose as before, that there is a standing (adult) evaluation function, representing the Batesonian compromise between imprinting and habituation. Assume further that there is a fluctuating threshold, which responds to such factors as length of time since sexual activity and physiological state.[32] If X encounters a person Y for whom $e_x(y) > t_x(y)$ then X feels sexual desire for Y. Whether this desire is translated into action depends on X's assessment of the situation. Instead of seeing sexual desire as automatically issuing in a sexual advance, we conceive of the "sexual desire module" as interacting with a "consequence assessment module".

Complex ontogenetic processes will issue in X's having a disposition to undertake (or to avoid) certain kinds of risks. Cultural transmission of lore about the appropriateness of various kinds of sexual relations will partially specify some of the perceived consequences. Features of the occurrent situation will also be relevant to the determination of those perceived consequences. Thus, whether X makes an incestuous advance may depend not only on X's sexual desire for a relative (itself the product of the standing evaluation function and the threshold function) but also on X's general willingness to take risks, on the social transmission of a taboo against incest, and on X's assessment of the likely response of the object of desire (and of the effects of that response on others). Foregoing abstraction for the moment, an elder brother with sexual desire for a younger sister may be restrained because of his unwillingness to tolerate even a slight chance of detection, or because the sanctions against incest (or perhaps, against sexual coercion) in his social group are very severe, or because he believes that his sister will complain to others and that her testimony will be credible. We can think of the assessment module as interposing barriers between the sexual desire and action expressing that desire: incest may occur when the checks are negligible or when the desire is so strong (that is, the difference between $e_x(y)$ and $t_x(y)$ is so great) that is overwhelms them.[33]

8. Glimpses Beyond

"This is all very well," the human sociobiologist may say, "but how can we formulate this approach so as to obtain testable hypotheses?" I have some sympathy for the question. My reply comes in two parts.

First, if my arguments are correct, it's highly likely that the measurements and correlations that figure in traditional human sociobiology are meaningless. The ways in which ecological constraints impinge on human behavior simply cannot be appreciated without some conception of the ways in which forms of behavior are bound together by proximate mechanisms and developmental processes. Second, and more important, the *general* model sketched in the last section can be filled out by proposing specific hypotheses about the forms of the functions and the processes that affect the transformation of the threshold. From these hypotheses one can generate expectations about human sexual behavior under various conditions. Furthermore, such hypotheses can also be tested by designing variations on standard types of psychological experiments, for example, observations of the differential reactions of subjects to various stimulus objects.

I don't doubt that the process of testing will be difficult, and that there will be familiar kinds of controversies about the interpretation of experimental results. But I take it as a positive virtue of my approach that it suggests many questions that have been virtually ignored in the vast anthropological and clinical literature on incest and incest-avoidance. What is the frequency with which, in postpubertal individuals, the evaluation function takes values that are higher than those of the threshold function for arguments that correspond to family members? How often are incestuous advances made? How often would they be made if the person moved by desire were not afraid of the consequences? Does the process of habituation play a more general role than that of modifying the attractiveness evaluation? Even a preliminary model helps us achieve more precise characterizations of the phenomena that are worth investigating.

We can proceed to study incest-avoidance in several different ways. First, and most obviously, one can attempt to make the model I have sketched more concrete by linking it to the available clinical literature on incest occurrences. The character of the underlying processes is partially revealed by seeing where they break down.[34] Second, one can inquire how the treatment in terms of evaluation functions and thresholds relates to psychological variables adduced in other contexts or to details of neurophysiology.[35] Third, one can pursue functional questions in the context presented by the emerging story of proximate and developmental mechanisms, inquiring not what is the function of incest-avoidance but what function is served by the process of imprinting or by some hypothetical transformation of the threshold function.[36] None of these ways of proceeding has priority over the others, and each should be responsive to the findings of the others. That is the moral of my story.

Let me close by returning to the abstract perspective with which we began. Once human behavioral ecology has rid itself of the errors of genetic determinism, and once it has vowed to emulate the genuine achievements of nonhuman behavioral ecology, the large obstacle that must be overcome is the identification of those aspects of the behavioral phenotype that actually have functional significance. The task is difficult because we need to refine our evolutionary modeling in light of theories of development and mechanism in psychology (or neurophysiology) that we simply do not have. I have suggested that we address that task by trying to integrate developmental and evolutionary studies, assembling what constraints we can as we go, and using each field of investigation to shed light on the other. There is no guarantee that this can be done successfully, but I hope that the example of incest-avoidance does indicate that we can introduce considerable order into vague and apparently inimical hypotheses by thinking clearly about both the developmental and the functional aspects of the situation.[37] If developmental psychology, neurobiology, and evolutionary analyses are undertaken together, then the grain of truth in orthodox human sociobiology, the claim that we, like other species, are products of the evolutionary process, might flower into something significant.

Notes

Ancestors of parts of this paper were presented to audiences at the Indiana University of Pennsylvania interdisciplinary conference on mind and brain, the 1986 meeting of the British Society for the Philosophy of Science, the 1986 meeting of the Philosophy of Science Association, the Institute for Child Development at the University of Minnesota, and the UCLA Behavioral Biology Seminar. I am grateful to many people in these audiences for their valuable comments, and, in particular, to Rob Boyd, Robert Hinde, Jerry Fodor (twice), and Kim Sterelny. Conversations and correspondence with Bill Charlesworth, Alan Sroufe, and Donald Symons have also enlightened me on many points, and I am extremely grateful to Michael Dietrich for his research assistance. Above all, I want to thank Patrick Bateson for much advice and encouragement and Lisa Hirschman for her gentle guidance through the eddies of the clinical literature on incest.

1. P. S. Kitcher, *Vaulting Ambition: Sociobiology and the Quest for Human Nature* (Cambridge, Mass.: MIT Press, 1985).

2. See the different enterprises of E. O. Wilson, *On Human Nature* (Cambridge, Mass.: Harvard University Press, 1978); R. Alexander, *Darwinism and Human Affairs* (Seattle: University of Washington Press, 1979); and the critique in Kitcher, *Vaulting Ambition*, chapters 1, 4, 6, and 9.

3. Kitcher, *Vaulting Ambition*.

4. P. S. Kitcher, "Précis of *Vaulting Ambition*, with Peer Review and Reply to Commentators," *Behavioral and Brain Sciences* 10 (1987): 61–99; Kitcher, "The Transformation of Human Sociobiology," in Arthur Fine and Peter Machamer, eds., *PSA 1986* (East Lansing, Mich.: Philosophy of Science Association, 1987), 63–74; Kitcher, "Imitating Selection," in Sidney Fox and Mae-Wan Ho, eds., *Metaphors and Models in Evolutionary Theory* (Chichester: John Wiley, 1987), 295–318.

5. S. J. Gould, "Cardboard Darwinism," in Gould, *The Urchin in the Storm* (New York: Norton, 1987); originally published in the *New York Review of Books*, October 1986.

6. N. Tinbergen, "On War and Peace in Animals and Man," *Science* 160 (1968): 1411–1418; reprinted in Arthur Caplan, ed., *The Sociobiology Debate* (New York: Harper and Row, 1978).

7. G. Woolfenden and J. Fitzpatrick, *The Florida Scrub Jay: Ecology of a Cooperative Breeding Bird* (Princeton, N.J.: Princeton University Press, 1985); J. R. Krebs and N. B. Davies, *Behavioral Ecology: An Evolutionary Approach* (Sunderland, Mass.: Sinauer, 1984).

8. R. Boyd and P. J. Richerson, *Culture and the Evolutionary Process* (Chicago: University of Chicago Press, 1985).

9. I have offered a sketch of how this larger project might go in Kitcher, "Imitating Selection."

10. Alexander seems to me to be correct when he points out the difficulty of seeking for developmental constraints in advance of making any decision about functional description (*Darwinism*, 101), but the correct conclusion to draw is not that human behavioral ecology can operate independently of proximo-developmental studies. Rather *primitive* ideas about functionality must be refined in the light of *equally primitive* views about development, and conversely. The hope is that iterations of such mutual refinement will lead us to defensible views in both areas of inquiry.

11. See N. Bischof, "The Biological Foundations of the Incest Taboo," *Social Sciences Information* 11 (1972): 7–36; P. van den Berghe, "Human Inbreeding Avoidance: Culture in Nature," *Behavioral and Brain Sciences* 6 (1983): 91–123; J. Shepher, *Incest: A Biosocial View* (New York: Academic Press, 1983).

12. P. Bateson, "Optimal Outbreeding and the Development of Sexual Preferences in Japanese Quail," *Zeitschrift für Tierpsychologie* 53 (1980): 231–244; Bateson, "Preferences for Cousins in Japanese Quail," *Nature* 295 (1982): 236–237; A. Pusey, "Inbreeding Avoidance in Chimpanzees," *Animal Behavior* 28 (1980): 543–552.

13. A. Wolf and C. Huang, *Marriage and Adoption in China 1845–1945* (Stanford, Calif.: Stanford University Press, 1980).

14. E. Seemanova, "A Study of Children of Incestuous Matings," *Human Heredity* 21 (1971): 108–128.

15. See M. Kaffman, "Sexual Standards and Behavior of the Kibbutz Adolescent," *American Journal of Orthopsychiatry* 47 (1977): 207–217. There are hints of similar pressures in Y. Talmon, "Mate Selection on Collective Settlements," *American Sociological Review* 29 (1964): 491–508, esp. p. 503.

16. J. Hartung, "Review of Shepher 1983," *American Journal of Physical Anthropology* 67 (1985): 169–171.

17. See B. Bengtsson, "Avoiding Inbreeding: At What Cost?" *Journal of Theoretical Biology* 73 (1978): 439–444; R. May, "When to Be Incestuous," *Nature* 279 (1979): 192–194.

18. For a forceful recent restatement of the problem, see R. C. Lewontin, S. Rose, and L. Kamin, *Not in Our Genes* (New York: Pantheon, 1984).

19. See M. Spiro, *Oedipus in the Trobriands* (Chicago: University of Chicago Press, 1982).

20. I should note that the clinical literature of the father-daughter incest is relatively rich and detailed by comparison with that on sibling incest. This may suggest that father-daughter incest occurs more frequently than sibling incest, although as K. C. Meiselman (*Incest* [San Francisco: Jossey-Bass, 1979]) lucidly points out, some rather basic facts about reproduction suggest that there will be more opportunities for father-daughter incest than for sibling incest (every daughter has a father, and most have fathers or father-surrogates who are present in the household; not every daughter has a sibling of appropriate age and sex to be a potential sexual partner). The situation is further complicated by the fact that studies of convicted incest offenders (for example, S. Weinberg, *Incest Behavior* [New York: Citadel, 1955]), as well as clinical studies in which reports of incest emerge in the course of therapy (Meiselman, *Incest*) are more likely to involve episodes in which one more powerful individual has coerced a weaker person into sexual activity. Sibling incest may be underreported both because it has less traumatic effects (assuming that it does) and because parents do not want to risk losing their children by notifying authorities about such incidents.

In a pioneering study, David Finkelhor (*Sexually Victimized Children* [New York: Free Press, 1979]) administered a questionnaire to a sample of college students and obtained interesting results about the frequency and type of childhood sexual experiences. Approximately one-quarter of the respondents reported a childhood sexual experience with a relative or near-relative, and for each sex, approximately one-half of the instances involved siblings. Father-daughter instances were significantly less frequent than brother-sister cases — and even than sister-sister cases (see ibid., 87).

Perhaps the clearest point to emerge from the literature is that mother-son incest is far less common than father-daughter incest. A well-argued attempt to explain the asymmetry is J. Herman and L. Hirschman, "Father-Daughter Incest," *Signs* 2 (1977): 735–756.

21. See W. Laymon, "Homosexuality, Sexual Dysfunction, and Incest in Male Identical Twins," *Canadian Journal of Psychiatry* 27 (1982): 144–147.

22. For major examples, see E. Tylor, "On a Method of Investigating the Development of Instituttions: Applied to Laws of Marriage and Descent," *Journal of the Royal Anthropological Institute* 18 (1888): 245–269; and C. Levi-Strauss, *The Elementary Structures of Kinship* (Boston: Beacon, 1969).

23. R. Fox, *The Red Lamp of Incest* (London: Hutchinson, 1980), chapter 1.

24. Thus Finkelhor's findings (*Sexually Victimized Children*; Finkelhor, *Child Sexual Abuse: New Theory and Research* [New York: Free Press, 1984]) are based on a far more inclusive conception of sibling incest than that employed by the human sociobiologists. For explanations and defense of the concepts used by clinicians, see Finkelhor, *Sexually Victimized Children*, 54; E. Macfarlane et al., *Sexual Abuse of Young Children* [New York: Guilford, 1986], chapter 1; and J. Herman, *Father-Daughter Incest* [Cambridge, Mass.: Harvard University Press, 1981], 70). If the Westermarckian approach is to be understood in terms of the extinguishing of feelings of sexual attraction among children raised together, it cannot be protected against the mass of evidence of a variety of expressions of such feelings that emerges in recent clinical studies, simply through the stratagem of refusing to apply the label "incest."

25. Herman, *Father-Daughter Incest*, 70.

26. Plainly this is not a sharp dichotomy, but the categories mark two ends of a continuum. Extreme instances of consensual incest would be examples in which the partners were both equally attracted to one another and in which the sexual relations are as unproblematic retrospectively as sexual relations normally are. Extreme instances of coercive incest would be examples of forcible incestuous rape (possibly repeated on many occasions) with continued resistance on the part of the victim. The studies of Herman and Hirschman, "Father-Daughter Incest"; Meiselman, *Incest*; and Herman, *Father-Daughter Incest* all show that women who unsuccessfully resist the advances of more powerful male relatives are typically better able to overcome the effects of the incident—perhaps because they do not perceive themselves as in any way responsible for what occurred.) I should emphasize that, even in those cases where there is no overt resistance, the power asymmetries in many incestuous situations support the clinicians' classifications of them as examples of victimization. Finkelhor's survey makes plain the fact that less coercion is felt when the participants are closer in age, and the relationships that develop among peers seem significantly less problematic than those involving crossing of generations. It is remarkable that there are at least some approximations to the consensual extreme that have been recorded. See for example Case 11 in Meiselman. *Incest*, 62, 265, 279; Case 3 in S. Kubo, "Researches and Studies on Incest in Japan," *Hiroshima Journal of Medical Sciences* 8 (1959): 99–139, 117–118; and numerous instances in R. E. L. Masters, *Patterns of Incest* (New York: Julian, 1963). (I am more skeptical of the data supplied by Masters and cite it only because, in this instance, there is independent evidence of the same phenomenon.)

27. See, for example, chapter 5 of Finkelhor, *Child Sexual Abuse*.

28. As I have pointed out elsewhere (Kitcher, *Vaulting Ambition*, 367–368), ventures in human sociobiology err in supposing that if there is a probability p that individuals reared together will not acquire the (Westermarckian) disposition to avoid sexual relations with one another, then the frequency of sibling incest should be (approximately) p. If we suppose that sibling incest is consensual, then the frequency of sibling incest should be (approximately) p^2. If we suppose that there is an uncorrelated trait of being able and willing to coerce sexual activity, and that this occurs with small probability q, then the frequency should be (approximately) $p^2 (1 - q) + pq$. In either event, using frequencies of consummate incest as indicators of frequencies of incestuous desire will underrepresent the frequency with which siblings are attracted to one another.

29. The studies of Finkelhor (*Sexually Victimized Children* and *Child Sexual Abuse*), Meiselman (*Incest*), and Herman (*Father-Daughter Incest*) show very clearly that relative age is extremely important in affecting the attitudes of participants in incestuous relationships. Herman, in particular, gives a compelling presentation of the damage inflicted on daughters who are victimized by their fathers, and of the sense of powerlessness that those daugh-

ters have. By contrast, the respondents to Finkelhor's questionnaire who were involved in incestuous relationships with peers seem to have been far *less* troubled by these relationships—which is not to maintain, as members of some "sexual liberation" groups hope to do, that some cases of incest are entirely free from victimization.

30. A number of writers have discussed the "frustration hypothesis" as an account of how initial attraction for the familiar is transformed into sexual indifference. See, for example, Fox, *Red Lamp of Incest*, 50; and P. Bateson, "Optimal Outbreeding," in Bateson, ed., *Mate Choice* (Cambridge: Cambridge University Press, 1983), 270. Commenting on the results that he has already achieved with quail, Bateson suggests that the hypothesis is not very likely, although it cannot be discounted. Although the hypothesis might easily be developed into a neat synthesis of Freudian and Westermarckian ideas, I don't think that the evidence on human sexual development is particularly favorable to it. For, as Kinsey originally documented and some recent writers on child sexuality have emphasized with enthusiasm (see F. Martinson, "Sexual Responses of Children," in L. Constantine and F. Martinson, eds., *Children and Sex* [Boston: Little, Brown, 1982], 24–25), there is little basis for the once popular claim that infantile sexual feelings inevitably—or even usually—are frustrated by adult responses to them. Moreover, studies of development of sexual preference indicate that the sex of childhood sexual partners has little impact on postpubertal sexual preference (A. Bell, M. Weinberg, and S. Hammersmith, *Sexual Preference: Its Development among Men and Women* [Bloomington: Indiana University Press, 1981]).

31. Note that the simple machinery I have introduced allows us to make sense of the idea that incest-avoidance is a bio-cultural phenomenon. We assume that there are certain ontogenetic processes that may be relatively imperturbable by the cultural input but that these processes interact with others that are highly sensitive to patterns of behavior and teaching within the family and within the broader culture. As Bateson has repeatedly emphasized, simplistic dichotomies about biological and cultural factors need to be overcome, and his own work on the ontogeny of mate preference in Japanese quail should not be interpreted as underwriting the idea that human incest-avoidance can be completely understood in biological terms.

32. One of the obvious points for further investigation is study of the effects of alcohol in the initiation of incest. Much of the literature on father-daughter incest relates incestuous advances to heavy drinking—see, for example, Finkelhor, *Sexually Victimized Children*, 22, 207; Macfarlane et al., *Sexual Abuse of Young Children*, 76, 112; and Weinberg, *Incest Behavior*. In terms of the account I offer in the text, it is natural to propose either that alcohol intake causes a generalized lowering of the threshold for sexual desire or that it eases the translation of sexual desire into overt behavior (of course, it is possible that both effects are present). If one could provide a neural substructure for the type of psychological account that I offer, it would be interesting to try to link the clinical evidence to ongoing research on the effects of alcohol on the brain.

33. The distinction between sexual desire and action that expresses sexual desire, which I have drawn in a preliminary way in this section, seems to me to be crucial in understanding one important facet of examples of incest. Clinical studies make it relatively clear that some daughters feel sexual attraction toward their incestuous fathers. Herman and Hirschman write: "Half of the women acknowledged, however, that they had felt some degree of pleasure in the sexual contact, a feeling which had only increased their sense of guilt and confusion" ("Father-Daughter Incest," 747). The point is elaborated with explicit citations of reports to therapists, including one woman's statement that she was "in love" with her father and another's description of herself as "very attracted" to her father (ibid., 747–748). *It should not be inferred from these reports that the women in question were not victimized in the relationship*. For, in such cases, there is (typically, and probably almost always) either a

reluctance to translate the sexual desire into action or an inability to foresee consequences of action, a lack that is exploited by the more knowledgeable father. Note that the presence of desires of this kind seems incompatible with a generalizd version of the Westermarck hypothesis, although it can readily be understood in terms of the model I've sketched: we can think of the women concerned as acquiring the standard attractiveness curve but as having a threshold function that is locally depressed in the region corresponding to the father, partly as a result of their isolation from other potentially attractive males and partly as a result of the father's previous attentions.

34. Much of the anthropological literature on incest-avoidance has, I think, failed to see the importance of identifying mechanisms by trying to understand where they don't work. Given the approach that I've proposed, the class of interesting cases is greatly expanded: we should no longer be looking just at instances in which incest is consummated, but at occasions on which people *feel incestuous desires*. A natural extension of Finkelhor's questionnaire would enable us to gather information about such occasions—although the methodological circumspection that Finkelhor himself expresses about his survey would be even more appropriate as we move from asking subjects to report overt behavior to inquiring, retrospectively, about their feelings.

35. Another obvious extension of the model I've sketched is to attempt to offer a connectionist articulation of the claims about thresholds and evaluation functions, along lines similar to the studies in J. McClelland and D. Rumelhart, *Parallel Distributed Processing*, 2 vols. (Cambridge, Mass.: MIT Press, 1986).

36. Given the scruples advanced in my *Vaueting Ambitions* about the dangers of casual attributions of function, I am loth to offer *qualitative* claims about the possible advantages of certain types of psychological mechanisms. However, in case the central point of the paper is lost in the details of the example, let me indicate how the approach I recommend might modify our functionalist hypotheses. Instead of looking for the function of a disposition to refrain from copulating with those with whom one has been reared, our attention should now shift to the (possible) function of the ontogenetic processes that give rise to the initial attractiveness curve, to its modification under habituation, and to the transformations of the threshold. The function of such processes may only be specifiable in terms of their effects *across the entire range of sexual behavior*, and it is possible that their consequences for survival and reproduction are mediated by *nonsexual* behavior as well. Thus an important issue for the synthesis of functional and developmental ideas is whether there is a single attractiveness evaluation that guides all social interactions, or whether there are separate evaluations for sexual and nonsexual relationships.

The challenge will obviously be to bring to the psychological level the same kind of precision in modeling and detail of observation that has begun to emerge in the behavioral ecological study of mate-choice *behavior* in nonhuman animals. See, in particular, Bateson, "Optimal Outbreeding and the Development of Sexual Preferences," "Preferences for Cousins," and "Optimal Outbreeding"; L. Partridge, "Non-Random Mating and Offspring Fitness, in P. Bateson, ed., *Mate Choice* (Cambridge: Cambridge University Press, 1983); and F. Cooke and J. Davies, "Mating in Snow Geese," in P. Bateson, ed., *Mate Choice* (Cambridge: Cambridge University Press, 1983).

37. Of course, it is eminently possible that the study of incest-avoidance is the wrong case to begin with. I use it here as an illustration for several reasons. First, it is an instance in which traditional human sociobiology has often taken pride, so that the differences between the approach that I recommend and the sociobiological tradition can appear more starkly in the discussion of this parade case. Second, thanks to the work of Bateson and others, there is a growing collection of illuminating studies of nonhuman organisms and their choices of mates. Third, the clinical studies of Meiselman, Herman, Hirschman and others,

together with Finkelhor's important studies, yield a body of psychological literature and data that can *in principle* be linked with biological investigations. For all this, it may turn out that the approach I have sketched is entirely off-target, and that human behavioral ecology ought to pursue some more tractable problem. (The moral of the history of genetics, in which those who began with fruit flies came far further than those who hoped to tackle the exciting problems of human heredity directly, is, I trust, obvious.) Let me emphasize that the treatment of incest sketched here is an *illustration*, and that, whatever its merits as an explanation of the particular phenomena, it is the general approach and the *style* of explanation that matters.

15

Four Ways of "Biologicizing" Ethics (1993)

I

In 1975, E. O. Wilson invited his readers to consider "the possibility that the time has come for ethics to be removed temporarily from the hands of the philosophers and biologicized."[1] There should be no doubting Wilson's seriousness of purpose.[2] His writings from 1975 to the present demonstrate his conviction that nonscientific, humanistic approaches to moral questions are indecisive and uninformed, that these questions are too important for scholars to neglect, and that biology, particularly the branches of evolutionary theory and neuroscience that Wilson hopes to bring under a sociobiological umbrella, can provide much-needed guidance. Nevertheless, I believe that Wilson's discussions of ethics, those that he has ventured alone and those undertaken in collaboration first with the mathematical physicist Charles Lumsden and later with the philosopher Michael Ruse, are deeply confused through failure to distinguish a number of quite different projects. My aim in this chapter is to separate those projects, showing how Wilson and his co-workers slide from uncontroversial truisms to provocative falsehoods.

Ideas about "biologicizing" ethics are by no means new, nor are Wilson's suggestions the only proposals that attract contemporary attention.[3] By the same token, the distinctions that I shall offer are related to categories that many of those philosophers Wilson seeks to enlighten will find very familiar. Nonetheless, by developing the distinctions in the context of Wilson's discussions of ethics, I hope to formulate a map on which would-be sociobiological ethicists can locate themselves and to identify questions that they would do well to answer.

II

How do you "biologicize" ethics? There appear to be four possible endeavors:

1. Sociobiology has the task of explaining how people have come to acquire ethical concepts, to make ethical judgments about themselves and others, and to formulate systems of ethical principles.

2. Sociobiology can teach us facts about human beings that, in conjunction with moral principles that we already accept, can be used to derive normative principles that we had not yet appreciated.

3. Sociobiology can explain what ethics is all about and can settle traditional questions about the objectivity of ethics. In short, sociobiology is the key to metaethics.

4. Sociobiology can lead us to revise our system of ethical principles, not simply by leading us to accept new derivative statements—as in number 2 above—but by teaching us new fundamental normative principles. In short, sociobiology is not just a source of facts but a source of norms.

Wilson appears to accept all four projects, with his sense of urgency that ethics is too important to be left to the "merely wise,"[4] giving special prominence to endeavor 4. (Endeavors 2 and 4 have the most direct impact on human concerns, with endeavor 4 the more important because of its potential for fundamental changes in prevailing moral attitudes. The possibility of such changes seems to lie behind the closing sentences of the essay by Ruse and Wilson.[5]) With respect to some of these projects, the evolutionary parts of sociobiology appear most pertinent; in other instances, neurophysiological investigations, particularly the exploration of the limbic system, come to the fore.

Relatives of endeavors 1 and 2 have long been recognized as legitimate tasks. Human ethical practices have histories, and it is perfectly appropriate to inquire about the details of those histories. Presumably, if we could trace the history sufficiently far back into the past, we would discern the coevolution of genes and culture, the framing of social institutions, and the introduction of norms. It is quite possible, however, that evolutionary biology would play only a very limited role in the story. All that natural selection may have done is to equip us with the capacity for various social arrangemets and the capacity to understand and to formulate ethical rules. Recognizing that not every trait we care to focus on need have been the target of natural selection, we shall no longer be tempted to argue that any respectable history of our ethical behavior must identify some selective advantage for those beings who first adopted a system of ethical precepts. Perhaps the history of ethical thinking instantiates one of those coevolutionary models that show cultural selection's interfering with natural selection.[6] Perhaps what is selected is some very general capacity for learning and acting that is manifested in various aspects of human behavior.[7]

Nothing is wrong with endeavor 1, so long as it is not articulated in too simplistic a fashion and so long as it is not overinterpreted. The reminders of the last paragraph are intended to forestall the crudest forms of neo-Darwinian development of this endeavor. The dangers of overinterpretation, however, need more derailed charting. There is a recurrent tendency in Wilson's writings to draw unwarranted conclusions from the uncontroversial premise that our ability to make ethical judgments has a history, including, ultimately, an evolutionary history. After announcing that "everything human, including the mind and culture, has a material base and originated during the evolution of the human genetic constitution and its interaction with the environment,"[7] the authors assert that "accumulating empir-

ical knowledge" of human evolution "has profound consequences for moral philosophy." For that knowledge "renders increasingly less tenable the hypothesis that
ethical truths are extrasomatic, in other words divinely placed within the brain or
else outside the brain awaiting revelation." Ruse and Wilson thus seem to conclude
that the legitimacy of endeavor 1 dooms the idea of moral objectivity.[8]

That this reasoning is fallacious is evident once we consider other systems of
human belief. Plainly, we have capacities for making judgments in mathematics,
physics, biology, and other areas of inquiry. These capacities, too, have historical
explanations, including, ultimately, evolutionary components. Reasoning in parallel fashion to Ruse and Wilson, we could thus infer that objective truth in mathematics, physics, and biology is a delusion and that we cannot do *any* science without
"knowledge of the brain, the human organ where all decisions . . . are made."[9]

What motivates Wilson (and his collaborators Ruse and Lumsden) is, I think,
a sense that ethics is different from arithmetic or statics. In the latter instances, we
could think of history (including our evolutionary history) bequeathing to us a
capacity to learn. That capacity is activated in our encounters with nature, and we
arrive at objectively true beliefs about what nature is like. Since they do not see
how a similar account could work in the case of moral belief, Wilson, Ruse, and
Lumsden suppose that their argument does not generalize to a denunciation of the
possibility of objective knowledge. This particular type of skepticism about the possibility of objectivity in ethics is revealed in the following passage: "But the philosophers and theologians have not yet shown us how the final ethical truths will be
recognized as things apart from the idiosyncratic development of the human
mind."[10]

There is an important challenge to those who maintain the objectivity of ethics,
a challenge that begins by questioning how we obtain ethical knowledge. Evaluating that challenge is a complex matter I shall take up in connection with project
3. However, unless Wilson has independent arguments for resolving questions in
metaethics, the simple move from the legitimacy of endeavor 1 to the "profound
consequences for moral philosophy" is a blunder. The "profound consequences"
result not from any novel information provided by recent evolutionary theory but
from arguments that deny the possibility of assimilating moral beliefs to other kinds
of judgments.

III

Like endeavor 1, endeavor 2 does not demand the removal of ethics from the hands
of the philosophers. Ethicists have long appreciated the idea that facts about human
beings, or about other parts of nature, might lead us to elaborate our fundamental
ethical principles in previously unanticipated ways. Card carrying Utilitarians who
defend the view that morally correct actions are those that promote the greatest happiness of the greatest number, who suppose that those to be counted are presently
existing human beings, and who identify happiness with states of physical and psychological well-being will derive concrete ethical precepts by learning how the maximization of happiness can actually be achieved. But sociobiology has no monopoly

here. Numerous types of empirical investigations might provide relevant information and might contribute to a profitable division of labor between philosophers and others.

Consider, for example, a family of problems with which Wilson, quite rightly, has been much concerned. There are numerous instances in which members of small communities will be able to feed, clothe, house, and educate themselves and their children far more successfully if a practice of degrading the natural environment is permitted. Empirical information of a variety of types is required for responsible ethical judgment. What alternative opportunities are open to members of the community if the practice is banned? What economic consequences would ensue? What are the ecological implications of the practice? All these are questions that have to be answered. Yet while amassing answers is a prerequisite for moral decision, there are also issues that apparently have to be resolved by pondering fundamental ethical principles. How should we assess the different kinds of value (unspoiled environments, flourishing families) that figure in this situation? Whose interests, rights, or well-being deserve to be counted?

Endeavors like the second one are already being pursued, especially by workers in medical ethics and in environmental ethics. It might be suggested that sociobiology has a particularly important contribution to make to this general enterprise, because it can reveal to us our deepest and most entrenched desires. By recognizing those desires, we can obtain a fuller understanding of human happiness and thus apply our fundamental ethical principles in a more enlightened way. Perhaps. However, as I have argued at great length, the most prominent sociobiological attempts to fathom the springs of human nature are deeply flawed, and remedying the deficiencies requires integrating evolutionary ideas with neuroscience, psychology, and various parts of social science.[11] In any event, recognizing the legitimacy of endeavor 2 underscores the need to evaluate the different desires and interests of different people (and, possibly, of other organisms), and we have so far found no reason to think that sociobiology can discharge that quintessentially moral task.

IV

Wilson's claims about the status of ethical statements are extremely hard to understand. It is plain that he rejects the notion that moral principles are objective because they encapsulate the desires or commands of a deity (a metaethical theory whose credentials have been doubtful ever since Plato's *Euthyphro*). Much of the time he writes as though sociobiology settled the issue of the objectivity of ethics negatively. An early formulation suggests a simple form of emotivism:

> Like everyone else, philosophers measure their personal emotional responses to various alternatives as though consulting a hidden oracle. That oracle resides within the deep emotional centers of the brain, most probably within the limbic system, a complex array of neurons and hormone-secreting cells located just below the "thinking" portion of the cerebral cortex. Human emotional responses and the

more general ethical practices based on them have been programmed to a substantial degree by natural selection over thousands of generations.[12]

Stripped of references to the neural machinery, the account Wilson adopts is a very
simple one. The content of ethical statements is exhausted by reformulating them
in terms of our emotional reactions. Those who assent to, "Killing innocent children is morally wrong," are doing no more than reporting on a feeling of repugnance, just as they might express gastronomic revulsion. The same type of
metaethics is suggested in more recent passages, for example, in the denial that
"ethical truths are extrasomatic" which I have already quoted.

Yet there are internal indications and explicit formulations that belie interpreting Wilson as a simple emotivist. Ruse and Wilson appear to support the
claim that "'killing is wrong' conveys more than merely 'I don't like killing.'"
Moreover, shortly after denying that ethical truths are extrasomatic, they suggest
that "our strongest feelings of right and wrong" will serve as "a foundation for
ethical codes," and their paper concludes with the visionary hope that study will
enable us to us to see "how our short-term moral insights fail our long-tem needs,
and how correctives can be applied to formulate more enduring moral codes."[13] As
I interpret them, they believe that some of our inclinations and disinclinations, and
the moral judgments in which they are embodied, betray our deepest desires and
needs and that the task of formulating an "objective" ("enduring," "corrected")
morality is to identify these desires and needs, embracing principles that express
them.

Even in Wilson's earlier writings, he sounds themes that clash with any simple
emotivist metaethics. For example, he acknowledges his commitment to different
sets of "moral standards" for different populations and different groups within the
same population.[14] Population variation raises obvious difficulties for emotivism.
On emotivist grounds, deviants who respond to the "limbic oracle" by wilfully tor-
turing children must be seen as akin to those who have bizarre gastronomic preferences. The rest of us may be revolted, and our revulsion may even lead us to
interfere. Yet if pressed to defend ourselves, emotivism forces us to concede that
there is no standpoint from which our actions can be judged as objectively more
worthy than the deeds we try to restrain. The deviants follow their hypothalamic
imperative, and we follow ours.

I suspect that Wilson (as well as Lumsden and Ruse) is genuinely torn between
two positions. One hews a hard line on ethical objectivity, drawing the "profound
consequence" that there is no "extrasomatic" source of ethical truth and accepting
an emotivist metaethics. Unfortunately, this position makes nonsense of Wilson's
project of using biological insights to fashion an improved moral code and also leads
to the unpalatable conclusion that there are no grounds for judging those whom
we see as morally perverse. The second position gives priority to certain desires,
which are to be uncovered through sociobiological investigation and are to be the
foundation of improved moral codes, but it fails to explain what normative standard gives these desires priority or how that standard is grounded in biology. In my
judgment, much of the confusion in Wilson's writings comes from oscillating
between these two positions.

I shall close this section with a brief look at the line of argument that seems to lurk behind Wilson's emotivist leanings. The challenge for anyone who advocates the objectivity of ethics is to explain in what this objectivity consists. Skeptics can reason as follows: If ethical maxims are to be objective, then they must be objectively true or objectively false. If they are objectively true or objectively false, then they must be true or false in virtue of their correspondence with (or failure to correspond with) the moral order, a realm of abstract objects (values) that persists apart from the natural order. Not only is it highly doubtful that there is any such order, but, even if there were, it is utterly mysterious how we might ever come to recognize it. Apparently we would be forced to posit some ethical intuition by means of which we become aware of the fundamental moral facts. It would then be necessary to explain how this intuition works, and we would also be required to fit the moral order and the ethical intuition into a naturalistic picture of ourselves.

The denial of "extrasomatic" sources of moral truth rests, I think, on this type of skeptical argument, an argument that threatens to drive a wedge between the acquisition of our ethical beliefs and the acquisition of beliefs about physics or biology (see the discussion of endeavor 1 above). Interestingly, an exactly parallel argument can be developed to question the objectivity of mathematics. Since few philosophers are willing to sacrifice the idea of mathematical objectivity, the philosophy of mathematics contains a number of resources for responding to that skeptical parallel. Extreme Platonists accept the skeptic's suggestion that objectivity requires an abstract mathematical order, and they try to show directly how access to this order is possible, even on naturalistic grounds. Others assert the objectivity of mathematics without claiming that mathematical statements are objectivity true or false. Yet others may develop an account of mathematical truth that does not presuppose the existence of abstract objects, and still others allow abstract objects but try to dispense with mathematical intuition.

Analogous moves are available in the ethical case. For example, we can sustain the idea that some statements are objectively justified without supposing that such statements are true. Or we can abandon the correspondence theory of truth for ethical statements in favor of the view that an ethical statement is true if it would be accepted by a rational being who proceeded in a particular way. Alternatively, it is possible to accept the thesis that there is a moral order but understand this moral order in naturalistic terms, proposing, for example, with the Utilitarians, that moral goodness is to be equated with the maximization of human happiness and that moral rightness consists in the promotion of the moral good. Yet another option is to claim that there are indeed nonnatural values but that these are accessible to us in a thoroughly familiar way—for example, through our perception of people and their actions. Finally, the defender of ethical objectivity may accept all the baggage that the skeptic assembles and try to give a naturalistic account of the phenomena that skeptics take to be incomprehensible.

I hope that even this brief outline of possibilities makes it clear how a quick argument for emotivist metaethics simply ignores a host of metaethical alternatives—indeed the main alternatives that the "merely wise" have canvassed in the history of ethical theory. Nothing in recent evolutionary biology or neuro-science forecloses these alternatives. Hence, if endeavor 3 rests on the idea that sociobiology yields a quick proof of emotivist metaethics, this project is utterly mistaken.

On the other hand, if Wilson and his co-workers intend to offer some rival metaethical theory, one that would accord with their suggestions that sociobiology might generate better ("more enduring") moral codes, then they must explain what this metaethical theory is and how it is supported by biological findings. In the absence of any such explanations, we should dismiss this endeavor as deeply confused.

V

In the search for new normative principles, project 4, it is not clear whether Wilson intends to promise or to deliver. His early writing sketches the improved morality that would emerge from biological analysis.

> In the beginning the new ethicists will want to ponder the cardinal value of the survival of human genes in the form of a common pool over generations. Few persons realize the true consequences of the dissolving action of sexual reproduc-tion and the corresponding unimportance of "lines" of descent. The DNA of an individual is made up of about equal contributions of all the ancestors in any given generation, and it will be divided about equally among all descendants at any future moment. . . . The individual is an evanescent combination of genes drawn from this pool, one whose hereditary material will soon be dissolved back into it.[15]

I interpret Wilson as claiming that there is a fundamental ethical principle, which we can formulate as follows:

> W: Human beings should do whatever is required to ensure the survival of a common gene pool for *Homo sapiens*.

He also maintains that this principle is not derived from any higher-level moral statement but is entirely jusified by certain facts about sexual reproduction. Wilson has little time for the view that there is a fallacy in inferring values from facts[16] or for the "absolute distinction between *is* and *ought*."[17] It appears, then, that there is supposed to be a good argument to W from a premise about the facts of sex:

> S: The DNA of any individual human being is derived from many people in earlier generations and, if the person reproduces, will be distributed among many people in future generations.

I shall consider both the argument from S to W and the correctness of W.

Plainly, one cannot deduce W from S. Almost as obviously, no standard type of inductive or statistical argument will sanction this transition. As a last resort, one might propose that W provides the best explanation for S and is therefore acceptable on the grounds of S, but the momentary charm of this idea vanishes once we recognize that S is explained by genetics, not by ethical theory.

There are numerous ways to add ethical premises so as to license the transition from S to W, but making these additions only support the uncontroversial enterprise 2, not the search for fundamental moral principles undertaken under the aegis of endeavor 4. Without the additions, the inference is so blatantly fallacious that we can only wonder why Wilson thinks that he can transcend traditional criticisms of the practice of inferring values from facts.

The faults of Wilson's mrthod are reflected in the character of the fundamental moral principle he identifies. That principle, W, enjoins actions that appear morally suspect (to say the least). Imagine a stereotypical postholocaust situation in which the survival of the human gene pool depends on copulation between two people. Suppose, for whatever reason, that one of the parties is unwilling to copulate with the other. (This might result from resentment at past cruel treatment, from recognition of the miserable lives that offspring would have to lead, from sickness, or whatever.) Under these circustances, W requires the willing party to coerce the unwilling person, using whatever extremes of force are necessary—perhaps even allowing for the murder of those who attempt to defend the reluctant one. There is an evident conflict between these conseqences of W and other ethical principles, particularly those that emphasize the rights and autonomy of individuals. Moreover, the scenario can be developed so as to entail enormous misery for future descendants of the critical pair, thus flouting utilitarian standards of moral correctness. Faced with such difficulties for W, there is little consolation in the thought that our DNA was derived from many people and will be dispersed among many people in whatever future generations there may be. At stake are the relative values of the right to existence of future generations (possibly under dreadful conditions) and the right to self-determination of those now living. The biological facts of reproduction do not give us any information about that relationship.

In his more recent writings, Wilson has been less forthright about the principles of "scientific ethics." Biological investigations promise improved moral codes for the future: "Only by penetrating to the physical basis of moral thought and considering its evolutionary meaning will people have the power to control their own lives. They will then be in a better position to choose ethical precepts and the forms of social regulation needed to maintain the precepts."[18] Ruse and Wilson are surprisingly reticent in expressing substantive moral principles, apparently preferring to disuss general features of human evolution and results about the perception of colors. Their one example of an ethical maxim is not explicitly formulated, although since it has to do with incest avoidance, it could presumably be stated as, "Do not copulate with your siblings!"[19] If this is a genuine moral principle at all, it is hardly a central one and is certainly not fundamental.

I believe that the deepest problems with the sociobiological ethics recommended by Wilson, Lumsden, and Ruse can be identified by considering how the most fundamental and the most difficult normative questions would be treated. If we focus attention, on the one hand, on John Rawls's principles of justice (proposals about fundamental questions) or on specific claims about the permissibilty of abortion (proposals about a very difficult moral question), we discover the need to evaluate the rights, interests, and responsibilities of different parties. Nothing in sociobiological ethics speaks to the issue of how these potentially conflicting sets of rights, interests, and responsibilities are to be weighed. Even if we were confident that sociobiology could expose the deepest human desires, thus showing how the enduring happiness *of a single individual* could be achieved, there would remain the fundamental task of evaluating the competing needs and plans of different people. Sociobiological ethics has a vast hole at its core—a hole that appears as soon as we reflect on the implications of doomsday scenarios for Wilson's principle (W). Nothing in the later writings of Wilson, Lumsden, and Ruse addresses the deficiency.

The gap could easily be plugged by retreating from project 4 to the uncontroversial project 2. Were Wilson a Utilitarian, he could address the question of evaluating competing claims by declaring that the moral good consists in maximizing total human happiness, conceding that this fundamental moral principle stands outside sociobigical ethics but contending that sociobiology, by revealing our evolved desires, shows us the nature of human happiness. As noted above in connection with project 2, there are grounds for wondering if sociobiology can deliver insights about our "deepest desires." In any case, the grafting of sociobiology onto utilitarianism hardly amounts to the fully naturalistic ethics proclaimed in Wilson's rhetoric.

If we try to develop what I take to be Wilson's strongest motivating idea, the appeal to some extrasociobiological principle is forced upon us. Contrasting our "short-term moral problems" with our "long-term needs," Ruse and Wilson hold out the hope that biological investigations, by providing a clearer picture of ourselves, may help us to reform our moral systems.[20] Such reforms would have to be carried out under the guidance of some principle that evaluated the satisfaction of different desires within the life of an individual. Why is the satisfaction of long-term needs preferable to the palliation of the desires of the moment? Standard philosophical answers to this question often presuppose that the correct course is to maximize the total life happiness of the individual, subject perhaps to some system of future discounting. Whether any of those answers is adequate or not, Wilson needs some principle that will play the same evaluative role if his vision of reforming morality is to make sense. Wilson's wrtings offer no reason for thinking of project 4 as anything other than a blunder, and Wilson's own program of moral reform presupposes the nonbiological ethics whose poverty he so frequently decries.

VI

Having surveyed four ways of "biologicizing" ethics, I shall conclude by posing some questions for the aspiring sociobiological ethicist. The first task for any sociobiologcal ethics is to be completely clear about which project (or projects) are to be undertaken. Genuine interchange between biology and moral philosophy will be achieved only when eminent biologists take pains to specify what they mean by the "biologicizations" of ethics, using the elementary categories I have delineated here.

Project 1 is relatively close to enterprises that are currently being pursued by biolologists and anthropologists. Human capacites for moral reflection are phenotypic traits into whose histories we can reasonably inquire. However, those who seek to construct such histories would do well to ask themselves if they are employing the most sophisticated machinery for articulating coevolutionary processes and whether they are avoiding the adaptationist pitfalls of vulgar Darwinism.

Project 2 is continuous with much valuable work done in normative ethics over the last decades. Using empirical information, philosophers and collaborators from other disciplines have articulated various types of moral theory to address urgent concrete problems. If sociobiological ethicists intend to contribute to this enterprise, they must explicitly acknowledge the need to draw on ectrabiological moral pinciples. They must also reflect on what ethical problems sociobiological information can help to illuminate and on whether human sociobiology is in any position to deliver such information. Although project 2 is a far more modest enterprise than that which Wilson and his collaborators envisage, I am very doubtful that human sociobiology is up to it.[21]

Variants of the refrain that "there is no morality apart from biology" lead sociobiologists into the more ambitious project 3. Here it is necessary for the aspiring ethicists to ask themselves if they believe that some moral statements are true, others false. If they do believe in moral truth and falsity, they should be prepared to specify what grounds such truth and falsity. Those who think that moral statements simply record the momentary impulses of the person making the statement should explain how they cope with people who have deviant impulses. On the other hand, if it is supposed that morality consists in the expression of the "deepest" human desires, then it must be shown how, *without appeal to extrabiological moral principles*, certain desires of an individual are taken to be privileged and how the conflicting desires of different individuals are adjudicated.

Finally, those who undertake project 4, seeing biology as the source of fundamental normative principles, can best make their case by identifying such principles, by formulating the biological evidence for them, and by revealing clearly the character of the inferences from facts to values. In the absence of commitment to any specific moral principles, pleas that "the naturalistic fallacy has lost a great deal of its force in the last few years"[22] will ring hollow unless the type of argument leading from biology to morality is plainly identified. What kinds of premises will be used? What species of inference leads from those premises to the intended normative conclusion?

It would be folly for any philosopher to conclude that sociobiology can contribute nothing to ethics. The history of science is full of reminders that initially unpromising ideas sometimes pay off (but there are even more unpromising ideas that earn the right to oblivion). However, if success is to be won, criticisms must be addressed, not ignored. Those inspired by Wilson's vision of a moral code reformed by biology have a great deal of work to do.

Notes

1. Edward O. Wilson, *Sociobiology: The New Synthesis* (Cambridge, Mass.: Harvard University Press, 1975), 562.

2. Some of Wilson's critics portray him as a frivolous defender of reactionary conservatism; see, for example, Richard Lewontin, Stephen Rose, and Leon Kamin, *Not in Our Genes* (New York: Pantheon, 1984). While I agree with several of the substantive points that these critics make against Wilson's version of human sociobiology, I dissent from their assessment of Wilson's motives and commitments. I make the point explicit because some readers of my *Vaulting Ambition: Sociobiology and the Quest for Human Nature* (Cambridge, Mass.: MIT Press, 1985) have mistaken the sometimes scathing tone of that book for a questioning of Wilson's intellectual honesty or of his seriousness. As my title was intended to suggest, I view Wilson and other eminent scientists who have entered into human sociobiology as treating important questions in a ham-fisted way because they lack crucial intellectual tools and because they desert the standards of rigor and clarity that are found in their more narrowly scientific work. The tone of *Vaulting Ambition* stems from the fact that the issues are so important and the treatment of them often so bungled.

3. For historical discussion, see R. Richards, *Darwin and the Emergence of Evolutionary Theories of Mind and Behavior* (Chicago: University of Chicago Press, 1987). Richard Alexander, *The Morality of Biological Systems* (Chicago: Aldine, 1987), offers an alternative version of sociobiological ethics, while Michael Ruse, *Taking Darwin Seriously* (London: Routledge, 1986), develops a position that is closer to that espoused in Wilson's later writings (particularly in Ruse and Edward O. Wilson, "Moral Philosophy as Applied Science," *Philosophy* 61 [1986]: 173–192).

4. E. O. Wilson, *On Human Nature* (Cambridge, Mass.: Harvard University Press, 1978), 7.

5. Ruse and Wilson, "Moral Philosophy," 192.

6. Robert Boyd and Peter Richerson, *Culture and the Evolutionary Process* (Chicago: University of Chicago Press, 1985).

7. Philip Kitcher, "Developmental Decomposition and the Future of Human Behavioral Ecology," *Philosophy of Science* 57 (1990): 96–117; reprinted as chapter 14 of this book.

8. Ruse and Wilson, "Moral Philosophy," 173–174.

9. Ibid., 173.

10. Charles Lumsden and Wilson, *Promethean Fire* (Cambridge, Mass.: Harvard University Press, 1983), 182–183.

11. Kitcher, *Vaulting Ambition*; see also Kitcher "Precis of *Vaulting Ambition* and Reply to Twenty-Two Commentators ('Confessions of a Curmudgeon')," *Behavioral and Brain Sciences* 10 (1987): 61–100; Kitcher, "The Transformation of Human Sociobiology," in A. Fine and P. Machamer, eds., *PSA 1986, Proceedings of the Philosophy of Science Association* (East Lansing, Mich.: Philosophy of Science Association, 1987), 63–74; Kitcher, "Imitating Selection," in Sidney Fox and Mae-Wan Ho, eds., *Metaphors in the New Evolutionary Paradigm* (Chichester: Wiley, 1988), 295–318; Kitcher, "Developmental Decomposition."

12. Wilson, *On Human Nature*, 6.

13. Ruse and Wilson, "Moral Philosophy," 178, 173, 192, respectively.

14. Wilson, *Sociobiology*, 564.

15. Wilson, *On Human Nature*, 196–197.

16. Wilson, "The Relation of Science to Theology," *Zygon* 15 (1980): 425–434, at page 431; Wilson "Comparative Social Theory," in *The Tanner Lectures* (University of Utah Press).

17. Ruse and Wilson, "Moral Philosophy," 174.

18. Lumsden and Wilson, *Promethean Fire*, 183.

19. See Ruse and Wilson, "Moral Philosophy," 183–185; for discussion of human incest avoidance, see Kitcher, "Developmental Decomposition," reprinted as chapter 14 in this book.

20. Ruse and Wilson, "Moral Philosophy," 192.

21. For reasons given in Kitcher, *Vaulting Ambition*, and Kitcher "Developmental Decomposition," the latter reprinted as chapter 14 in this book.

22. Wilson, "Relation of Science to Theology," 431.

16

Pop Sociobiology Reborn

The Evolutionary Psychology of Sex and Violence (2002)

COAUTHORED WITH A. LEAH VICKERS

1. Introduction: A Dismal History

Here's a recipe for winning fame and fortune as an architect of the new-and-improved human sciences. First, make a bundle of claims to the effect that certain features are universal among human beings, or among human males, or among human females. Next, couple each claim with a story of how the pertinent features were advantageous for primitive hominids, or males, or females, as they faced whatever challenges you take to have been prevalent in some lightly sketched savannah environment. (Don't worry that your knowledge of past environments is rather thin—Be creative!) Finally, announce that each feature in the bundle has been shaped by natural selection, and so corresponds to something very deep in human nature (male human nature, female human nature), something that may be overlain with a veneer of culture but that molds our behavior and the forms of our societies. Accompany everything with hymns to the genius of Darwin, broadsides against "blank slate" views of the human mind, and vigorous denunciations of the lack of rigor and clarity that has hitherto reigned in the human sciences.

In the second half of the twentieth century, three major movements tried to follow this recipe. First came animal ethology with stirring yarns about naked apes and territorial imperatives. These stories were recast by the second wave, as human sociobiology drew more systematically on the resources of contemporary evolutionary theory. In the 1960s and 1970s, the integration of mathematical models with field observations enabled students of animal behavior to advance, support, and refine detailed theories about caste structure in social insects, copulation in dungflies, and the mating structures of red deer.[1] Successes like these inspired the ambitious to propose that kindred insights could be achieved with respect to our own species: they claimed that human beings are, by nature, xenophobic and "absurdly easy to indoctrinate," that human societies are inevitably stratified by relations of power and domination, that men are fated to be fickle and women to be coy, that human altruism is an illusion and that we can't hope to achieve genuine sexual equality.[2] Pop sociobiology was born.

By the mid-1980s, the movement had attracted a barrage of criticism. Skeptics pointed out that, by contrast with the careful studies of nonhuman animals, the suggestions about universals of human behavior (or male behavior, or female behavior) rested on anecdotal evidence. Furthermore, pop sociobiology contented itself with telling informal stories about advantages, instead of putting to work the mathematical tools of evolutionary theory, painstakingly deployed by workers on deer and dungflies. Careful work on the evolution of behavior had appreciated, from the beginning, the need to consider alternative hypotheses and to discriminate among them using data from evolutionary genetics, experiments, comparative observation, or mathematical modelling, but no such pains were taken by the leading proponents of pop sociobiology.[3] Nor was there, to begin with, any appreciation of the possibility that cultural transmission might affect the traits of human beings, and when, belatedly, pop sociobiology came to terms with this issue, its attempts to show that "the genes hold culture on a leash" depended on arbitrary assignments of values to crucial parameters.[4]

Yet perhaps the most important defect lay in the conclusions, often announced with commendable regret, that certain unpleasant features were so deeply ingrained in human nature as to be unmodifiable. Critics noted that such conclusions cannot validly be derived from the kinds of evolutionary scenarios presented.[5] The most those scenarios could reveal is that there are pieces of DNA that, in the particular environments encountered by our hominid ancestors, give rise to characteristics— competitiveness, coyness, xenophobia, whatever—that proved beneficial in those environments; the scenarios have no bearing on whether, under different regimes of development, those traits would be bound to arise (nor whether they would be advantageous in these rival circumstances).

Would-be Darwinian reformers of the human sciences adopted a strategy for coping with these criticisms. "Indeed," they explained, "some sociobiologists have made unwarranted claims; but our approach should not be dismissed; we are aware of the criticisms; we have made them ourselves; we are reformed; we have abandoned the idea that genes are destiny; we are evolutionary psychologists, who aim to use Darwinian insights to fathom human tendencies." Some of them continued to insist on the importance of the enterprise in indicating to us how we might amend unwanted forms of behavior. In the late 1980s, when evolutionary psychology kept its claims modest and its head down, charity commended giving the new movement the benefit of the doubt. But the publication of a rousing revival of the pop favorites of the past[6] made it apparent that the old mistakes haven't lost their allure. Evolutionary psychology turns out to be pop sociobiology with a fig leaf.

2. The Pop Sociobiology Revival: An Overview

We'll try to substantiate this last accusation by looking at two of the most prominent exhibits in the Pop Revival, David Buss's proposals about male and female sexual attractiveness and the hypotheses of Randy Thornhill and Craig Palmer[7] on rape (Thornhill and Palmer draw on Buss's efforts, so our critique of Buss will extend to their program). First, however, we'll offer a more general view of the evolutionary psychology movement.

The principal advance evolutionary psychologists take themselves to have made consists in recognizing that natural selection doesn't shape human behavior directly, but rather the psychological mechanisms underlying behavior. Bad old pop sociobiology supposed that natural selection would favor males who were fickle and promiscuous. Thoroughly modern Darwinian analyses recognize the need to integrate biology with the right approach to psychology, to wit the view that the mind consists of lots of special-purpose devices (modules[8]) that prompt different forms of behavior. Evolutionary psychology reflects on the problems and challenges faced by our hominid ancestors, generating hypotheses about the kinds of psychological traits natural selection has bequeathed to us. These hypotheses are evaluated by collecting evidence from human subjects who report their feelings and preferences in actual or imagined situations, or by studying human behavior. Support for a psychological claim is supposed to come from juxtaposing contemporary data with an *independent* Darwinian expectation about what kinds of ancestral tendencies would have contributed to reproductive success.

If this is to be successful, then *both* the evidence collected and the Darwinian theorizing have to satisfy important constraints. Let's start with the evidence. Whether or not this consists of responses to questionnaires or statistical patterns of behavior, it will have probative force with respect to a hypothesis about a psychological mechanism only if that hypothesis can be integrated with other claims about the psychology of human subjects to generate expectations about what should be observed in the pertinent experimental or natural situations. When the mind is conceived as a bundle of psychological capacities and dispositions that interact with one another and that are causally affected by external cues, the psychological account has to tell us enough about the nature of the interactions and the responses to the cues so that we can derive specific claims about human actions. A claim about a single trait, in splendid isolation, leaves entirely open what sorts of behavior are to be expected—since the activity of other mechanisms could override, suppress, amplify, or redirect whatever tendency is hypothesized—and, in consequence, loose associations between hypothesized psychological tendencies and a pattern of behavior should impress nobody.

It would, of course, be unfair to ask any evolutionary psychologist to provide us with a complete, detailed psychology. Yet if the psychological account provided introduces a collection of capacities that might easily prompt an agent to incompatible forms of behavior—as for example when we're told that people are attracted to different characteristics that regularly turn up in different locations—then we can't tell much about what typical subjects will do. Consider preferences for various types of food. It's a familiar fact that someone's actual diet may not reflect her craving for a particular food, precisely because what she chooses to eat is a function of several underlying psychological dispositions. So we could "protect" an evolutionary story about universal gustatory yearnings by supposing that the underlying tendencies are inhibited by other mechanisms. Or, to put the point differently, the hypothesis that human beings have evolved to crave large hunks of red meat (say) issues no definite predictions about the frequency of carnivorous displays in any human population.

Turning now to the specifically Darwinian part of the enterprise, we should recognize an important point often made by John Maynard Smith: model-building requires attention to the details, and mathematical modeling uncovers and refines hidden presuppositions. (Maynard Smith and W. D. Hamilton are pioneering figures in evolutionary theory, on whose work sociobiology has drawn; the illuminating work of people like Eric Charnov, Geoffrey Parker, Peter Harvey, John Krebs, and many others shows the salutary influence of Maynard Smith and Hamilton.) Mathematical models aren't always necessary in evolutionary work: sometimes alternative hypotheses can be screened out by considerations drawn from genetics, or careful experiments, or detailed cross-species comparisons. In human sociobiology, however, where rival hypotheses can easily be multiplied, where genetic ignorance is the order of the day, where many of the experiments that might clear up controversy are rightly forbidden as unethical, and where cross-specific comparisons are vulnerable to worries about salient differences, it's crucial that the proposals about histories of natural selection should be formulated clearly and precisely. Pop sociobiology often substituted casual stories about selective advantages for rigorous models of selective pressures. To do better, one must know enough about the alleged environment in which the selection process occurred to formulate defensible claims about reproductive costs and benefits.

In the human case (and, quite possibly, in investigations of other species) it's also important to recognize the possibility of cultural transmission. Since the important work of Robert Boyd and Peter Richerson,[9] everyone interested in Darwinizing the human sciences should have known that a population under the joint influence of natural selection and cultural transmission can exhibit characteristics different from those of a population under the influence of natural selection alone, and that the modes of cultural selection generating this type of deviation can themselves be sustained under natural selection.[10] Hence even when one works out the precise details of a hypothesis about the *natural* selection of some trait, it will always be pertinent to wonder if that characteristic would have emerged under the *joint* influence of natural selection and cultural transmission. In short, then, the models that reformed pop sociobiologists are going to use have to be *more* elaborate than those used by their counterparts pursuing nonhuman studies.

There are two theoretical points that add further difficulties to pursuing a serious Darwinian psychology. As many leading Darwinians have declared repeatedly, Darwin replaced the notion of a species as a *type* with an emphasis on intraspecific variability. Perhaps, then, evolutionary psychology's commitment to a universal human nature is suspect. Even though there are surely some traits that are found (almost) universally across our species, it's important not to suppose that universal fixation is the norm. One can't reply that natural selection is a homogenizing force, for, although there are some circumstances—when the underlying genetics is free from well-known complications and there's an optimal form of a particular trait— in which natural selection would be expected to make one variant virtually universal, the necessary hedges can't be disregarded. Sometimes the genetic details make it impossible that the optimal form of a trait should be fixed (a simple example is when the optimal trait is coded by a heterozygote), and there are other instances in which natural selection is expected to generate a polymorphic equilibrium (a classic case is the Hawk-Dove polymorphism from elementary evolutionary game theory).

The idea of individually selected psychological capacities should also be carefully scrutinized. For all their shortcomings, earlier pop sociobiologists did recognize that evolution has something to do with genes, and they were frequently chastised for naïve assumptions that there were genes available to direct females to be coy or human beings in general to be xenophobic. The error, here, as we've already remarked, was to introduce a form of genetic determinism: if the underlying genotype generated the pertinent trait in the ancestral environment, then, it was assumed, it would yield the trait in all environments. Recent pop sociobiologists, by contrast, don't like to talk about genes. For all their reticence, however, they can't avoid advancing genetic hypotheses. After all, without a genetic basis for a trait—that is, a tendency for the underlying genotype to yield a particular phenotype *in the selective environment*—there can be no natural selection. To suppose that there's a naturally selected psychological mechanism for this or that—cheater detection, say, or directing young women to swoon at the prospect of powerful older men—is to claim that there's been genetic variation in some ancestral population pertinent to the propensity to perform such narrowly defined tasks. Although they don't say as much, they must think that there are two alleles—call them A and B—associated in the primeval environment (or range of environments), with a greater or lesser ability to carry out the appointed task (detect cheats or swoon appropriately).

Let's take a deep breath at this point. It's worth reminding ourselves of what genes do. Genes encode proteins. So A and B encode different proteins, and, on a simple version, it seems that evolutionary psychologists are committed to saying that these differences amount to solely and precisely a difference in cheat-spotting-acuity or swoonability. We're prepared to concede that differences in proteins might show up in alternative forms of neural chemistry, evident in psychological changes—it's not incredible that a modified neural receptor protein might make a mouse, or a human, more or less good at remembering things, or slower or faster to learn. What's highly implausible is that changing a protein could leave all our psychological tendencies untouched while fine-tuning the talent for cheat-spotting or weakness at the knees at the thought of a mate with status, power, and wealth. Until we are offered some plausible idea about mechanisms, we ought to dismiss these suggestions as vague speculation. The overreaching is hidden only because the latest Darwinizers have learned from the demise of old-style pop sociobiology: Be cagey about genetic hypothesizing!

This is surely simplistic, and evolutionary psychologists ought to repudiate the words we've put into their mouths. A better suggestion would be that the pertinent proteins have lots of different phenotypic consequences, but *the one that matters* concerns the narrowly specified psychological disposition (spotting cheats, swooning appropriately). The claim, then, is that the rival genotypes give rise to phenotypes that differ in lots of ways, but *only the evolutionary psychologist's favorite disposition* makes a serious difference to reproductive success—the rest is a wash.

The fitness contribution of the chosen trait swamps any correlated effects. But, lacking any hints about the underlying genotypes, how their differences might make neural—and therefore psychological—differences, and what impact such *overall* differences might have, there's just no reason to believe that claim. Why should a priori guesses about the nonexistence of correlations with selective significance serve as the basis for evolutionary analysis?

Let's put the point more positively. Forget the fine-grained psychological dispositions for the moment, and ask how natural selection might shape human psychology. Absent revolutionary proposals, the obvious answer is that different genotypes might encode proteins that participate differently in the reactions that underlie neural development, in the formation or pruning of synapses, in the sensitivity to various molecular signals, or in the speed of processes of transmission. It doesn't follow that selective modification of genotypes would affect all aspects of our psychology. But these considerations do suggest the real possibility that psychological phenomena are genetically linked in ways about which we're currently ignorant, so that a particular genetic modification would produce a spectrum of psychological responses, increasing some aspects of human performance and diminishing others. If so, then hunting for the ways in which selection has shaped such fine-grained psychological traits as a disposition to detect cheats is an unpromising strategy, and one can't do any serious Darwinian psychological analysis until there's much greater knowledge of the intricacies of neurodevelopment. Many evolutionary psychologists naively posit their favorite psychological atoms, each under individual selective control and thus each associated with some locus that affects nothing else. This is myth-making, not serious science.

We anticipate a response: "We have to start somewhere. Science must always begin from ignorance, so to demand knowledge at the beginning is antiscience."[11] We acknowledge that no investigation begins from complete knowledge; so much is truism. But well-planned investigations recognize which forms of current ignorance matter and endeavor to ameliorate them, rather than whistling away the complications and hoping that they won't prove significant.

Our review of general issues is intended to highlight the mistakes that attend the recent pop sociobiology of sex and violence. We now turn to the details.

3. Savannah Yearnings: A Romance

The sun is setting, casting a soft bronze glow on the meadow. You, Primeval Pru, realize that you face the hardest decision of your life as a hunter-gatherer: It is time to choose your man. Two stand before you. On the left is a younger man whose deep-set eyes are framed by rich black lashes. His body is unscarred, suggesting that he has not exerted himself much in close encounters with beast or man. But you find it hard to turn your gaze from his warm smile. On the right is an older, balding fellow with plain features and a commanding manner. He gestures to his impressive hut and his collection of animal skins. Whom should you pick?

David Buss knows. He has a theory of evolved mate selection in humans—his "Sexual Strategies Theory"—which informs us as to what Primeval Pru and her contemporary descendants will do (or, more exactly, what Primeval Pru would have done if she has a lot of contemporary descendants). This "theory" is best conceived as an amalgam of claims about mate selection, all of which rely on the same few fundamental tenets. The basic principle from which Buss generates his conclusions (as Thornhill and Palmer after him) is that "the sexes will differ in precisely those domains in which women and men have faced different sorts of adaptive problems."[12] The pertinent evolutionary pressures are supposed to have operated during the "environment of evolutionary adaptedness" (EEA), apparently the Pleistocene, when our ancestors lived as hunter-gatherer groups.

Here's the story. Men's and women's roles in reproduction are asymmetrical in three different ways. First men, but not women, face "parental uncertainty." Second, women are fertile for a smaller portion of their lives than are men. Third, women invest considerably more in reproduction than do men. Following many other pop sociobiologists, Buss waxes lyrical about the contrast between the roughly 450 nutrient-loaded gametes that a woman will produce in a lifetime and the millions of tiny mobile gametes in a single male ejaculate (replenished, as he points out, at a rate of about twelve million an hour). After conception, a woman is also committed to nine months of pregnancy, and after birth, only she can lactate and thus provide milk for the offspring.

These asymmetries create three adaptive problems for men and women. Men will need to increase the probability of paternity and to identify female reproductive value (which peaks in a woman's mid-teens when she has all of her fertile years before her).[13] Women will need to find men who can provide them with resources and defend them and their children against predators and human aggressors. Natural selection will thus select for psychological dispositions that incline men to sexual jealousy, that will prompt them to take advantage of whatever opportunities they have for a quick copulation on the side, and that lead them to be attracted to women with the signs of peak reproductive value—full lips, clear eyes, lustrous hair, a bouncy gait (all these figure in Buss's catalogue, as does a waist-hip ratio of roughly 0.7). Similarly, selection will favor women whose psychological dispositions lead them to be attracted to older men (men with power and resources) and that make them less inclined to wander.

So much for the Darwinian "expectations." Now for the data. To his credit, Buss has carried out an extensive survey in which questionnaires were administered to members of 37 cultures in 33 countries. Besides asking for biographical information (age, sex, religion, etc.) the questionnaires contain queries about mate preferences, first in the form of open-ended questions and then by means of rating and ranking tasks. The open-ended part requires the subject to state the age at which he or she wishes to marry, the age difference the subject would prefer to exist between the subject and the subject's spouse, and the number of children desired. The second part of the first instrument requires respondents to rate 18 characteristics (such as earning capacity, ambition/industriousness, youth, physical attractiveness, and chastity) based on how "important or desirable" each would be in choosing a mate. The respondent must give a numerical rating on a scale from 0 to 3, ranging from "irrelevant or unimportant" (0) to indispensable (3). The second instrument asks subjects to rank 13 characteristics, based on their desirability in a mate. Ten thousand and forty-seven (10,047) subjects were included in the study.[14]

Buss reports that the results accord with his Darwinian expectations. For 36 of 37 samples, there's a statistically significant difference showing that women rate "good financial prospect" higher than do men. In 29 of 37 samples, there's a statistically significant difference with respect to ambition/industriousness (women rating it more highly), and in 34 samples there's a statistically significant difference with respect to physical attractiveness (men rating it as more important). Averaged over all samples, women responded that they prefer men who are 3.42 years older than themselves, while men answered that they prefer women who are 2.66 years younger.[15]

Although his study is the centerpiece of his evidence, Buss defends his "Sexual Strategies Theory" with other considerations more squarely in the pop sociobiological tradition.

A comparison of the statistics derived from personal advertisements in newspapers reveals that a man's age has a strong effect on his preferences. As men get older, they prefer as mates women who are increasingly younger than they are. Men in their thirties prefer women who are roughly five years younger, whereas men in their fifties prefer women ten to twenty years younger. He also reminds us of the familiar male pride in "conquests" and "notches on the belt," which he views as signaling an adaptation to brief sexual encounters.[16] A favorite tale of the differences in "short-term mating strategies" stems from an experiment conducted on a college campus: an "attractive person" approaches a member of the opposite sex and issues a sexual invitation; 100 percent of the women declined, 75 percent of the men accepted.[17]

So there's a clear message for Primeval Pru. Avert your gaze. Forget that smile. Snuggle down with the animal skins.

We disagree. We don't think we know enough to offer Pru any advice at all. In line with the general conclusions drawn in the previous section, we find Buss's claims about the operation of selection naïve and his alleged empirical support questionable. Let's start with the data.

What exactly does Buss's questionnaire measure? Consider first the issue of whether the responses accord with respondents' preferences. Subjects may have beliefs about how they should respond to the questionnnaire, or how those who distribute the questionnaire want them to respond. Although Buss notes that his research assistants did not know his hypotheses, any concordance between his predictions and the stereotypes prevalent in a culture will leave his results vulnerable to bias, whatever the ignorance of his subjects and those who administer the instruments. Furthermore, even if we neglect possibilities that responses will reflect widespread cultural values, Buss must assume that people have access to their own preferences. Interestingly, he emphasizes that "sexual strategies do not require conscious planning or awareness," so that his faith in the questionnaire has to rest on a nice distinction in typical human levels of awareness: we know our preferences but we don't recognize why we have them.[18] As we'll note shortly, inquiring what subjects would say in explaining their responses might well prove illuminating. An even more fundamental assumption is that there are such things as stable preferences that endure beyond the situation of answering the questionnaire into the contexts in which people actually make their decisions. A significant tradition of psychological research—pioneered by Walter Mischel over a period of three decades—has produced convincing evidence that many personality traits are situation-specific, and recent data suggests that the same may apply to preferences.[19]

Yet even if we grant that Buss is measuring genuine stable preferences, uncontaminated by cultural norms, the most important question concerns the *content* of these preferences. The connection between "mate choice"—the topic of the various questions and tasks—and sexual attraction needs scrutiny. Choosing a mate typically means more than picking a sexual partner (or even a reproductive partner), and in many, if not all, of the cultures that figure in Buss's survey, the consequences of mate choice affect many dimensions of the parties' lives. Recall a point from the last section: actual behavior results from the interaction among psychological mechanisms. Assuming that there are such mechanisms, it's only the most simplistic psychology that takes mate choice to reflect the pure operation of the "sexual attraction" mechanism(s). Can we seriously believe that, in societies in which virtually all of a woman's aspirations will be affected by the economic status of the man she marries, the response to questions about "mates" will be unaffected by nonsexual considerations? Buss's brief attempt to confront one instance of this point—his discussion of the hypothesis that women like men with resources because they are cut off from acquiring such resources for themselves—fails to appreciate both the force and the scope of the challenge. Data indicating that successful women have a strong preference for men with resources do not forestall the obvious concern that such women can attain their nonsexual goals, in the kinds of societies in which they live, only by following the culturally approved course for their less fortunate sisters and cousins. Furthermore, the general point is that in all cases libido may run one way and socioeconomic considerations quite another. Indeed, Buss might have found this out had he probed why his respondents gave the answers they did, for their explanations might have shown the various life dimensions along which they viewed mate choice. Perhaps, as Mae West unfortunately did not say, sex has nothing to do with it.

The point we've been developing extends to a broader criticism of Buss's "theory" by exposing its psychological poverty. As we noted above, in any attempt to link hypothetical psychological traits to behavior—even to the relatively special behavior of filling out a questionnaire—one must know how the traits interact and how they are affected by environmental cues. Imagine Buss's hero, Savannah Sam, with wonderfully refined dispositions to react to waist-hip ratio, hair lustre, bounciness in gait, and so forth. If Primeval Pru sets all the sensitivities aquiver, then, provided that no nonsexual disposition interferes (a large assumption), we can expect Sam to court (if that's the right verb) Pru. Sam's alternatives are not likely to be Pru, on the one hand, and Geriatric Georgina on the other. Maybe one of the women Sam confronts is ahead on bounciness and fullness of lips, but another wins on hair lustre and waist-hip ratio. What should the poor lad do? Buss doesn't tell us what the mate choice should be, and this is typical of the looseness of the amalgam of claims he offers. You can predict just about anything you want to from his hypotheses by adjusting the relative strength of the sexual attraction dispositions or by invoking interference from other parts of the psyche.

Does this matter? One might think that Buss has done enough by describing a bundle of psychological traits and that he can leave it to future researchers to decide how these traits interact to produce behavior. Recall, however, that the point of the enterprise was to connect human psychology with evolution under natural selection, and natural selection will presumably discriminate our primeval players on the basis of their behavior. Until we have some idea of how the traits posited will issue in behavior, we can't make any judgment about their selective impact.

The elasticity of the connection between claims and evidence can be illustrated by returning to the proposition-in-the-quad. On the face of it, there's a striking asymmetry in male and female responses to the opportunity for a spot of recreational sex. But what accounts for the difference? Just the firing of the "sexual attraction" disposition in the men and its inhibition in the women? We agree with Natalie Angier's suggestion that the evidence may have more to tell us about women's fears than about their sexual yearnings.[20] Depending on how you adjust the relative strengths of the "attraction disposition" and the "fear disposition" you can predict the data from any hypothesis you choose about asymmetries in male-female sexual desire. Buss's favorite has no special privilege.

Even though we think that Buss's arguments from the data he assembles have the flaws to which we've pointed, we see his search for empirical evidence as an improvement in the customs of pop sociobiology. We can't be so positive about his Darwinizing. Consider his claim that "over a one-year period, an ancestral man who managed to have short-term sexual encounters with dozens of women would likely have caused many pregnancies."[21] A little sober physiology will show that there's a 1 to 2 percent chance of producing offspring per copulation. If Savannah Sam manages one-shot sex with one hundred different women, he may produce two offspring. His enduring evolutionary contribution will, of course, depend on whether these children survive (with whose support, exactly?). Even though one might wonder just what the expected reproductive success might be, it's important to recall that significant evolutionary change can occur when selection pressures are very small (of the order of 0.001, for example). So Sam's modest chances may make a crucial difference.

At just this point, however, the EEA fades into a rosy blur. Sam is supposed to be competing with other aggressive males for the chance to copulate. Some of his female targets may have long-term mates, primed (we recall) to be on the watch for lowered paternity certainty. The females themselves (we remember) are supposed to be less-than-completely interested in casual sex, so Sam is going to have to do a fair bit of talking before they go off with him for a romp in the bushes (but stay tuned! late-breaking news from Thornhill and Palmer suggests that talk may not be needed!). So let's ask the obvious questions: How big is the population to which Sam belongs? To what extent is it possible for his rendezvous to go undetected by others? In what percentage of the pregnancies he brings about will the child receive biparental support? What's the chance of surviving to sexual maturity without biparental support? It may spoil the fun to raise these questions, but until they've been answered there's no way of telling whether Sam's ventures in sperm-spreading will prove selectively advantageous (or disastrous). To put it bluntly, we have to do some delicate accounting to decide if the expected increase in reproductive success is outweighed by the expected effects on Sam of the reactions of those around him to his activities. Any serious exploration of the operations of natural selection must make definite assumptions about what strategies are available to the organisms involved and what ecological constraints affect the reproductive payoffs.

One fundamental oversight of many misadventures in pop sociobiology (and its recent offshoots) is their neglect of within-group differences in strategies. Back to Primeval Pru. If (as Buss and others suggest) ancestral societies were pyramidal with a few men in power and many more scrambling underneath, it's not entirely obvious that being attracted to the Big Man with the Resources is a good female strategy. Maybe there's too much competition there, and Pru would do better to latch on to Mid-Level Mel. (Similarly, if all the males are drooling over Pru, Sam may do better to respond to the maternal promise of Plain Jane across the watering hole.) Pru needs enough to support herself and the kids, but that doesn't mean she'll be at an advantage if she goes for power, age, and the big bucks. If she's good at spotting talent, then Energetic Ernie—nothing but promise but nothing but promise!—would be a better bet. These are only *possibilities*, but they are rival accounts of selection that must be explored, not simply neglected. We leave as exercises to the reader the construction of formal models that will yield any number of different "Darwinian expectations,"[22] although we're prepared to concede to Buss the banal point that in none of these will Pru find Doddering Dan the Deadbeat the lodestone of her life.

We'll close our critique of Buss by pointing out how his conclusions, allegedly generated from Darwinian analyses of life in the EEA are, in fact, used as premises in ameliorating his ignorance about ancestral environments and their demands. Consider the following claims that are typical of Buss's efforts in evolutionary analysis:

> Women over evolutionary history could often garner far more resources for their children through a single spouse than through several temporary sex partners.[23]
>
> A lone woman in ancestral environments may have been susceptible to food deprivation. She may also become a target for aggressive men.[24]

The second is cagey enough, but he quickly slides from the cautious "may" in order to argue that ancestral women would need the protection and support of mates. So in both instances we have definite pronouncements about the challenges of the EEA. Intriguing and informative pronouncements.

In fact, current researchers know very little about the EEA—or even whether there's some privileged time period on which we should concentrate in understanding the evolutionary origins of human psychological tendencies. Should we even be concerned with selection on our hunter-gatherer ancestors rather than considering primate evolution on the one hand, and more recent gene-culture coevolution on the other? But Buss has a simple way of overcoming his ignorance. Consider his defense of the idea that paternity uncertainty was a problem for ancestral men: "Behavioral, physiological, and psychological clues point powerfully to a human evolutionary history in which paternity uncertainty was an adaptive problem for men."[25] So here's the argument. We know that current preferences and propensities are actually adaptations because we can identify them as selectively advantageous in the EEA. And we recognize the selective advantages by drawing conclusions about the EEA on the basis of our knowledge that those current preferences and propensities are really adaptations. The analysis is viciously circular.

4. The Slavering Beast Within: A Gothic Novella

The most substantial part of Thornhill and Palmer's A *Natural History of Rape* is its second chapter, in which the authors draw on earlier pop sociobiological discussions of asymmetries in sexual strategies, particularly the work of David Buss. The authors aim to build on those discussions to advance an account of how natural selection underlies many aspects of rape. Thornhill and Palmer are particularly interested in three main points, advanced in the writings we've just reviewed. First, the appropriate female strategy is to be choosy about potential mates. Second, the appropriate male strategy is to try to copulate as much as possible. Third, males have been selected to worry about issues of paternity. From these three points, Thornhill and Palmer draw their central conclusions. Rape should be especially painful to females because their attempts to choose their mates have been subverted. Males should be more inclined to rape because they are primed to copulate even when females are not interested, and, of course, they should be especially tempted by those females who exhibit the signs of high reproductive value (the young with bouncy gait, lustrous hair, and so forth). Males have also evolved to be suspicious of female claims that they have been coerced into copulating (more specifically: men have evolved to suspect the claims made by their mates), and that is why rape laws have taken the historical forms that they have.

So there we have it. An explanation of the principal features of rape by applying sound Darwinian principles. Add on a denunciation of that feminist canard that rape isn't a sexual act—what nonsense!—and we're done.

Well, not quite. What exactly are the Darwinian explanations supposed to be? Let's begin with the fundamental phenomenon. Some men rape women, and sometimes, men rape other men. Why do these acts occur and why do they occur in the contexts they do with a certain distribution of types of victims? Critics of previous sociobiological stories about rape have pointed out that many instances of rape involve as victims girls who haven't yet reached menarche or women who are past menopause. Thornhill and Palmer reply that "younger women are greatly overrepresented and that girls and older women greatly underrepresented in the data on victims of rape."[24] Waiving some concerns that will occupy us later, we note that this evidence seems relevant only to the kinds of questions that occupy Buss: the most it can show is something about the women rapists find most attractive (and, of course, we don't think it shows much about that). The question has been subtly shifted. Given that some men rape—for whatever reasons—why do they tend to rape young women? Answer: men are more likely to be attracted to young women, so whatever it is that impels them to sexual coercion, young women are more likely to be the victims.

We are concerned with two features of this answer. First, we want to note that there's a controversial assumption that the psychology of rape parallels that of consensual sex. The rapist's behavior is seen as the product of a disposition to be attracted toward certain kinds of people, whether or not they are willing, and a disposition to force sex on a particular occasion. There's an obvious alternative psychological hypothesis, one that not only corresponds to many people's introspective awareness but also seems to permeate the folk tales, poetry, dramas, and stories of almost every culture, that views reciprocity as a central feature of sexual attraction. If that alternative hypothesis is right, then the strategy of seeing the rapist as someone whose tendencies to sexual attraction are just like those of any one else of the same sex, with something extra added on, is misguided. We don't know that the hypothesis is true—indeed, we recommend psychological exploration of it—but we don't think it should simply be dismissed without careful consideration.

We'll spend more time on a second issue. In our view, the major question about rape concerns the causes of coercion. At risk of being pedantic, let's aim for maximal clarity on this point. Imagine two stylized situations. In the first, a man (Adam) is attracted to a woman (Eve) and makes her a sexual proposition. Eve demurely declines. Adam does not force her (he may try to persuade, but he doesn't coerce). In the second, another man (Tarquin) is attracted to a different woman (Lucretia). Like Eve, Lucretia says "No." Tarquin presses on and eventually forces Lucretia to couple with him. Surely the centerpiece of a Darwinian account of rape should not be a story (a bad story, we've argued) about why Eve and Lucretia are found attractive, but rather an explanation of the difference between Adam and Tarquin. What is it about Adam that makes him hold back when Tarquin uses force?

Thornhill and Palmer don't offer any clear answer to this question. Whether this is because they don't have the issues in focus or because they haven't made up their minds we don't presume to judge. They do tell their readers that there are two different ways to apply Darwinian ideas to the study of rape. The *direct* approach supposes that there are "psychological mechanisms designed specifically to influence males to rape in ways that would have produced a net reproductive benefit in the past" (p. 59). The *by-product* approach proposes that there are a number of psychological mechanisms that have been shaped by natural selection that sometimes combine to trigger an act of rape. In a version of this approach that the authors draw from Donald Symons,[27] the mechanisms hypothesized are "the human male's greater visual sexual arousal, greater autonomous sex drive, reduced ability to abstain from sexual activity, much greater desire for sexual variety per se, greater willingness to engage in impersonal sex, and less discriminating criteria for sexual partners."[28] For reasons we've offered in earlier sections, we doubt that these hypothetical characteristics have been targets of natural selection, but the example does have the virtue of exposing Thornhill and Palmer's intended contrast. On the by-product approach, there's no commitment to supposing that acts of rape enhance (or once enhanced) the reproductive success of the rapist. Maybe there are all these adapted psychological dispositions that sometimes combine in ways that are unfortunate for the rapist (as well as being terrible for the victim).

Thornhill and Palmer don't advance any definite hypotheses about the Adam/Tarquin difference. We'll try to do better. Start with the direct approach. There are two possibilities. Either the adaptation is almost universal among human males or it isn't. On the former assumption, the rape disposition is present in just about every human being with a Y chromosome, and the fact that a lot of men don't engage in rape must be explained by invoking some combination of contextual cues and the inhibiting activity of other psychological dispositions. Plainly there's not going to be a lot of direct data to support this hypothesis until we've been told a lot more about possible cues and interactions. But maybe we can get some clues by thinking about the past action of natural selection.

Here's the simplest story. Males have been programmed to rape when they have a chance for copulating with a potentially fertile female and they can get away with it. If there were genetic variation in some savannah population with respect to the disposition to use force, so that most of the male population never engaged in sexual coercion while occasional mutants would rape fertile females only under conditions in which they incurred no costs, then the mutants would have slightly higher expected reproductive success (alternatively, we might suppose a disposition to use force only when the expected costs are lower than the expected reproductive benefits). At this point, everything depends on the details. As we noted in the last section, the chance that a copulation will lead to a birth is 1–2 percent (a figure with which Thornhill and Palmer[29] seem to agree), and this figure has to be discounted by the chance that the child will be abandoned, die before attaining puberty, or simply be ill-prepared for a successful reproductive future. Equally, we need a sober evaluation of the potential costs of an act of rape. Under what conditions, if any, in the savannah environment, could a rapist be expected to recognize that the chances of physical injury from other hominids were sufficiently low that the small benefit of forcing a copulation outweighed the expected costs? Again, we leave to the reader the exercise of constructing formal models that show rampant rape, a low incidence of rape, or no possibilities for the aspiring rapist. Hint: it's simply a matter of adjusting group size, daily habits, social structures, and aggressive tendencies.

The natural selection of the rape disposition is, of course, mediated by that remarkable mutant genotype that expresses itself in just the tendency to coerce copulation in the face of female reluctance when the circumstances are right (or whose effects on fitness are only so mediated). We harbor doubts about that genotype just as we are doubtful that some (or all) of us carry a genotype that enabled our Pleistocene ancestors to stand firm and pick an extra berry or two just when a lion was sufficiently far off to let them garner a small nutritive benefit without cost.

As we acknowledged, the story we've been telling is the simplest version of the universal variant of the direct approach. One embarrassing feature of our tale is that it fails to account for the difference between Adam and Tarquin—there are many Adams who seem to pass up opportunities that Tarquins exploit. Plainly, we need some epicycles, another psychological disposition or two to explain Adam's undue reticence or Tarquin's lack of proper caution. We'll also have to face up to the fact that rape victims are sometimes young girls or older women, so there'll have to be other causal factors that make the tendency to rape misfire. Of course, as we build these in, we'll have to be very careful that we don't subvert whatever story we've been telling about the advantages in the ancestral environment; it will, for example, be disastrous if the sources of inhibition or excitation might have led our ancestors to actions that incurred great risks of injury (like the mythical Pleistocene berry-picker who tarries an instant too long).

Maybe we can do better by switching to the polymorphic variant of the direct approach. Now we suppose that some men develop the rape disposition and others don't. No problem now with explaining the difference between Adam and Tarquin: Tarquin has it, Adam doesn't. The challenge this time is to conjure up a plausible tale about the way in which natural selection on our ancestors produced this poly-morphism. Here's one way to try. Suppose that all males share a conditional disposition: if one experiences one type of developmental environment the rape dis-position develops, if one experiences a different type of developmental environment it doesn't. Back now to Savannah Sam, first bearer of the mutant allele associated with this conditional disposition. Sam is going to have to have some reproductive edge. If this fails to involve any act of rape on his part, then it's hard to see why the allele should persist in the population. But if Sam's Darwinian advantage is a con-sequence of his developing in the pertinent environment, acquiring the rape dis-position, and going in for a rape or two, then it's hard to see why a *fixed* disposition to acquire the rape disposition, come what may, wouldn't have been equally good. Once again, we urge readers to be imaginative and to construct evolutionary models for their favorite outcomes.

Perhaps the indirect approach will fare better. Indeed, there's a reading of Thornhill and Palmer on which the indirect approach must succeed if the direct approach fails. For, unfortunately, rape happens. The people who commit rape belong to a species that has evolved under natural selection. So, when an act of rape occurs, some combination of psychological features that humans have evolved to have must combine with environmental stimuli to prompt it. A triumph for the Darwinian approach to the human sciences?

Not really. The interpretation we've offered is banal, and would go through equally well whatever human activity—chopstick use or needlepoint, say—we were to consider. If the indirect approach is to vindicate Thornhill and Palmer's adver-tisement that evolutionary theory will guide "the scientific study of life in general and of humans in particular to fruitful ends of deep knowledge,"[30] then it will have to provide something more substantive than the vacuous suggestion that human actions draw on evolved psychological mechanisms. Something more like the version Thornhill and Palmer reconstruct from Symons, perhaps.

Let's assume for the time being that the asymmetries celebrated by Symons, Buss, and Thornhill and Palmer are genuine: males are more inclined to want casual sex than females and so forth.[31] Somehow these differences are supposed to be parlayed into an account of why rape sometimes occurs. So far as we can tell, there's just one option that will serve Thornhill and Palmer's turn. From time to time some men get so overstimulated that they just can't hold back, even though what they go on to do may be maladaptive (as well, of course, as being traumatic for their victims).

It doesn't take much thought to see why so simple a proposal won't do. Without further elaborate psychological hypotheses, we have no reason to reject the apparent evidence that a fair number of men who are as sexually stimulated as those who rape manage to accept a woman's refusal. On the face of it, the difference between Adams and Tarquins isn't simply one of the strength of sexual desire. If Thornhill and Palmer want to argue that appearances are deceptive, then they have a lot of work to do—they would have to show that there is some psychological (or neurophysiological) measure of level of sexual arousal that distinguishes all the rapists from all those men who accept rejection.

So what exactly is the difference between those males who behave like Tarquin and those, equally ardent, who emulate Adam? The obvious suggestion is that there are inhibitory mechanisms whose strength varies between the cases. Can we find any Darwinian clues about what such mechanisms might be? Thornhill and Palmer seem to believe we can. They cite work by "the evolutionary psychologist Neil Malamuth" on reduced sexual restraint. Malamuth, and others, have found that certain kinds of developmental experiences are correlated with an apparent "sexual impulsiveness and risk taking." Apparently "reduced parental investment (resulting from poverty or the absence of the father)" leads to "a male's perception of rejection by potential mates." Allegedly, "men emerge from this background with a perception of reduced ability to invest in women, an expectation of brief sexual relationships with women, a reduced ability to form enduring relationships, a coercive sexual attitude toward women, and an acceptance of aggression as a tactic for obtaining desired goals."[32]

The Darwinian language in the passage from which we have quoted is entirely gratuitous. What the studies reveal is that boys who are brought up in poor environments without a father have a higher tendency to harbor certain attitudes toward women and toward sexual relationships, attitudes that increase the chances that they will force sex. There's no warrant whatsoever for suggesting that this has a lot to do with parental investment or the young men's investment in potential mates. You don't need an evolutionary perspective to discover these attitudes and you don't require an evolutionary perspective to interpret them. The basic point is that there do seem to be variations among males in the mechanisms that inhibit the expression of sexual desire in the face of female reluctance, and by standard psychological studies of rapists, one can find correlations between the relative strength of the inhibitory mechanisms and characteristics of the developmental environment.

Once we've come this far, it's not hard to see that the insistent Darwinizing is at best irrelevant and at worst an obstacle. The fundamental question concerns the complex of psychological attitudes that inhibit, or fail to inhibit, the forcing of sex. If we consider the entire spectrum of rapes, including the rape of children and postmenopausal women, which Thornhill and Palmer consistently downplay, we can reasonably conjecture that the rapist's attitude often fails to acknowledge the victim as a person and sometimes even embodies a deliberate intention to demonstrate that the victim is the object of hostility or contempt. Adam holds back, even in the grip of intense desire because he acknowledges Eve's right to say "No." Tarquin, by contrast, sees Lucretia as less than fully human, or wishes to show his dominance of her, or intends that his rape will serve as an act of revenge. The critical task for a theory of rape is to be able to characterize these attitudes as precisely as possible and to understand how they come about. We are prepared to believe that poverty can breed frustration, that a father's absence and the lack of parental affection can engender tendencies to see others as utensils rather than people. Exploring these psychological issues and the causal relationships they involve is not advanced by the speculative invocations of Darwin that Thornhill and Palmer favor.

But wait! Don't Thornhill and Palmer have a reply to the charges we've leveled? After all, they devote an entire chapter to attacking "the social science explanation of rape," in which they consider, and take themselves to demolish, arguments to the effect that rape is about hostility, dominance, punishment, and the desire for control. Consider the following typical passage.

> Brownmiller (1975) sees rape in large-scale war as stemming in part from the frenzied state of affairs and the great excitement of men who have just forcefully dominated the enemy. That hypothesis predicts that soldier rapists would be indiscriminate about the age of the victims. But they are not; they prefer young women.[33]

The second sentence we've quoted is, we believe, unwarranted. Brownmiller's position, as we would reconstruct it, can be developed as a pair of claims:

1. For whatever reasons (not necessarily the Darwinian tales Thornhill and Palmer borrow from Buss), men are typically more attracted to young women.
2. The coercive expression of sexual desire is the result of a failure in an inhibitory mechanism that can be caused by hostility toward the victim.

So Brownmiller (at least on our reconstruction) would predict both that the frequency of rape would be greater in a situation of war, in which soldiers express hostility towards the victims (and, very probably, their desire to show dominance), and that the distribution of rape victims would be skewed towards younger women.

The logical mistake evident here is common to T&P's other discussions of social scientific hypotheses about rape in general and of feminist proposals in particular. They claim that all kinds of confusions flow from viewing rape "as an act of violence."[34] But the confusions are all Thornhill and Palmer's. Rape is not just about violence: there's a difference between the rapist and the batterer. In our judgement, however, rape isn't just about sex either. If Thornhill and Palmer had seen clearly that they need to account for the difference between Adam and Tarquin, they'd have recognized that other psychological mechanisms and attitudes come into play and have appreciated the obvious possibility that, in most instances of rape, motives of aggression and dominance are also present. Further they might have seen that general characteristics of societies are pertinent to the attitudes that adult human beings have toward one another, and in particular to the attitudes that men have toward women. They might then have acknowledged that broad social tendencies can permeate psychological development and lead men to acknowledge women as full persons—or not. The feminist authors who have suggested that prevalent cultural images of women are relevant to how a woman's refusal is heard have a genuine point.[35]

We'll be completely explicit. When rape occurs, there's a sexual dimension to the event. When sexual intercourse is forced, there are typically nonsexual dimensions to the event. The attitudes that lead to the coercive sex often involve intentions to hurt, dominate, humiliate, and obtain revenge. Those attitudes are themselves often present because of a complex developmental history, one that may involve not just details of individual ontogenies (lack of parental affection, for example) but also more general cultural influences that lead men not to see women as full people (but, for example, as collections of salient body parts—genitals, breasts, buttocks, lustrous hair, full lips, and so on).

Let's sum up the discussion of this section. We've examined the two variants of the direct adaptation approach and found that the task of working out a coherent Darwinian model that will fit the evidence is, to say the least, challenging; the challenge is not taken up by Thornhill and Palmer. The by-product approach leads fairly quickly to the sensible proposal that rape occurs when certain inhibitory mechanisms are weakened. Despite their attempts to drag in Darwinian language, T&P fail to show how evolutionary psychology can illuminate the character of these inhibitory mechanisms. Further pursuit of the sensible proposal seems to require research in developmental psychology, and quite possibly elaborations of the social science hypotheses that Thornhill and Palmer deride.

We'll spare the reader an equally extensive treatment of Thornhill and Palmer's two other major claims, the thesis that rape is especially hurtful to women because it subverts their preferred mating strategy and the idea that rape laws reflect male concern with paternity certainty. The analysis of these proposals would proceed on similar lines. Once again, we'd ask just what the selective advantage of intense female pain is supposed to be. Is this a psychological adaptation shared with other primates, or is it part of a female tactic for reassuring Mr. Big Bucks with his refined paternity uncertainties? We'd invite consideration of the hypothesis that people have a general tendencies to feel hurt when they have been used and to expect tenderness and the expression of affection in sexual contact. Similarly, it would be appropriate to ask exactly why attitudes of suspicion toward female testimony are supposed to be adaptive, and to consider the precise costs and benefits of reacting to rape in different ways.

We have offered only hints. Any serious evolutionary account is going to have to advance definite claims about the character of the adaptation, the set of available strategies, and the environment in which selection is alleged to have taken place. This, of course, is what evolutionary theorists do. But Thornhill and Palmer do not live up to the standards of the discipline. Their identification of adaptations is entirely elusive, and there's not a shred of discussion of available strategies (let alone of potential genetic bases for them!) or of the environmental details.

These are harsh words, and we anticipate protests. Surely Thornhill and Palmer do appeal to broad and familiar features of evolution on sexual species, the sexual asymmetries, paternity worries, and so forth that they treat as cardinal dogmas of general evolutionary theory. Isn't it enough to rely on the work of others and to consider ways in which the challenges of natural and sexual selection might be met? No. To make progress in understanding the springs of human behavior, it's necessary to be far clearer about the nature of the selection pressures, the consequences of the allegedly favored strategy and the possible rivals. Thornhill and Palmer tell us nothing specific about the problems that might be addressed by a tendency to rape or by a disposition to feel intense pain at being raped. All their readers get are vague gestures. Such insubstantial suggestions would not be taken seriously in other areas of evolutionary studies. Workers on social insects or sage grouse don't simply talk vaguely about the requirements of obtaining food or avoiding predators; they explore the ecological parameters they take to be significant; they engage in studies to discover the kinds of strategies their organisms can employ; they collect data on reproductive rates. We appreciate the difficulties of meeting such high standards in the study of our own species. But, when the gap between standards and practice is as vast as it is in this discussion of human rape, it's simply false advertising to claim to be in the same business.

5. Conclusion: In Defense of Irreverence

We believe that the studies we have reviewed are scientifically shoddy. But there's surely a fair amount of bad work in the world. Why should people become so upset with the evolutionary psychology of sex and violence, as practiced by Buss, Thornhill, and Palmer? We'll close with a brief attempt at explanation.

It's not incumbent on scientific researchers to offer policy suggestions, but some recent pop sociobiologists—including Thornhill and Palmer—have defended their proposals about human nature by declaring that they can help resolve urgent social issues. Even though we concede that they have good intentions, that they want to help decrease the incidence of rape, it's hard to avoid the judgment that Thornhill and Palmer's suggestions, where not banal, will do little good. Given the speculative character of their Darwinizing and the elusiveness of their proposals, even their inability to recognize crucial issues, policies influenced by their text might well make matters worse.

Consider, for example, their suggestions about educational programs. They begin with a program for boys, agreeing "with social scientists that males should be educated not to use force or the threat of force to obtain sex."[36] No problem so far, but we didn't need any Darwinizing to arrive at this judgment. Keen to show the fecundity of their ideas, Thornhill and Palmer continue with two disastrous further suggestions. First, they propose that educators should explain the differences between male and female sexuality. As we pointed out repeatedly in the last section, even granting the pop sociobiological claims about these differences, the crucial question is why some men (Adams) hold back from forcing women to their desires and others (Tarquins) don't. Any program based on stating "the evolutionary reasons why a young man can get an erection just by looking at a photo of a naked woman."[37] is pointing in the wrong direction and encouraging a view of the springs of rape that may encourage young men to downplay its importance ("Well, it's only human nature after all!"). The critical part of the education, as so many feminists and their social scientific allies have insisted, should be to teach young men that "No" means No, and to help them overcome the kinds of hostility, dominance, and desires for power that are so frequently part of the psychological cause of rape.

A misguided program for boys is bad enough. But Thornhill and Palmer also want a parallel program for girls, pointing out to them the True Nature of the Slavering Beasts with whom they are doomed to reproduce. Young women "should be made aware of the costs associated with attractiveness."[38] Not only is this vulnerable to just the criticisms we directed at the tutorial for boys, but its social consequence is likely to be a continued perception that women are partly responsible for rape ("She was asking for it").[39] Any sensible approach to rape education should be freed from suggestions of female responsibility or complicity, directed toward correcting a problem in male attitudes, clearly demarcated from the expression of some hypothetically universal male sexuality, and firmly linked to a failure in inhibiting mechanisms. Thornhill and Palmer seem to be suggesting an educational program that will reinforce attitudes that ought to be extinguished.

No wonder, then, that they arouse such ire. But we still have told only part of the story. If, as many scholars believe, individual ontogenies are affected by stereotypes in the broader culture, so that male views of women are sometimes shaped by a widespread tendency to reduce them to sexual playthings, then pop sociobiologists don't just ignore crucial causal factors. In their style of analysis, their tendentious talk of "reproductive potential," "investment," "paternity certainty," and so forth, they dehumanize the complex activity of human courtship, love and marriage, embodying in their prose just those images of women as bundles of sexually pertinent body parts—genitals, breasts, lustrous hair, and the rest—that are taken to contribute to the devaluation of women and the incidence of rape. Buss, Thornhill and Palmer and their colleagues give academic respectability to ways of regarding women and of viewing sexual relations that many people see as profoundly damaging, and they do so by using an idiom that portrays women as resources and sex as commerce.

There are self-pitying moments in A Natural History of Rape in which the authors wonder why their work inspires hostile reactions. No prizes for guessing their preferred explanation: they stand in a line of thinkers that extends back to Galileo, a line of fearless revolutionaries dedicated to science and truth. We offer a harsher alternative. They pretend to scientific rigor when they have none; they misunderstand the positions of those whom they lambast; they blunder into sensitive issues, self-righteously offering proposals that it's reasonable to fear will be counterproductive; and they employ language and images that reinforce just those social tendencies their opponents view as crucial factors in producing pain and humiliation for women.

Just as we think the comparison with Galileo inappropriate, we don't recommend that pop sociobiologists be shown the instruments of torture. We think instead that what Thornhill and Palmer, and others of their ilk, merit is a thorough irreverence, born of recognizing that the dignity of academic prose is not in order here. In short, the Bronx cheer.

Notes

We would like to thank Allan Gibbard for helpful conversations, although we are not persuaded by his more positive view of evolutionary psychology; we are also grateful to Patricia Kitcher for some extremely constructive advice about an earlier draft. Jerry Coyne and Richard Lewontin supplied extensive written comments on the penultimate version and have helped us to improve it in a large number of ways; we are deeply indebted to them.

1. Respectively: G. Oster and E. O. Wilson, *Caste and Ecology in the Social Insects* (Princeton, N.J.: Princeton University Press, 1978); G. Parker, "Searching for Mates," in J. R. Krebs and N. Davies, eds., *Behavioral Ecology: An Evolutionary Approach* (Oxford: Blackwell, 1978); T. Clutton-Brock et al., *Red Deer* (Chicago: University of Chicago Press, 1981).

2. E. O. Wilson, *Sociobiology: The New Synthesis* (Cambridge, Mass.: Harvard University Press, 1975); Wilson, *On Human Nature* (Cambridge, Mass.: Harvard University Press, 1978); David Barash, *The Whisperings Within* (London: Penguin, 1979); Pierre van den Berghe, *Human Family Systems* (New York: Elsevier, 1979).

3. For critique, see Richard Lewontin, Steven Rose, and Leon Kamin, *Not in Our Genes* (New York: Pantheon, 1984); Philip Kitcher, *Vaulting Ambition: Sociobiology and the Quest for Human Nature* (Cambridge, Mass.: MIT Press, 1985).

4. Charles Lumsden and E. O. Wilson, *Genes, Minds, and Culture* (Cambridge, Mass.: Harvard University Press, 1981); John Maynard Smith and N. Warren, "Review of *Genes, Minds, and Culture,*" *Evolution* 36 (1982): 620–627; Kitcher, *Vaulting Ambition*, chapter 10.

5. Lewontin et al., *Not in Our Genes*; Kitcher, *Vaulting Ambition*.

6. Randy Thornhill and Nancy Thornhill, "The Evolutionary Psychology of Men's Sexual Coercion," *Behavioral and Brain Sciences* 15 (1992): 365–375.

7. Randy Thornhill and Craig Palmer, *A Natural History of Rape* (Cambridge, Mass.: MIT Press, 2000).

8. A classic source of the modular approach to the mind is Jerry Fodor, *The Modularity of Mind* (Cambridge, Mass.: MIT Press, 1981). Whether Fodor would recognize the use that evolutionary psychologists make of his ideas is quite another matter. But many of the most influential writings in evolutionary psychology, particularly the articles of Leda Cosmides and John Tooby, do champion the Fodorian notion of module as an "informationally encapsulated psychological subsystem." The terminology is much less evident in the authors whose views we discuss here, although they share the common evolutionary psychological strategy of atomizing the mind into parts that are taken to be under independent selective control. We'll henceforth avoid the technical term *module*.

9. Robert Boyd and Peter Richerson, *Culture and the Evolutionary Process* (Chicago: University of Chicago Press, 1985).

10. In a rather uninformed discussion of culture and its impact on behavior, Thornhill and Palmer (see *Natural History of Rape*, 27) show that they do not really understand the work of Boyd and Richerson. They show a similar lack of comprehension in lumping the recent group selectionist proposals of Elliott Sober and David Sloan Wilson with oder views that have been decisively discredited (ibid., 6). It strikes us as odd that authors who are so keen to introduce an evolutionary perspective into the social sciences should be so superficially informed about theoretical issues pertaining to evolution.

11. See Thornhill and Thornhill, "Evolutionary Psychology," 405.

12. David Buss, "Psychological Sex Differences: Origins through Sexual Selection," *American Psychologist* 50 (1995): 164–168, at page 164.

13. David Buss, "Sex Differences in Human Mate Preferences: Evolutionary Hypotheses Tested in 37 Cultures," *Behavioral and Brain Sciences* 12 (1989): 1–49.

14. Ibid.

15. Ibid.

16. David Buss, *The Evolution of Desire* (New York: Basic Books, 1994), 52, 77.

17. David Buss, *Evolutionary Psychology: The New Science of the Mind* (Boston: Allyn and Bacon, 1999), 161. The observant reader will note that there's a slight problem in Buss's coopting this experiment for his own purposes, since the point of his investigations is to *discover* what kinds of people men and women find attractive. The experiment was, however, carried out (by Clarke and Hartfield) on the basis of a prior estimate of attractiveness. But we let this pass.

18. See Buss, "Mate Preferences Mechanisms: Consequences for Partner Choice and Intrasexual Competition," in J. Barkow et al., eds., *The Adapted Mind* (New York: Oxford University Press, 1992), 249–266, at p. 253.

19. Walter Mischel, *Personality and Assessment* (New York: Wiley, 1968); D. A. Moore, "Order Effects in Preference Judgments: Evidence for Context Dependence in the Generation of Preferences," *Organizational Behavior and Human Decision Processes* 78 (1999): 146–165.

20. Natalie Angier, *Woman: An Intimate Geography* (Boston: Houghton Mifflin, 1999).

21. Buss, *Evolutionary Psychology*, 162.

22. See Kitcher, *Vaulting Ambitions*, 170–171, for some straightforward ways of replacing casual speculations about sexual strategies with the kinds of models that are constructed in competent evolutionary studies.

23. Buss, *Evolution of Desire*, 23.

24. David Buss, "The Psychology of Human Mate Selection," in C. B. Crawford and D. L. Krebs, eds., *Handbook of Evolutionary Psychology* (Mahwah, N.J.: Erlbaum, 1998), 405–429, at page 416.

25. David Buss, "Paternity Uncertainty and the Complex Repertoire of Human Mating Strategies," *American Psychologist* 51 (1996): 161–162, at page 161.

26. Thornhill and Palmer, *Natural History of Rape*, 72, drawing on Randy Thornhill and Nancy Thornhill, "Human Rape: An Evolutionary Analysis," *Ethology and Sociobiology* 4 (1983): 137–173.

27. Donald Symons, *The Evolution of Human Sexuality* (New York: Oxford University Press, 1979), 264–267.

28. Thornhill and Palmer, *Natural History of Rape*, 62.

29. Ibid., 100.

30. Ibid., 3

31. Ibid., 62.

32. Ibid., 68–69.

33. Ibid., 134. Citation to Susan Brownmiller, *Against Our Will: Men, Women, and Rape* (New York: Simon and Schuster, 1975).

34. Thornhill and Palmer, *Natural History of Rape*, 136ff.

35. Perhaps there's a more charitable interpretation of Thornhill and Palmer, one that sees them as recognizing the fact that rape isn't only about sex or only about aggression (power, dominance, etc.). Perhaps Thornhill and Palmer and the feminists they criticize can agree on rejecting both polar positions (rape is a matter of sex alone, rape is a matter of aggression alone). We think that the constant emphasis on sexual strategies shaped by selection, and the failure to distinguish the question of explaining the characteristics of rape victims from the question of distinguishing between Adam and Tarquin makes any such interpretation unlikely. Authors with the more charitable interpretation clearly in view would have written a very different book.

36. Thornhill and Palmer, *Natural History of Rape*, 171.

37. Ibid., 179.

38. Ibid., 181.

39. As Dick Lewontin pointed out to us, this phrase needs careful consideration. Sometimes women do dress in ways that they hope will lead men to find them desirable. But surely
these women do not want the male desires to lead to sexual coercion. Educational programs
should surely be very clear about the difference between the desire to be desired and the
desire to be attacked.

17

Born-Again Creationism (2002)

1. The Creationist Reformation

In the beginning, Creationists believed that the world was young. But Creation "science" was without form and void. A deluge of objections drowned the idea that major kinds of plants and animals had been fashioned a few thousand years ago, and hardly modified since. Then the spirit of piety brooded on the waters and brought forth something new. "Let there be design!" exclaimed the reformers—and lo! there was Born-Again Creationism.

Out in Santee, California, about twenty miles from where I used to live, the old movement, dedicated to the possibility of interpreting Genesis literally, continues to ply its wares. Its spokesmen still peddle the familiar fallacies, their misunderstandings of the second law of thermodynamics, their curious views about radiometric dating with apparently revolutionary implications for microphysics, the plundering of debates in evolutionary theory for lines that can be usefully separated from their context, and so forth. But the most prominent Creationists on the current intellectual scene are a new species, much smoother and more savvy. Not for them the commitment to a literal interpretation of Genesis with all the attendant difficulties. Some of them even veer close to accepting the so-called "fact of evolution," the claim, adopted by most scientists within a dozen years of the publication of Darwin's *Origin*, that living things are related and that the history of life has been a process of descent with modification. The sticking point for the Born-Again Creationists, as it was for many late-nineteenth-century thinkers, is the mechanism of evolutionary change. They want to argue that natural selection is inadequate, indeed that no natural process could have produced the diversity of organisms, and thus that there must be some designing agent, who didn't just start the process but who has intervened throughout the history of life.

From the viewpoint of religious fundamentalists the Creationist Reformation is something of a cop-out. Yet for many believers, the new movement delivers everything they want—particularly the vision of a personal God who supervises the history of life and nudges it to fulfil His purposes—and even militant evangelicals may come to appreciate the virtues of discretion. Moreover, the high priests of the Reformation are clad in academic respectability, professors of law at UC/Berkeley and of biochemistry at Lehigh, and two of the movement's main cheerleaders are highly respected philosophers who teach at Notre Dame. Creationism is no longer hick, but *chic*.

2. Why Literalism Failed

In understanding the motivations for, and the shortcomings of, Born-Again Creationism, it's helpful to begin by seeing why the movement had to retreat. The early days of the old-style "creation science" campaign were highly successful. Duane Gish, debating champion for the original movement, crafted a brilliant strategy. He threw together a smorgasbord of apparent problems for evolutionary biology, displayed them very quickly before his audiences, and challenged his opponents to respond. At first, the biologists who debated him laboriously offered details to show that one or two of the problems Gish had raised could be solved, but then their time would run out and the audience would leave thinking that most of the objections were unanswerable. In the mid-1980s, however, two important changes took place: first, defenders of evolutionary theory began to take the same care in formulating answers as Gish had given to posing the problems, and there were quick, and elegant, ways of responding to the commonly reiterated challenges; second, and more important, debaters began to fight back, asking how the observable features of the distribution and characteristics of plants and organisms, both those alive and those fossilized, could be rendered compatible with a literal interpretation of Genesis.

Suppose that the earth really was created about ten thousand years ago, with the major kinds fashioned then, and diversifying only a little since. How are we to account for the distributions of isotopes in the earth's crust? How are we to explain the regular, worldwide ordering of the fossils? The only Creationist response to the latter question has been to invoke the Noachian deluge: the order is as it is because of the relative positions of the organisms at the time the flood struck. Take this suggestion seriously, and you face some obvious puzzles: sharks and dolphins are found at the same depths, but, of course, the sharks occur much, much lower in the fossil record; pine trees, fir trees, and deciduous trees are mixed in forests around the globe, and yet the deciduous trees are latecomers in the worldwide fossil record. Maybe we should suppose that the oaks and beeches saw the waters rising and outran their evergreen rivals?

Far from being a solution to Creationism's problems, the flood is a real disaster. Consider biogeography. The ark lands on Ararat, say eight thousand years ago, and out pop the animals (let's be kind and forget the plants). We now have eight thousand years for the marsupials to find their way to Australia, crossing several large bodies of water in the process. Perhaps you can imagine a few energetic kangaroos making it—but the wombats? Moreover, Creationists think that, while the animals were sorting themselves out, there was diversification of species within the "basic kinds"; jackals, coyotes, foxes, and dogs descend, so the story goes, from a common "dog kind." Now despite all the sarcasm that they have lavished on orthodox evolutionary theory's allegedly high rates of speciation, a simple calculation shows that the rates of speciation "creation science" would require to manage the supposed amount of species diversification are truly breathtaking, orders of magnitude greater than any that have been dreamed of in evolutionary theory. Finally, to touch on just one more problem, Creationists have to account for the survival of thousands of parasites that are specific to our species. During the days on the ark, these would have had to be carried by less than ten people. One can only speculate about the degree of ill-health that Noah and his crew must have suffered.

A major difficulty for old-style Creationism has always been the fact that very similar anatomical structures are co-opted to different ends in species whose ways of life diverge radically. Moles, bats, whales, and dogs have forelimbs based on the same bone architecture that has to be adapted to their methods of locomotion. Not only is it highly implausible that the common blueprint reflects an especially bright idea from a designer who saw the best ways to fashion a burrowing tool, a wing, a flipper, and a leg, but the obvious explanation is that shared bone structure reflects shared ancestry. That explanation has only been deepened as studies of chromosome banding patterns have revealed common patterns among species evolutionists take to be related, as comparisons of proteins have exposed common sequences of amino acids, and, most recently, as genomic sequencing has shown the affinities in the ordering of bases in the DNA of organisms. Two points are especially noteworthy. First, like the anatomical residues of previously functional structures (such as the rudimentary pelvis found in whales), parts of our junk DNA have an uncanny resemblance to truncated, or mutilated, versions of genes found in other mammals, other vertebrates, or other animals. Second, the genetic kinship even among distantly related organisms is so great that a human sequence was identified as implicated in colon cancer by recognizing its similarity to a gene coding for a DNA repair enzyme in yeast. The evidence for common ancestry is so overwhelming that even the Born-Again Creationist Michael Behe is moved to admit that it is "fairly convincing" and that he has "no particular reason to doubt it."[1] (Notice that Behe doesn't quite commit himself here—in fact, to use an example from Richard Dawkins that Behe and others have discussed, there's an obvious line to describe Behe's phraseology: METHINKS IT IS A WEASEL.)

Imagine Creationists becoming aware, at some level, of this little piece of history, and retreating to the bunker in which they plot strategy. What would they come up with? First, the familiar idea that the best defense is a good offense: they need to return to the tried-and-true, give-'em-hell, Duane Gish fire-and-brimstone attack on evolutionary theory. Second, they need to expose less to counterattack, and that means giving up on the disastrous "Creation model" with all the absurdities that Genesis-as-literal-truth brings in its train; better to make biology safe for the central tenets of religion by talking about a design model so softly focused that nobody can raise nasty questions about parasites on the ark or the wombats' dash for the Antipodes. Third, they should do something to mute the evolutionists' most successful arguments, those that draw on the vast number of cross-species comparisons at all levels to establish common descent; this last is a matter of some delicacy, since too blatant a commitment to descent with modification might seem incompatible with Creative Design; so the best tactic here is a carefully choreographed waltz—advance a little toward accepting the "fact of evolution" here, back away there; as we shall see, some protagonists have an exquisite mastery of the steps.

Surprise, surprise. Born-Again Creationism has arrived at just this strategy. I'm going to look at the two most influential versions.

3. The Hedgehog and the Fox

Isaiah Berlin's famous division that contrasts hedgehogs (people with one big idea) and foxes (people with lots of little ideas) applies not only to thinkers but to Creationists as well. The two most prominent figures on the Neo-Creo scene are Michael Behe (a hedgehog) and Philip Johnson (a fox), both of whom receive plaudits from such distinguished philosophers as Alvin Plantinga and Peter van Inwagen. (Since Plantinga and van Inwagen have displayed considerable skill in articulating and analyzing philosophical arguments, the only charitable interpretation of their fulsome blurbs is that a combination of *Schwärmerei* for Creationist doctrine and profound ignorance of relevant bits of biology has induced them to put their brains in cold storage.) Johnson, a lawyer by training, is a far more subtle rhetorician than Gish, and he moves from topic to topic smoothly, discreetly making up the rules of evidence to suit his case as he goes. Many of his attack strategies refine those of country-bumpkin Creationism, although, like the White Knight in *Alice*, he has a few masterpieces of his own invention.

Behe, by contrast, mounts his case for Born-Again Creationism by taking one large problem, and posing it again and again. The problem isn't particularly new: it's the old issue of "complex organs" that Darwin tried to confront in the *Origin*. Behe gives it a new twist by drawing on his background as a biochemist and describing the minute details of mechanisms in organisms so as to make it seem impossible that they could ever have emerged from a stepwise natural process.

4. Behe's Big Idea

Here's the general form of the problem. Given our increased knowledge of the molecular structures in cells and the chemical reactions that go on within and among cells, it's possible to describe structures and processes in exceptionally fine detail. Many structures have large numbers of constituent molecules and the precise details of their fit together are essential for them to fulfil their functions. Similarly, many biochemical pathways require numerous enzymes to interact with one another, in appropriate relative concentrations, so that some important process can occur. Faced with either of these situations, you can pose an obvious question: how could organisms with the pertinent structures or processes have evolved from organisms that lacked them? That question is an explicit invitation to describe an ancestral sequence of organisms that culminated in one with the structures or processes at the end, where each change in the sequence is supposed to carry some selective advantage. If you now pose the question many times over, canvass various possibilities, and conclude that not only has no evolutionist proposed any satisfactory sequences, but that there are systematic reasons for thinking that the structure or process could not have been built up gradually, you have an attack strategy that appears very convincing.

That, in outline, is Behe's big idea. Here's a typical passage, summarizing his quite lucid and accessible description of the structures of cilia and flagella:

As biochemists have begun to examine apparently simple structures like cilia and flagella, they have discovered staggering complexity, with dozens or even hundreds of precisely tailored parts. It is very likely that many of the parts we have not considered here are required for any cilium to function in a cell. As the number of required parts increases, the difficulty of gradually putting the system together skyrockets, and the likelihood of indirect scenarios plummets. Darwin looks more and more forlorn.[2]

This sounds like a completely recalcitrant problem for evolutionists, but it's worth asking just why precisely Darwin should look more and more forlorn.

Notice first that lots of sciences face all sorts of unresolved questions. To take an example close to hand, Behe's own discussions of cilia frankly acknowledge that there's a lot still to learn about molecular structure and its contributions to function. So the fact that evolutionary biologists haven't yet come up with a sequence of organisms culminating in bacteria with flagella or cilia might be regarded as signaling a need for further research on the important open problem of how such bacteria evolved. Not so!, declares Behe. We have here "irreducible complexity," and it's just impossible to imagine a sequence of organisms adding component molecules to build the structures up gradually.

What does this mean? Is Behe supposing that his examples point to a failure of natural selection as a mechanism for evolution? If so, then perhaps he believes that there was a sequence of organisms that ended up with a bacterium with a flagellum (say), but that the intermediates in this sequence added molecules to no immediate purpose, presumably being at a selective disadvantage because of this. (Maybe the Good Lord tempers the wind to the shorn bacterium.) Or does he just dispense with intermediates entirely, thinking that the Creator simply introduced all the right molecules de novo? In that case, despite his claims, he really does doubt common descent. Behe's actual position is impossible to discern because he has learned Duane Gish's lesson (Always attack! Never explain!). I'll return at the very end to the cloudiness of Behe's account of the history of life.

Clearly, Behe thinks that Darwinian evolutionary theory requires some sequence of precursors for bacteria with flagella and that no appropriate sequence could exist. But why does he believe this? Here's a simple-minded version of the argument. Assume that the flagellum needs 137 proteins. Then Darwinians are required to produce a sequence of 138 organisms, the first having none of the proteins and each one having one more protein than its predecessor. Now, we're supposed to be moved by the plight of organisms numbers 2–137, each of which contains proteins that can't serve any function, and is therefore, presumably, a target of selection. Only number 1, the ancestor, and number 138, in which all the protein constituents come together to form the flagellum, have just what it takes to function. The intermediates would wither in the struggle for existence. Hence, evolution under natural selection couldn't have brought the bacterium from there to here.[3]

But this story is just plain silly, and Darwinians ought to disavow any commitment to it. After all, it's a common theme of evolutionary biology that constituents of a cell, a tissue, or an organism are put to new uses because of some modification of the genotype. So maybe the immediate precursor of the proud possessor of the flagellum is a bacterium in which all the protein constituents were already present, but in which some other feature of the cell chemistry interferes with the reaction that builds the flagellum. A genetic change removes the interference (maybe a protein assumes a slightly different configuration, binding to something that would have bound to one of the constituents of the flagellum, preventing the assembly). "But, Professor Kitcher [Creos always try to be polite], do you have any evidence for this scenario?" Of course not. That is to shift the question. We were offered a proof of the impossibility of a particular sequence, and when one tries to show that the proof is invalid by inventing possible instances, it's not pertinent to ask for reasons to think that those instances exist. If they genuinely reveal that what was declared to be impossible isn't, then we no longer have a claim that the Darwinian sequence couldn't have occurred, but simply an open problem of the kind that spurs scientists in any field to engage in research.

Behe has made it look as though there's something more here by inviting us to think about the sequence of precursors in a very particular way. He doesn't actually say that proteins have to be added one at a time—he surely knows very well that that would provoke the reaction I've offered—but his defense of the idea that there just couldn't be a sequence of organisms leading up to bacteria with flagella insinuates, again and again, that the problem is that the alleged intermediates would have to have lots of the components lying around like so many monkey-wrenches in the intracellular works. This strategy is hardly unprecedented. Country-bumpkin Creos offered a cruder version when they dictated to evolutionists what fossil intermediates would have to be like: the transitional forms on the way to birds would have to have had half-scales and half-feathers, halfway wings—or so we are told.[4] Behe has made up his own ideas about what transitional organisms must have been like, and then argued that such organisms couldn't have existed.

In fact, we don't need to compare my guesswork with his. What Darwinism is committed to (at most) is the idea that modifications of DNA sequence (insertions, deletions, base changes, translocations) could yield a sequence of organisms culminating in a bacterium with a flagellum, with selective advantages for the later member of each adjacent pair. To work out what the members of this sequence of organisms might have been like, our ideas should be educated by the details of how the flagellum is actually assembled and the loci in the bacterial genome that are involved. Until we know these things, it's quite likely that any efforts to describe precursors or intermediates will be whistling in the dark. Behe's examples cunningly exploit our ability to give a molecular analysis of the end product and our ignorance of the molecular details of how it is produced.

Throughout his book, Behe repeats the same story. He describes, often charmingly, the complexities of molecular structures and processes. There would be nothing to complain of if he stopped here and said: "Here are some interesting problems for molecularly minded evolutionists to work on, and in a few decades time, perhaps, in light of increased knowledge of how development works at the molecular level, we may be able to see what the precursors were like." But he doesn't. He tries to argue that precursors and intermediates required by Darwinian evolutionary theory couldn't have existed. This strategy has to fail because Behe himself is just as ignorant about the molecular basis of development as his

Darwinian opponents. Hence he hasn't a clue what kinds of precursors and inter-mediates the Darwinian account is actually committed to—so it's impossible to demonstrate that the commitment can't be honored. However, again and again, Behe disguises his ignorance by suggesting to the reader that the Darwinian story must take a very particular form—that it has to consist in something like the simple addition of components, for example—and on that basis he can manufacture the illusion of giving an impossibility proof.

Although this is the main rhetorical trick of the book, there are some impor-tant subsidiary bits of legerdemain. Like pre-Reformation Creationists, Behe loves to flash probability calculations, offering spurious precision to his criticisms. Here's his attack on a scenario for the evolution of a blood-clotting mechanism, tentatively proposed by Russell Doolittle:

> Let's do our own quick calculation. Consider that animals with blood-clotting cas-cades have roughly 10,000 genes, each of which is divided into an average of three pieces. This gives a total of about 30,000 gene pieces. TPA [Tissue Plasminogen Activator] has four different types of domains. By "variously shuffling," the odds of getting those four domains together is 30,000 to the fourth power, which is approx-imately one-tenth to the eighteenth power. Now, if the Irish Sweepstakes had odds of winning of one-tenth to the eighteenth power, and if a million people played the lottery each year, it would take an average of about a thousand billion years before *anyone* (not just a particular person) won the lottery. . . . Doolittle appar-ently needs to shuffle and deal himself a number of perfect bridge hands to win the game.[5]

This sounds quite powerful, and Behe drives home the point by noting that Doolittle provides no quantitative estimates, adding that "without numbers, there is no science"[6]—presumably to emphasize that Born-Again Creationists are better scientists than the distinguished figures they attack. But consider a humdrum phe-nomenon suggested by Behe's analogy to bridge. Imagine that you take a standard deck of cards and deal yourself thirteen. What's the probability that you got exactly those cards in exactly that order? The answer is 1 in 4×10^{21}. Suppose you repeat this process ten times. You'll now have received ten standard bridge hands, ten sets of thirteen cards, each one delivered in a particular order. The chance of getting just those cards in just that order is 1 in $4^{10} \times 10^{210}$. This is approximately 1 in 10^{222}. Notice that the denominator is far larger than that of Behe's trifling 10^{18}. So it must be *really* improbable that you (or anyone else) would ever receive just those cards in just that order in the entire history of the universe. But, whatever the cards were, you did.

What my analogy shows is that, if you describe events that actually occur from a particular perspective, you can make them look improbable. Thus, given a description of the steps in Doolittle's scenario for the evolution of TPA, the fact that you can make the probability look small doesn't mean that that isn't (or couldn't) have been the way things happened. One possibility is that the evolution of blood-clotting was genuinely improbable. But there are others.

Return to your experiment with the deck of cards. Let's suppose that all the hands you were dealt were pretty mundane—fairly evenly distributed among the suits, with a scattering of high cards in each. If you calculated the probability of receiving ten mundane hands in succession, it would of course be much higher than the priority of being dealt those very particular mundane hands with the cards arriving in just that sequence (although it wouldn't be as large as you might expect). There might be an analogue for blood-clotting, depending on how many candidates there are among the 3,000 "gene pieces" to which Behe alludes that would yield a protein product able to play the necessary role. Suppose that there are a hundred acceptable candidates for each position. That means that the chance of success on any particular draw is $(1/30)^4$, which is about 1 in 2.5 million. Now, if there were 10,000 tries per year, it would take, on average, two or three centuries to arrive at the right combination, a flicker of an instant in evolutionary time.

Of course, neither Behe nor I knows how tolerant the blood-clotting system is, how many different molecular ways it allows to get the job done. Thus we can't say if the right way to look at the problem is to think of the situation as the analogue to being dealt a very particular sequence of cards in a very particular order, or whether the right comparison is with cases in which a more general type of sequence occurs. But these two suggestions don't exhaust the relevant cases.

Suppose you knew the exact order of cards in the deck prior to each deal. Then the probability that the particular sequence would occur would be extremely high (barring fumbling or sleight of hand, the probability would be 1). The sequence only *looks* improbable because we don't know the order. Perhaps that's true for the Doolittle shuffling process as well. Given the initial distribution of pieces of DNA, plus the details of the biochemical milieu, principles of chemical recombination might actually make it very probable that the cascade Doolittle hypothesizes would ensue. Once again, nobody knows whether this is so. Behe simply assumes that it isn't.

Let me sum up. There are two questions to pose: What is the probability that the Doolittle sequence would occur? What is the significance of a low value for that probability? The answer to the first question is that we haven't a clue: it might be close to 1, it might be small but significant enough to make it likely that the sequence would occur in a flicker of evolutionary time, it might be truly tiny (as Behe suggests). The answer to the second question is that genuinely improbable things sometimes happen, and one shouldn't confuse improbability with impossibility. Once these points are recognized, it's clear that, for all its rhetorical force, Behe's appeal to numbers smacks more of numerology than of science. As with his main line of argument, it turns out to be an attempt to parlay ignorance of molecular details into an impossibility proof.

I postpone until the very end another fundamental difficulty with Behe's argument for design, to wit his fuzzy faith that appeal to a Creator will make all these "difficulties" evaporate. As we shall see, both he and Johnson try to hide any positive views. With good reason.

5. Johnson's Kangaroo Court

Darwin on Trial is a bravura performance by a formidable prosecutor, able to assemble nuggets of evidence and to present them in the most damning fashion. The defense lawyer isn't even court-appointed, simply absent or asleep. I'm going to argue that, when the defense actually shows up, Johnson's apparently devastating attacks turn out to be slick versions of old sophisms.

Unlike Behe, who officially admits the universal relatedness of organisms, Johnson takes some trouble to blur the distinction between the process of descent with modification (the "fact of evolution") and the mechanism that drives the process. Here are some typical passages:

> The arguments among the experts are said to be matters of detail, such as the precise timescale and mechanism of evolutionary transformations. These disagreements are signs not of crisis but of healthy creative ferment within the field, and in any case there is no room for doubt about something called the "fact" of evolution.
>
> But consider Colin Patterson's point that a fact of evolution is vacuous unless it comes with a supporting theory. Absent an explanation of how fundamental transitions can occur, the bare statement that "humans evolved from fish" is not impressive. What makes the fish story impressive, and credible, is that scientists think they know how a fish can be changed into a human without miraculous intervention.[7]

> We observe directly that apples fall when dropped, but we do not observe a common ancestor for modern apes and humans. What we *do* observe is that apes and humans are physically and biochemically more like each other than they are like rabbits, snakes, or trees. The ape-like common ancestor is a hypothesis in a *theory*, which purports to explain how these greater and lesser similarities came about. The theory is plausible, especially to a philosophical materialist, but it may nonetheless be false. The true explanation for natural relationships may be something much more mysterious.[8]

> Paleontologists now report that a *Basilosaurus* skeleton recently discovered in Egypt has appendages which appear to be vestigial hind legs and feet. The function these could have served is obscure. They are too small even to have been much assistance in swimming, and could not conceivably have supported the huge body on land.[9]

Here, and in other places, Johnson confuses the question of whether the history of life shows a process of descent with modification with problems about evolutionary mechanisms, as well as cleverly raising the standards of evidence appropriate for calling something a "fact."

Contemporary evolutionary theorists, notably Stephen Jay Gould, have wanted to distinguish the "fact" of evolution (the universal relatedness of life, the process of descent with modification) from theories about the mechanisms of evolutionary change, precisely because Creationists have exploited debate on the latter issue to cast doubt on the former. The distinction was already clear in the late nineteenth century, when the claim that organisms are related by descent with modification became virtually universally accepted, even though naturalists continued to debate Darwin's preferred account of the causes of evolutionary change. Johnson wants to turn the clock back. His first sally charges that facts are "vacuous" unless they come with supporting theories—and, of course, there's an appeal to authority, thrown in. The word choice is interesting. Does Johnson think that the claim that organisms are related isn't true? Or that it's equivalent to some elementary logical truth (such as "All fish are fish")? The latter is completely implausible. Of course, Johnson would like to say that the claim of descent with modification is incorrect, but, since he can't defend that, he insinuates it by using a negative term.

In fact, scientific claims are often made without "supporting theories." Consider Kepler's laws about planetary orbits. Prior to the articulation of Newtonian theory, were these "vacuous"? Were chemists' proposals about chemical composition "vacuous" before we had detailed accounts of molecules and valences? Or Mendelian claims about hereditary factors in the absence of knowledge that genes are made of DNA? The point derived from Patterson seems straightforwardly false, since it often seems a scientific advance to establish *that* something is the case without being able to say why or how it is so. But Johnson cleverly buttresses his argument by misformulating the claim of common descent—instead of "Humans evolved from fishes", we should have "All living things in the history of life on earth are related through a process of descent with modification."

The next step is to offer an appraisal. Johnson's opines that the doctrine he dislikes isn't "impressive." Again, it's not obvious that the ability to wow Johnson or his Creo friends is the appropriate criterion—shouldn't we be concerned with whether or not the doctrine is *true*? But, of course, it's been made to *seem* less impressive because of the pathetically reduced formulation. Only as an afterthought, does Johnson link the irrelevant "impressiveness" to the pertinent criterion, credibility, and then he garbles the relations of evidence. What makes claims about common descent credible is a wide variety of evidence drawn from comparative anatomy and physiology, comparative embryology, and biogeography—the kind of evidence that clinched the case in the post-Darwinian decades, and that has been extended ever since (most notably in recent biochemical studies)—not any embedding in a theory about the causes of evolutionary change. Precisely the point made by defenders of evolutionary theory like Gould is that we have overwhelming evidence for common descent even though we may debate the mechanisms of evolutionary change.

On to the second version. Here Johnson starts by pouncing on an analogy used by Gould, the comparison of relations of descent to the falling of apples. Of course, it's perfectly correct to point out that we don't *observe* all the intermediates hypothesized by the claim that all organisms are related through descent with modification. However, the fundamental point was to differentiate between parts of science that are so firmly accepted that they are classified as "facts" from parts of science that are more controversial. It's not obvious to me that the fact/theory terminology is the best way of marking this distinction—so Johnson may be justified in criticizing the rhetoric of his opponents. But that's just a preliminary point, and the main issue is whether the evidence for claims of descent is much stronger than that for causal explanations of the processes of modification.

Of course, there are plenty of parts of science that are not directly established by observation in the way that statements about falling apples are, but which nevertheless are counted as so firmly in place that scientists see themselves as building on them, rather than disputing them. Consider the claim that water molecules consist of two hydrogen atoms and one oxygen atom, or the identification of DNA as the molecular basis of heredity. Gould's characterization of common descent as a "fact" is meant to assimilate the thesis of universal relatedness to these scientific claims, to point out that its status is equally secure. If Johnson means to dispute this point, it's useless to note that one can't observe hypothetical intermediates—observation is just as inept to confirm molecular composition as it is to disclose ancestral organisms. What must be done is to show that the evidence in favor of common descent is far flimsier than evolutionary theorists have taken it to be, that it is nowhere near as strong as the support that has been garnered for the proposal that water is H_2O. To do that he has to explain why all those anatomical and physiological similarities, ranging from matters of gross morphology all the way down to molecular minutiae and including the apparently useless and nonfunctional residues of past structures, have been misinterpreted or overinterpreted by the defenders of evolution.

Johnson's half-hearted attempts to do just that are typified by the third passage I've quoted. As he notes, *Basilosaurus* is a sea-dwelling mammal related to whales, and it appears to retain rudimentary limbs. Evolutionary theorists account for the presence of these limbs by supposing that *Basilosaurus* is a modified descendant of some land-dwelling mammal, in whom the limbs were functional. The genetic changes that have taken place along the lineage have modified the body considerably, but the developmental program continues to produce vestigial versions of the structures present in the ancestors: they proclaim the animal's relatedness to land-dwelling forebears.

What Neo-Creos have to do at this point is explain that the vestiges don't signal any relationship to other mammals. So why are they there? What's the nonevolutionary explanation? Johnson doesn't tell us. Instead, he changes the subject, pointing out that the limbs aren't functional. But that wasn't the point at issue—indeed, the nonfunctionality was an indication that the limbs had been carried over from ancestral forms! Johnson has let the argument evolve from a dispute about descent with modification to a debate about the *causes* of evolutionary change, and he irrelevantly chides his opponents for not being able to tell a Darwinian selectionist story for these particular features of *Basilosaurus* and its immediate ancestors.[10]

In the end, then, Johnson's attempt to dispute the "fact of evolution" is an exercise in evasion. When the rhetorical tricks are unmasked, it's clear that he's failed to answer the big question: if organisms aren't related by descent with modification, what's the explanation for all the detailed similarities we find among living things? Yes, indeed, the true explanation for observed relationships might be "more mysterious"—as might the true explanation for the data from which chemists justify their views about the composition of water or of genes—but the mysteries are, apparently, to remain the strict property of Johnson and his cronies.

As a lawyer, Johnson has an excellent understanding of ways in which burdens of proof can be shifted and standards of evidence raised. Here are some samples of his skill:

> The question I want to investigate is whether Darwinism is based upon a fair assessment of the scientific evidence, or whether it is another kind of fundamentalism.
> Do we really know for certain that there exists some natural process by which human beings and all other living beings could have evolved from microbial ancestors, and eventually from nonliving matter?[11]

Archaeopteryx [*sic*] is on the whole a point for the Darwinists, but how important is it? Persons who come to the fossil evidence as convinced Darwinists will see a stunning confirmation, but skeptics will see only a lonely exception to a consistent pattern of fossil disconfirmation. If we are testing Darwinism rather than merely looking for a confirming example or two, then a single good candidate for ancestor status is not enough to save a theory that posits a worldwide history of continual evolutionary transformation.[12]

The first passage frames the issues so as to impose unnecessarily stringent requirements on defenders of evolutionary theory. We start with two options: either the acceptance of evolutionary theory rests on "a fair assessment of the evidence" or it's a "kind of fundamentalism." Strictly speaking, that doesn't exhaust the possibilities, but let that pass. In the very next sentence Johnson transmutes the first option, reformulating it as the requirement that we know *for certain* that some natural process produced people out of microbes. Now, of course, this is focused directly on the issue of the mechanism of evolutionary change and explicitly demands knowledge of the mechanisms that have operated over the entire sweep of evolutionary history, but the most glaring distortion occurs in the talk of certainty. In effect, the choices have been reduced to knowing all the details with certainty and being a fundamentalist, so that no space is left for the thoughtful evolutionary theorist who wants to say "The evidence for the universal relatedness of life is compelling. Further we know of a number of natural processes that have produced evolutionary change. We can't always say for sure which of these has been operative at which stage of the history of life, nor do we know that our inventory of possible mechanisms is complete, but, on the evidence we have, there's no reason to think that any supernatural process was needed in the evolution of organisms." That type of response is analogous to that of the chemist who declares "The evidence for our views about the kinds of bonds that occur between molecules in a vast number of substances is compelling. Further, we know in principle how the distributions of electrons in bonds result from basic principles of quantum mechanics. But we don't know how to solve the Schrödinger equation for any complex molecule, and it may be that our understanding of the microphysics may be limited in various respects. Given the evidence we have, however, there's no reason to think that supernatural processes are needed to keep the constituents of large molecules together." In chemistry, as in evolutionary biology, there are open problems, and while some parts of the science are quite firmly established (on the basis of compelling evidence), the idea that we should claim certainty *überhaupt* is as absurd as the thought that, if we can't do so, we've relapsed into fundamentalism.

The second passage occurs in the middle of a discussion of the fossil record (a discussion I'll treat from a different perspective shortly). After clouding the issues about the reptile-bird transition—mainly by claiming that evolutionists ought to produce fine-grained transitional sequences linking ancestral organisms to all the different species of birds—Johnson concedes, grudgingly, that the existence of *Archeopteryx* is "on the whole a point for the Darwinists." Indeed, since the explicit challenge was to find transitional forms linking major groups, it's hard to see how the production of an intermediate, such as *Archeopteryx*, could fail to meet the challenge: Johnson's strategy is like that of the child who bets his friend that she can't juggle three balls for a minute and then, when she does it successfully, welshes on the bet on the grounds that she didn't do it with her eyes shut. But, after magnanimously conceding that *Archeopteryx* is confirming evidence for the view that reptiles and birds are related by descent, he pooh-poohs the significance of this by suggesting that it's a "lonely exception to a consistent pattern of fossil disconfirmation." Let's formulate Johnson's implicit requirement explicitly: it's the demand that the fossil record would only confirm evolutionary theory if we could discover intermediate forms for every major transition (with Johnson reserving the right to decide which transitions count as "major" and also to demand the fineness of grain of the intermediate sequences). This is as arrogant as a counterdemand to be shown the fingerprint of the Creator in specified domains of the living world. As Darwin well knew, and as our improved understanding of the physics and chemistry of fossilization has shown us ever more clearly, the chances that any given species will be represented in the fossil record is extremely low. Our estimates of those chances are not, as Johnson likes to insinuate, specially cooked to favor evolutionary theory; they are based on independent parts of science. Given those estimates, we'd expect that for many major transitions the hypothesized intermediates would not be found in the fossil record, but, when the transitional fossils do occur, they provide striking confirmation of the claim of descent with modification because, if that claim were not true, the existence of such fossils would be highly improbable.

To see this more clearly, consider an analogy. In building the case against the notorious Moriarty, specifically in order to justify the conclusion that Moriarty visited the scene of the crime, the prosecution appeals to the fact that he was observed, just before the crime was committed, halfway between his lair and the crime scene. The defense responds that there has been no evidence of Moriarty's footprints on the pavement throughout the hundred-yard walk, that nobody saw Moriarty within ten yards of the crime scene, and so forth. The defense lawyer is a studious disciple of Phillip Johnson.

In fact, Johnson is sufficiently uneasy about the fossil evidence to go to considerable lengths to respond to examples on which evolutionary theorists (rightly) place special emphasis. He cites the reptile-mammal transition as the "crown jewel of the fossil evidence for Darwinism."[13] He continues with one of his most accurate condensations of the biology:

> At the boundary, fossil reptiles and mammals are difficult to tell apart. The usual criterion is that a fossil is considered reptile if its jaw contains several bones, of which one, the articular, connects to the quadrate bone of the skull. If the lower jaw consists of a single dentary bone, connecting to the squamosal bone of the skull, the fossil is classified as a mammal.[14]

It might initially appear very difficult for an animal to be "intermediate" between reptiles and mammals, given this criterion in terms of jaw morphology. However, there's a very rich set of fossils showing reduction of the reptilian features and development of the mammalian traits; particularly remarkable are fossils, most famously *Diarthrognathus*, in which both types of jaw-joint are present.

After quoting Stephen Jay Gould's description of the advanced mammallike reptiles, distinguished by the reduction in the quadrate and articular, Johnson comments:

We may concede Gould's narrow point, but his more general claim that the mammal-reptile transition is thereby established is another matter. Creatures have existed with a skull bone structure intermediate between that of reptiles and mammals, and so the transition with respect to this feature is possible. On the other hand, there are many important features by which mammals differ from reptiles besides the jaw and ear bones, including the all-important reproductive systems.[15]

Well, when you can't argue the facts, argue the law. The existence of *Diarthrognathus* and friends shows that transitional forms with respect to jaw morphology *actually appeared* (not just [sniff!] that they were "possible"). So Johnson has to contend that these are irrelevant to the case. He reaches into the Creationist bag of debating tricks for a well-known tactic, that of specifying just the intermediate forms that would satisfy him—he wants to see transitions in the "reproductive system." Clever indeed! For what he wants are the soft bits, the parts that don't have a prayer of being represented in the fossil record.

Now in fact, as Johnson ought to know, there isn't a single mammalian reproductive system—there are monotremes (egg-laying mammals), marsupials, and, most familiar, the placental mammals. So what is actually happening is that the question is being shifted. Instead of asking for an account of the reptile-mammal transition, Johnson is making a much more sweeping demand for a fine-grained sequence of transitional forms *within* mammalia to show the gradual emergence of the placental mammals. What Gould actually claimed to be able to do was to show how a feature shared by all mammals (monotremes and marsupials as well as placentals), the structure of the jaw joint, emerges in the fossil record in a fine-grained transition from the structure found in living and extinct reptiles. However much Johnson might like to invoke the character of the reproductive system as a way of separating the mammals from the reptiles, the criterion of jaw morphology is taxonomically fundamental. mammals are animals that have one jaw structure, reptiles are animals that have a different jaw structure, and the mammallike reptiles are those in which the bones involved in the reptilian jaw are being reduced, the most advanced of them being double-jointed. These criteria aren't pressed into service to save evolutionary theory. They are demanded by the diversity of the mammals with respect to other features. Johnson's revisionary taxonomy would sweep away some of the antipodean mammals.

There are signs that Johnson recognizes that all is not well with his first line of argument, for he follows it up with an alternative. After noting that the fossil record for the reptile mammal transition is so rich that a prominent evolutionary biologist (Douglas Futuyma) suggests that it's impossible to tell which species were the ancestors of modern mammals, Johnson continues

> But large numbers of eligible candidates are a plus only to the extent that they can be placed in a single line of descent that could conceivably lead from a particular reptile species to a particular early mammal species. The presence of similarities in many different species that are outside of any possible ancestral line only draws attention to the fact that skeletal similarities do not necessarily imply ancestry. The notion that mammals-in-general evolved from reptiles-in-general through a broad clump of diverse therapsid lines is not Darwimsm. Darwinian transformation requires a single line of ancestral descent.[16]

The claim of common descent is, apparently, to be defeated by an *embarras de richesse*.

Plainly, Johnson hasn't been reading contemporary evolutionary theory very carefully, for he seems to have overlooked the modern emphasis on a theme (already present in Darwin) that the tree of life turns out to be a bush. Well-documented cases of anagenesis (in which a single lineage is gradually transformed) are quite rare. Paleontological reconstructions typically show modifications associated with (branching) speciation. Hence, in studying the mammallike reptiles, there's no suprise in finding lots of closely related species, not just parents and daughters but sisters and cousins and aunts. Futuyma's point is that there are so many relatives in this family that it's hard to sort out the relationships and, in particular, hard to tell which mammallike reptiles are ancestral to the mammals.

So Johnson is quite wrong in thinking that there has to be a linear sequence linking all these fossils by ancestor-descendant relations: evolutionists would be quite surprised if that were so. But the rhetoric of his case depends on a skillful ambiguity. He insinuates doubts about whether jaw morphology is a reliable guide to relationship by talking of "species outside of any possible ancestral line." The suggestion, of course, is that evolutionists are committed to thinking of some of these species as *unrelated* and that this undermines their claims that anatomical features (like the size and positioning of bones) are a good indicator of evolutionary relationships. But that's completely false. Those who study this transition don't believe that the fossils can be fitted into a single line of ancestors and descendants, but they think of all of them as related. To repeat, there are daughters and sisters and cousins and aunts. The difficulty lies in assigning particular fossils particular degrees of relationship. But that difficulty doesn't interfere with the enterprise of revealing the reptile-mammal transition in the fossil record. Once again, a legal analogy may prove helpful. If the defense denies that any member of the Crebozo gang could have done the dirty deed, and the prosecution shows how Phil, Al, Pete, and Mike (Crebozos all) had motive, means, and access, the general claim that one of the Crebozos is guilty may be established without the prosecution's being able to tell which of the individual thugs delivered the decisive blow.

Johnson's entire book is filled with the sophistries I've been exposing, as he distorts the positions he opposes, shifts standards of evidence, quotes people out of context, and uses ambiguity to cover his argumentative gaps. Any well-trained philosopher with no particular axe to grind and a modest knowledge about evolutionary biology could hardly fail to see that rhetoric substitutes for sound argument on virtually every page—which is why the endorsements of Johnson by Plantinga and van Inwagen are so revealing. I'll close this part of the discussion by looking at one example that ought to strike philosophers as especially egregious, Johnson's invocation of Popper and the famous (infamous?) falsifiability criterion.

Creos old and new love Sir Karl. The country-bumpkin appeal to the falsifiability criterion proposed that the theory of evolution reduced to some principle of natural selection and that this principle turned out to be a tautology ("Those who survive, survive"—to be sung to the tune of *Che sera, sera*). That line of objection has been decisively refuted, both by pointing out that it misunderstands the principle of natural selection and, far more importantly in my opinion, by showing how absurd it is to reduce the theory of evolution to the principle of natural selection. Johnson tries a different tack. He starts from the idea that the theory of evolution is akin to Marx's account of history or Freud's psychoanalysis in making no genuine predictions at all.

There are well-known philosophical objections to so simple-minded an invocation of the falsifiability criterion. Both Freud and Marx offered sufficiently rich bodies of doctrine that they could make predictions about individual events. When their predictions went wrong, they modified their *total* doctrines to preserve their central principles. Johnson points this out and immediately cries "Foul!"[17] He fails to appreciate the fact that almost all scientists spend large portions of their time behaving in similar ways: they set up an experiment in light of their theoretical understanding, find that it fails to work, tinker a bit, revise their views about the situation, and in some instances, make modifications of parts of the underlying theory to protect the central principles. To cite a familiar example, when the orbit of Uranus wasn't as predicted, astronomers didn't abandon Newton's gravitational theory (in favor of what? one wonders), but hunted for a new planet. Popper himself understood the nuances of testing far better than those whom he has inspired, and in the wake of appeals to falsifiability, philosophers of science have shown how the refutation of a theory with broad scope proceeds, not by single decisive experiments but by the accumulation of cases which challenge defenders to find any way of supplementing the central principles.

The history of science is full of theories that came to grief because of the building up of difficulties. Darwin's theory of evolution isn't among the shipwrecks. Johnson charges that this is because "[t]he central Darwinist concept that later came to be called the "fact of evolution"—descent with modification—was thus from the start protected from empirical testing."[18] His false allegation rests on a confusion: the fact that evolutionary theory has survived doesn't entail that it was *bound* to succeed. If we had discovered the world to be different in different respects then evolutionary theory would have been given up because it faced insuperable difficulties.

Suppose we had found that the similarities in structure on which claims of evolutionary relationship are based didn't correlate with observed patterns of descent: we claim that two organisms are descended from a common ancestor because they share some feature and then discover that the feature is present in organisms we know to be unrelated; this happens again and again. Imagine further that, as we move from level to level, the groupings by similarity vary: anatomical and physiological similarities are underlain by very different tissue and cellular types; cellular similarities don't correlate with chromosomal differences; all of the groupings at higher levels are quite different from those we find at the molecular level—humans, for example, turn out to be biochemically very similar to frogs and palm trees and radically different from chimpanzees. Claims about biogeographical distribution consistently founder on the inability of organisms to make the journeys that are hypothesized (recall Darwin's concern with this problem, which drove him to do experiments on the transport of seeds, thus confuting Johnson's claim that he proposed "no daring experimental tests").[19] The fossil record turns out to be chaotic, with strata that are geologically ancient containing samples of all classes of organisms. Fundamental theories in physics imply that the earth is quite young. And so on and on.

My construction of this list is not particularly imaginative, for I've drawn on a large number of the ways in which Creos, including Phillip Johnson, have tried to falsify evolutionary theory. (The example about similarities between frog proteins and human proteins was once a staple of old-time Creationist entertainments; although it's been decisively rebutted, there may be a faint trace in Johnson's remarks about molecular variation in frogs.)[20] The point is that Darwin's evolutionary theory could have gone the way of phlogiston chemistry, the corpuscular theory of light, blending inheritance, the universal ether, stabilist theories of the continents, and many other discarded theories. It didn't, not because evolutionary theorists are stubborn ideologues but because the kinds of observations that would have discredited it (occasionally, but wrongly, hailed as "facts" in the Creationist literature) have not been made. Far from being "vacuous" or "unfalsifiable," evolutionary theory sticks its neck out again and again, denying the copresence of human and dinosaur footprints at Paluxy, predicting the morphology of ancestral ants (subsequently discovered by E. O. Wilson, W. L. Brown, and F. M. Carpenter), ruling out the possibility that the chicken genome is more similar to the human genome than the latter is to the chimpanzee genome, and in a host of further commitments.

Johnson's fondness for claiming that evolutionary theory doesn't belong in the Popperian Temple of Proper Science actually rests on his repeated contention that the theory is insulated against one particular type of problem, the absence of transitional forms in the fossil record. Here's a typical version of the charge.

> If Darwin had made risky predictions about what the fossil record would show after a century of exploration, he would not have predicted that a single "ancestral group" like the therapsids and a mosaic like *Archaeopteryx* would be practically the only evidence for macroevolution. Because Darwinists look only for confirmation, however, these exceptions look to them like proof.[21]

As I've already maintained, evolutionary theory does make a large number of predictions about the character of the fossil record, notably that fossils will occur in a particular order in strata world-wide. But what Johnson wants is a particular kind of prediction. If all these intermediates actually existed, he claims, we ought to find traces of them, and it's anti-Popperian dogmatic weaseling to be complacent when those traces aren't found. But this is simply to ignore an independently confirmed chunk of science, the geophysics and geochemistry of fossilization. If you couple evolutionary theory with an extremely bad theory, the Panglossian theory of fossilization, whose claim is that every organism that has ever existed has a chance greater than 1/100 of being fossilized, then, of course, evolutionary theory predicts that the fossil record should contain myriad intermediates that we don't find. But the fact that a theory conjoined with a very bad auxiliary theory predicts lots of falsehoods cuts no ice whatsoever. What Johnson ought to do is to show that evolutionary theory plus a realistic theory of fossilization predicts the existence of far more intermediate forms than we've yet found.

To the best of my knowledge, nobody has done that. In principle, one could use our understanding of the vicissitudes of fossilization to compute the chances of finding remains of various types, calculate estimates of population sizes, and then use our best account of phylogenies to arrive at probabilities that the fossil record will reveal a particular sort of distribution. Whether or not that can be done in practice is unclear, for the information currently available might be too limited to narrow the range of probabilities. But if this is the sort of prediction that Johnson demands (ignoring the other kinds I've mentioned), then he could try to work out the details in a serious and responsible fashion. Of course, it's a lot quicker—and probably more to his taste—to invoke the Panglossian theory and damn evolutionary theory unfairly.

I'll illustrate his tactic with another analogy. Consider the theory that the texts we print in volumes with "Holy Bible" on the spine have been produced by a historical process in which ancient manuscripts in various languages have been copied and recopied, translated and retranslated. Imagine a fringe sect that challenges this theory, holding that alleged similarities among texts are based on misunderstandings of the ancient languages, that the works we have are fictions composed in the late Middle Ages, and that there were no intermediates linking them back to the ancient world. Phillip Johnson, quite rightly, is moved to answer this criticism, but, in the ensuing debate, he finds himself facing the following objection. "Where are all those intermediate texts, Professor Johnson? Surely your theory predicts that the historical record should be full of copies showing the ways in which ancient versions have been successively amended. We know that it isn't. Are you going to desert your Popperian principles and ignore this falsification?" We can easily guess how Johnson would reply. He'd point out that libraries are often looted and burned, that texts are lost and thrown away, that vandalism was omnipresent in the ancient world, and so forth. He'd accuse his opponents of foisting on to him an absurd auxiliary theory about historical preservation. And he'd be right. Yet the sophistries of his challengers are exactly his own.

6. Where's the Beef?

I come at last to the most basic difficulty with the Neo-Creo attack, its dim suggestions that the scientific world needs a shot of supernaturalism. The Born Again Creationists tread different paths to a common destination. Whether hedgehogs or foxes, they conclude that evolutionary theory is beset by problems—one very deep and systematic problem for Michael Behe, a whole scatter of troubles for Phillip Johnson—and they portray the establishment as dogmatic in its insistence on excluding creative design: given that the going story of life and its history is such a shambles, why are these evolutionists so obstinate in thinking that some "purely naturalistic process" produced people? When this conclusion is made explicit, there's a natural question to pose to the Neo-Creos. How exactly is the appeal to creative design supposed to help?

I've been contending throughout that the charges of "insoluble problems" are wildly overblown. But let's play along for a bit. Consider the difficulties that Behe and Johnson cite and suppose that they really do need to be addressed. Why should we think that invoking creative design, with all its theological resonances, is just the ticket for solving them? Behe and Johnson don't say. They've learned from the failures of pre-Reformation Creationism, and they know much, much better than to put their literalist cards out on the table. Fine. But we ought to be a little curious about what sort of magic a creative design model might be able to work.

Let's start with Behe and concede to him that we haven't a clue about how you can produce the bacterial flagellum or the clotting cascade in small steps. We might think we'd get some clues once developmental molecular genetics has developed a bit, but maybe Behe has a plausible proposal that will save us the wait and the trouble. What could it be? Well, it has to involve creative design, so we can assume that the unbridgeable gaps between the bacteria sans flagella and their fully equipped successors are transcended through the activities of some Creator or "creative force." Continuing to be generous, let's give Behe the personalized version.

So what does the Creator do? Option 1: He (we'll throw in patriarchy as well) arranges the selection regime for the hapless intermediates, directs the mutations, and so forth; so, in accord with a doctrine Behe has "no particular reason to doubt," organisms are linked by descent, and the Creator's work is devoted to making sure that just the right mutations arise in the right order and that the organisms on the way to the complex final state are protected against the consequences of having lots of useless spare parts that will be assembled at some final stage. Option 2: the Creator dispenses with a lot of the intermediate steps by cunningly arranging for lots of mutations to happen at once; if 183 new proteins are needed for the new structure, then zap! He strikes the appropriate loci with his magical mutating finger; or maybe he does it in two goes of 92 and 91 (with a protective environmental regime for the halfway stages); or in three interventions of 61 mutations a trick. Here, again, organisms are related by descent with modification, although the "descent" and the "modifications" are a bit abnormal. Option 3: the Creator gives up on mutation and selection entirely, simply creating a bunch of organisms with the right molecular stuff de novo; of course, if Behe thinks that this is the way things worked, then he really does have doubts about descent with modification.

The first point to note is that there's absolutely no evidence in favor of any of these options—they are the kinds of things to which one would only be driven if one thought that Behe's Big Problem was so intractable that there was no alternative. But matters are actually much worse than that, as one can see by posing questions about the Creator's psychology. Why should anyone think that the kind of Creator for whom Behe and Johnson both want to make room would undertake any of these projects? In Option 1, we envisage a Creator with the power to direct mutations and contrive protective environments who prefers simulating natural selection with gerrymandered selection pressures to directing all the needed mutations at once. In Option 2, we envisage a Creator who has the power to create organisms, but who prefers to simulate descent by the magic of mass mutation rather than simply producing the kinds of organisms He wants (either successively or simultaneously). In Option 3, we envisage a Creator who creates all the kinds of organisms He wants, as He wants them, but equips them with the genomic junk found in organisms He's created earlier. I am no engineer, but these visions inspire me to echo Alfonso X on the complexities of the Ptolemaic account of the solar system—

had the Creator consulted me at the Creation, I think I could have given him useful advice.

Perhaps I am being unfair. Maybe the project of design looks ludicrous because I have selected the wrong options for Creative intervention. Behe could easily answer my concerns by coming up with an alternative, one that would explain how creative design has figured in the history of life on our planet and how that creative design is part of a project worthy of his favorite Creator. I'm inclined to think that he won't do that, that the silence in Neo-Creo positive proposals will continue to be deafening. After all, positive doctrines and explanations have always been Creationism's Achilles Heel.

Notice that the line of argument in which I'm now engaged isn't a defense of evolutionary theory. For the sake of argument, I've conceded that evolutionary theory faces deep and intractable problems, although I've spent most of my time arguing that that's totally false. To show that the problems alleged to face evolutionary theory can't be solved by appealing to creative design isn't to rehabilitate the theory, for one doesn't always have to adopt the better of two alternatives. But in demonstrating that evolutionary theory is clearly superior to the imaginable members of the Creationist family I ought to sap the motivation of those who are drawn to Creationism. Attacking evolutionary theory was supposed to make room for God, but, as we've seen, there's not much hope for an active role for the Deity in any successor to evolutionary theory.

Although it's hard to see just which of my three options Behe would choose, his position is less indefinite than Johnson's. What Johnson thinks actually happened in the history of life is deeply obscure. All he tells us is that the hallmark of Creationism is the idea that "a supernatural Creator not only initiated the process but in some meaningful sense *controls* it in furtherance of a purpose".[22] Controls it how and when? With what purpose? Johnson doesn't say. For one so enthusiastic about canonizing Sir Karl, Johnson's "creation model" is rather short on "risky predictions."

Suppose we were to concede Johnson's claims about the difficulties for evolutionary theory. It would be natural to expect, as his book puffs to its conclusion, that he would say something about how those difficulties vanish once one invokes the activities of the designing Creator. Consider those puzzles about the fossil record. What exactly do they indicate? Does Johnson believe that organisms whose fossilized remains are lower in the geological column were around long before those higher up? Which organisms does he think are related to which? And, if he denies the descent of major groups from the organisms evolutionary theorists identify as their ancestors, how does he think the later organisms were formed?

Here are the possibilities. Option 1: Johnson might claim that the fossil record is profoundly misleading, that there's been no succession of organisms; this would be to take over part of old-style Creationism, claiming that the major kinds of organisms were all formed at once, and have inhabited the earth since the Creation (except, of course, for those that have gone extinct); it would, however, remain uncommitted to whether or not the earth is old or young. Option 2: Johnson might claim that the fossil record really does show a sequence of organisms, with some appearing later in earth's history than others, claiming that the major kinds are created as they appear, and are not modified descendants of earlier organisms. Option 3: Johnson might propose that the history of life is one of descent with modification, but that the Creator has guided the processes (perhaps in the ways signaled in Behe's first two options, considered previously).

Now option 1 is highly problematic because it offers no account of the world-wide ordering of the fossils in geological strata and no account of the anatomical, physiological, developmental, and molecular affinities among organisms. The first of these is the familiar difficulty that led people early in the nineteenth century—including extremely devout naturalists—to abandon the idea of a single fixed creation in favor of a sequence of creations. Option 2 founders on the need to understand why later organisms take over features at all levels from earlier organisms, features that are often no longer functional. Are we to assume that the junk in our genomes and the vestigial bits and pieces of anatomy are just signs of the Creator's whimsy? Option 3 is, of course, an evolutionary account of life, one widely adopted in the later decades of the nineteenth century by theists who thought that there had to be a supernatural component to mechanisms of evolutionary change. It would require Johnson, like Behe, to explain just what it is that the Creator does, and why he does things that way. All three schematic Creation models face large and familiar problems, which is why all the detailed versions of all of them have been abandoned by thinkers whose knowledge and intellectual integrity greatly exceed Johnson's.

But wait! Maybe Johnson has some gleaming new version that will put the general worries to rest? Alas, any reader who expected that would be disappointed. Toward the end of his book, he confesses

> I am not interested in any claims that are based upon a literal reading of the Bible, nor do I understand the concept of creation as narrowly as Duane Gish does. If an omnipotent Creator exists He might have created things instantaneously in a single week or through gradual evolution over billions of years. He might have employed means wholly inaccessible to science, or mechanisms that are at least in part understandable through scientific investigation.
>
> The essential point of creation has nothing to do with the timing or the mechanism the Creator chose to employ, but with the element of design or purpose. In the broadest sense, a "creationist" is simply a person who believes that the world (and especially mankind) was *designed*, and exists for a *purpose*. With the issue defined that way, the question becomes: Is mainstream science opposed to the possibility that the natural world was designed by a Creator for a purpose? If so, on what basis?[23]

Not only is no creation model offered here, but the definition of "creationism" is modified to make it compatible with orthodox Darwinism!

Johnson's original formulation of the position required that the Creator not merely set things up and let 'em roll, but that He actively intervene in the history of life. In this later passage, the commitment to continued intervention has been abandoned. Some Darwinians would be prepared to allow that the Creator fixed the initial conditions for the universe, although they'd contend that everything that has occurred since can be understood as the outcome of natural processes. For those theistically inclined, a view like this has often seemed superior to one on which the universe requires continual janitorial work—Leibniz chided Newton for hypothesizing that the Creator might have to tinker with His handiwork. Of course, there's a residual cluster of worries centering around the motivations of an omnipotent Creator for proceeding in so indirect a fashion.

Johnson's official line is that we ought not second-guess the Almighty. Confronting concerns about the apparent policy of letting later organisms inherit the junk of their ancestors, he chides evolutionary theorists for "speculating" about what a "proper Creator" would do. But, if the creation model is to be taken seriously as an account of life and its history, the character of processes and products must be full of clues to the attitudes of the Creator, and on the basis of our observations, it's clear that the motivations of a Creator who let the evolutionary process unfold in the ways that it has in order to produce our own species are quite baffling. The more intimately the Creator becomes involved in the adjustments of the process, the greater the bafflement.

So Johnson leaves everything vague, hoping that nobody will notice that he's either committing himself to an extremely implausible hands-on Creator with purposes for which his means seem singularly ill-designed or a slightly more credible hands-off Creator who produced the current world in just the ways cosmologists and Darwinian evolutionary theorists suggest. But his final question reveals his blindness to the historical fate of his options. Unless he does come forward with a new proposal for understanding the role of the Creator in the history of life, we're entitled to suppose that the only ways of articulating a Creationist view are those that have been tried from the late eighteenth century to the present. Those are just the options I've canvassed, successively explored by ingenious, pious, but honest, thinkers who rejected them because they were at odds with the record of historical and contemporary life. Coffin nails are driven deeper with the advance of fossil discoveries, the dissection of molecular relationships, our increased understanding of biogeography, and all the rest. In the end, the only answer one can give to Johnson's question—presumably intended as rhetorical—is that the best mainstream science can allow him is a Creator who set things up and let them unfold by natural processes. Whether this does more than pay lip service to the yearning for purpose and design is a matter I leave to theists.

The Neo-Creo model factory is strikingly out of new resources. For all the fancy rhetoric, all the academic respectability, all the accusations and gesticulations, Born-Again Creationism is just what its country cousin was. A sham.

Notes

I am extremely grateful to Dan Dennett and Ed Curley for sharing with me their unpublished discussions of the Creationist writers I discuss here. I have also learned much from an illuminating essay by Niall Shanks and Karl Joplin, "Redundant Complexity: A Critical Analysis of Intelligent Design in Biochemistry," *Philosophy of Science* 66 (1999): 262–282. Finally. I'd like to thank Robert Pennock for his editorial encouragement and for the insights of his own excellent treatment of the Neo-Creos in *Tower of Babel*.

1. I'll be quoting extensively from two Creationist works, Michael Behe, *Darwin's Black Box: The Biochemical Challenge to Evolution* (New York: Free Press, 1996), and

Phillip Johnson, *Darwin on Trial* (Washington, D.C.: Regnery Gateway, 1993). Here: Behe, *Darwin's Black Box*, 5.

2. Behe, *Darwin's Black Box*, 73.

3. I borrow this pithy formulation from Dan Dennett.

4. For further discussion of this issue, see Philip Kitcher, *Abusing Science: The Case against Creationism*. (Cambridge, Mass.: MIT Press, 1982), 117.

5. Behe, *Darwin's Black box*, 94.

6. Ibid., 95.

7. Johnson, *Darwin on Trial*, 12.

8. Ibid., 47.

9. Ibid., 84.

10. Ibid., 85.

11. Ibid., 14.

12. Ibid., 79.

13. Ibid., 75.

14. Ibid., 75.

15. Ibid., 76.

16. Ibid.

17. Ibid., 146.

18. Ibid., 149.

19. Ibid., 149.

20. Ibid., 90.

21. Ibid., 153.

22. Ibid., 4, n.1.

23. Ibid., 113.

Index